NUMERICAL SOLUTION
OF
DIFFERENTIAL
EQUATIONS

WILLIAM EDMUND MILNE, Ph. D., D. Sc.

PROFESSOR EMERITUS, OREGON STATE UNIVERSITY

*Second Revised
and Enlarged Edition*

Dover Publications, Inc., New York

Published in Canada by General Publishing Company, Ltd., 30 Lesmill Road, Don Mills, Toronto, Ontario.
Published in the United Kingdom by Constable and Company, Ltd., 10 Orange Street, London WC 2.

This Dover edition, first published in 1970, is an unabridged and corrected republication of the work originally published by John Wiley & Sons, Inc., in 1953. The author has written a new preface for this edition, and has supplemented the original text with several new appendices and an updated bibliography. R. R. Reynolds of the Boeing Company has supplied a new bibliography on partial differential equations.

Standard Book Number: 486-62437-4
Library of Congress Catalog Card Number: 75-100546

Manufactured in the United States of America
Dover Publications, Inc.
180 Varick Street
New York, N.Y. 10014

To My Wife

PREFACE TO THE DOVER EDITION

Numerical Solution of Differential Equations was written at the very dawn of the computer era, before the author had a chance to test any of the methods on high speed machines. Problems of stability had not yet received the attention later devoted to them, and procedures were still under the limiting influence of paper and pencil arithmetic. Ease of computation was the first consideration.

Hence the author welcomes this opportunity to supplement the original Wiley text with several appendices and a more up-to-date bibliography in order to improve and modernize his treatment of the subject matter. Also a number of unfortunate blunders in formulas and in the text can now be corrected. Portions of Chapter 4 have been revised or replaced by Appendix E, and several articles of Chapter 6 have been superseded by the material of Appendix F. Machine computation, especially with higher-order formulas, now requires special methods for change of step length; this topic is briefly considered in Appendix G.

The Milne Method is no longer important if modern computers are available, but is still useful if they are not, since now its main fault, instability, can easily be controlled as explained in Appendix D. The simplicity of its formulas makes it handy for paper and pencil, or desk computer, arithmetic.

The limited treatment of partial differential equations in the original text is supplemented and modernized in Appendix I, written by Mr. R. R. Reynolds of the Boeing Company; he has included a special bibliography for this section.

Finally a bibliography covering roughly the period from 1950 to 1967 has been appended as Appendix H. The huge volume of papers written in this seventeen-year period has made anything beyond a selective sampling impossible.

It is a pleasure for the author to express his thanks to the *Journal of the Association for Computing Machinery* for permission to reprint material in Appendix D, to the National Science Foundation for timely grants, to Dr. A. T. Lonseth and Dr. Don Aufenkamp for use of the facilities of the Department of Mathematics and of the Computer Center respectively. Also, to R. R. Reynolds for help with the revision, to Richard Haefele and Walter Yungen for testing methods and computing tables on the CDC 3300, and finally to Dr. George E. Forsythe for a long list of corrections to the original edition.

Oregon State University, January 1969

W. E. MILNE

PREFACE TO THE FIRST EDITION

In the decade just passed a burst of interest and activity has animated the field of numerical methods in general and the numerical solution of differential equations in particular. At the same time progress in mechanical computation has opened up whole areas heretofore deemed inaccessible.

This book attempts to acquaint the reader with some of the principal techniques available for the numerical solution of ordinary and partial differential equations. It strives to present these techniques in an elementary manner and largely with the aid of examples worked out in minute detail. Clarity of presentation for the benefit of the practical computer is the primary objective, and, though mathematical rigor has been striven for within the scope of an elementary text, rigor is not an object of idolatry. It is also beyond the limitations of this text to cite all of the very many methods for numerical solution described in the literature; the effort rather is to treat adequately a few of those deemed most essential.

The multiplicity of procedures for numerical solution, some differing in basic concept, many differing only in methodology, makes difficult a clear and logical presentation. In an attempt to meet this difficulty the second chapter contains an elementary account, with a minimum of mathematics, of a simple way to solve equations numerically when no great accuracy is needed. The discussion is detailed because the ideas are basic and recur in later chapters. For convenience of reference and to avoid interruptions in the explanations of specific methods the necessary mathematical machinery for Part I has been mainly concentrated in Chapter 3. The remainder of Part I is devoted to the analysis of several representative methods selected from the large number available.

Since some of the processes described in Part I apply only in special situations or are becoming outmoded because of mechanical computing aids, the reader who wants just one good up-to-date method will welcome a guide. At the risk of arousing controversy the author recommends Methods II and VII for such a reader. Specifically Chapter 2, together with Articles 28, 29, 30, 37, 38, 39, and 41, will give the practical computer the essential machinery for handling almost any "decent" problem in ordinary differential equations.

viii   PREFACE TO THE FIRST EDITION

For the benefit of readers whose acquaintance with matrices may be limited it has seemed advisable to include Chapter 9 on Linear Equations and Matrices. In spite of the number of excellent texts on matrices it is believed that a brief treatment slanted specifically toward iterative methods of solution is justified.

In order to make the book more useful for classroom instruction a number of exercises have been inserted to provide the student with the practical experience so indispensable for the mastery of numerical methods. Answers to a number of these exercises have been computed.

I am happy to acknowledge my gratitude to Dr. I. S. Sokolnikoff, without whose encouragement the book would probably not have been written; to the National Bureau of Standards, for making the resources of its Institute for Numerical Analysis available to me; to the Office of Naval Research for financial support; to Doctors John H. Curtiss, George E. Forsythe, Magnus R. Hestenes, and Cornelius Lanczos of the National Bureau of Standards for advice and criticism; to Doctor ˙ William M. Stone of Oregon State College for criticism; to Doctor James F. Price of Oregon State College for carefully checking the manuscript; and to Elizabeth Harding, Gladys Franklin, Lindley Wilson, Nan Reynolds, and Bessie Milne, computers.

W. E. MILNE

Corvallis, January 1953

CONTENTS

PART I ORDINARY EQUATIONS

1 INTRODUCTION

6 SYSTEMS OF EQUATIONS. HIGHER-ORDER EQUATIONS

7 TWO-POINT BOUNDARY CONDITIONS

PART II PARTIAL EQUATIONS

8 EXPLICIT METHODS. PARABOLIC AND HYPERBOLIC EQUATIONS

9 LINEAR EQUATIONS AND MATRICES

PART I

Ordinary Equations

1

INTRODUCTION

When a practical problem in science or technology permits mathematical formulation, the chances are rather good that it leads to one or more differential equations. This is true certainly of the vast category of problems associated with force and motion, so that whether we want to know the future path of Jupiter in the heavens or the path of an electron in an electron microscope we resort to differential equations. The same is true for the study of phenomena in continuous media, propagation of waves, flow of heat, diffusion, static or dynamic electricity, etc., except that we here deal with partial differential equations.

Elementary courses in differential equations present a long list of clever devices by means of which one is supposed to be able to solve differential equations. It is therefore with a horrible sense of frustration and futility that one eventually discovers how few, relatively speaking, are the equations that these devices will crack, and how many are the problems that can be neatly formulated as differential equations but get no farther for lack of solutions. Textbooks on hydrodynamics, elasticity, quantum mechanics, and the like develop elegant systems of differential equations but limit the actual solutions to very simple cases.

Faced with this situation, some bold spirits decided to dispense with analytical solutions altogether and attempted to squeeze the desired information out of the differential equations directly. Hence it came about that various numerical methods were devised, not by professional mathematicians at all, but by workers in other fields, who were intent on another objective and to whom the method of solution was merely incidental.

Adams' method appeared in an article entitled "Theories of Capillary action" by Bashforth and Adams [4] in 1882.* Another method for the second-order equations of dynamics is found in an "Essay on

---

* Numbers in brackets refer to the Bibliography at the back of the book.

the Return of Halley's Comet" by Cowell and Crommelin [41]. Still other methods were formulated in connection with ballistic problems during World War I by the astronomer F. R. Moulton [142, 143, 94]. In fact, astronomical and ballistic problems have been especially productive of numerical methods.

Perhaps for this reason, perhaps because it involves much tedious calculation, the numerical solution of differential equations has never quite attained respectability among pure mathematicians. Otherwise surely the subject would have been much better standardized, the gaps in our present knowledge would have been filled, and arguments concerning which method is best would have been settled.

Actually several nearly similar methods have been discovered and published independently, and in many cases no completely adequate means for assessing accumulated error has been devised. For a subject that is really rather elementary the numerical solution of differential equations is still in a surprisingly crude state. Let us hope that the wave of fresh interest in numerical methods stirred up by large-scale digital computing machinery will attract the attention of able mathematicians to the problems still outstanding in this subject.

1. Differential equations

Equations involving variables and their derivatives are called differential equations. In the simplest case there are two variables x and y, one derivative dy/dx, denoted by y', and one equation

$$F(x, y, y') = 0.$$

This equation is commonly solved for the derivative if possible, and written in the form

(1.1) $$y' = f(x, y).$$

The case of two variables where derivatives of higher order occur leads to the equation

(1.2) $$F(x, y, y', \cdots, y^{(n)}) = 0.$$

Here n denotes the highest order among the derivatives. This is an ordinary differential equation of nth order.

An equation may contain several dependent variables $y_1, y_2, \cdots, y_m$, and their first-order derivatives $y_1', y_2', \cdots, y_m'$ with respect to a single independent variable x. The equation

$$F_1(y_1, \cdots, y_m, y_1', \cdots, y_m', x) = 0$$

represents such a situation.   In order that the m dependent variables
be determined as functions of x, there are required as many equations
as there are dependent variables.   It is customary to solve the equations for the derivatives as follows:

(1.3)
$$y_1' = f_1(y_1, \cdots, y_m, x),$$
$$y_2' = f_2(y_1, \cdots, y_m, x),$$
$$\cdots \cdots \cdots \cdots \cdots$$
$$y_m' = f_m(y_1, \cdots, y_m, x).$$

This is a system of differential equations of the mth order.

The nth-order differential equation 1.2 can be replaced by an
equivalent system of n equations of first order by the substitutions

$$y_1 = y, \quad y_2 = y', \quad y_3 = y'', \quad \cdots \quad y_n = y^{(n-1)}.$$

For, if equation 1.2 when solved for $y^{(n)}$ is

$$y^{(n)} = f(x, y, y', \cdots y^{(n-1)}),$$

it is apparent that an equivalent system is

$$y_1' = y_2,$$
$$y_2' = y_3,$$
$$\cdot$$
$$\cdot$$
$$\cdot$$
$$y_{n-1}' = y_n,$$
$$y_n' = f(x, y_1, y_2, \cdots, y_n).$$

Conversely, from a system of equations of mth order such as equation 1.3 it is possible, formally at least, to obtain a differential equation
of order not exceeding m and containing only one dependent variable.
It is often impossible, however, to perform the actual eliminations,
and hence this transformation is of theoretical rather than practical
interest.

Differential equations or systems of differential equations with
only one independent variable are called *ordinary* differential equations.   All those so far cited are ordinary.   Equations involving more
than one independent variable and the partial derivatives of the
dependent variables with respect to the independent variables are
called *partial* differential equations.

Here again there should be as many equations as there are dependent variables. For example, in the case of three independent variables x, y, z, and two dependent variables U, V, we may have the two first-order equations:

$$F_1(U, V, U_x, U_y, U_z, V_x, V_y, V_z) = 0,$$

$$F_2(U, V, U_x, U_y, U_z, V_x, V_y, V_z) = 0.$$

A general theory of partial differential equations scarcely exists, and interest mainly centers around particular equations such as the Cauchy-Riemann equations, the equations arising in hydrodynamics, elasticity, static and dynamic electricity, heat flow, and wave motion.

2. Solution of a differential equation. Arbitrary constants. Arbitrary functions

We recall that the general solution of a differential equation of first order

(2.1) $$y' = f(x, y)$$

is in the form of an equation

$$\phi(x, y, C) = 0,$$

connecting x, y, and an arbitrary constant. The significance of the arbitrary constant becomes more evident if we replace the differential equation by a difference equation as follows: Let x_0, x_1, x_2, $\cdots$ denote a set of equally spaced values of x, with $x_{i+1} - x_i = h$, let y_0, y_1, y_2, $\cdots$ denote corresponding values of y, as yet undetermined, and note that

$$y_i' = \lim_{x_{i+1} \to x_i} \frac{y_{i+1} - y_i}{x_{i+1} - x_i} = \lim_{h \to 0} \frac{y_{i+1} - y_i}{h}.$$

In place of the differential equation we consider the difference equation

$$(y_{i+1} - y_i)/h = f(x_i, y_i)$$

or

(2.2) $$y_{i+1} = y_i + hf(x_i, y_i).$$

Letting i assume in succession the n values 0, 1, $\cdots$, $n - 1$, we have in equation 2.2 a set of n equations involving the $n + 1$ unknowns, y_0, y_1, $\cdots$, y_n. These n equations are not sufficient to determine the $n + 1$ unknowns. However, if one of the unknowns, say y_0, is regarded as arbitrary, then the remaining y's can all be determined by equation 2.2 in terms of this arbitrary y_0.

Example. The differential equation

$$y' = x + y$$

has the solution

$$y = Ce^x - x - 1$$

or

$$y = (y_0 + x_0 + 1)e^{x-x_0} - x - 1,$$

where $y = y_0$ when $x = x_0$. The associated difference equation is

$$y_{i+1} = y_i + h(x_i + y_i),$$

whence

$$y_1 = y_0 + h(x_0 + y_0),$$

$$y_2 = y_1 + h(x_1 + y_1),$$

$$y_3 = y_2 + h(x_2 + y_2),$$

$$\cdot\ \cdot\ \cdot\ \cdot\ \cdot\ \cdot\ \cdot\ \cdot\ \cdot\ \cdot\ \cdot\ \cdot$$

Successive eliminations give

$$y_1 = (1 + h)(y_0 + x_0 + 1) - x_1 - 1,$$

$$y_2 = (1 + h)^2(y_0 + x_0 + 1) - x_2 - 1,$$

$$y_3 = (1 + h)^3(y_0 + x_0 + 1) - x_3 - 1,$$

and, in general,

$$y_n = (1 + h)^n(y_0 + x_0 + 1) - x_n - 1.$$

From the relation $x_n - x_0 = nh$, we have $n = (x_n - x_0)/h$ and

$$y_n = (1 + h)^{(x_n-x_0)/h}(y_0 + x_0 + 1) - x_n - 1.$$

Since $\lim\limits_{h \to 0} (1 + h)^{(x_n-x_0)/h} = e^{x_n-x_0}$, this solution approaches the solution of the differential equation as h approaches zero.

For a differential equation of second order,

$$y'' = f(x, y, y'),$$

we proceed in the same manner, except that here it is convenient (though not necessary) to replace y' by $(y_i - y_{i-1})/h$. The second derivative is replaced by

$$(y_{i+1} - 2y_i + y_{i-1})/h^2$$

so that the difference equation becomes

$$y_{i+1} - 2y_i + y_{i-1} = h^2 f(x_i, y_i, (y_i - y_{i-1})/h).$$

Giving i the $n - 1$ values $1, 2, \cdots, n - 1$, we get $n - 1$ equations containing the $n + 1$ unknowns $y_0, y_2, \cdots, y_n$. Here evidently two additional conditions must be supplied in order to determine the y's completely. This corresponds to the fact that we have two arbitrary constants in the solution of a differential equation of second order.

The case of a system of simultaneous equations may be illustrated by

$$x' = X(x, y, z, t), \qquad y' = Y(x, y, z, t), \qquad z' = Z(x, y, z, t),$$

a system of three equations with three dependent variables x, y, z, and one independent variable t. As usual primes denote differentiation with respect to the independent variable. Proceeding as before we get the three difference equations

$$x_{i+1} = x_i + hX_i, \qquad y_{i+1} = y_i + hY_i, \qquad z_{i+1} = z_i + hZ_i,$$

and, setting $i = 0, 1, \cdots, n - 1$, we obtain $3n$ equations for the determination of the $3n + 3$ unknowns $x_0, \cdots x_n, y_0, \cdots, y_n, z_0, \cdots, z_n$. Here again the number of equations falls short of the number of unknowns by precisely the number of arbitrary constants in the general solution of the differential system.

These three illustrations should make clear the situation for the case of ordinary equations. We next consider partial differential equations.

For simplicity consider the case of two independent variables x and y, which we shall look upon as the rectangular coordinates of a point in the xy plane. We subdivide the plane into rectangles by vertical lines spaced at distance h and horizontal lines spaced at distance k, as shown in Fig. 1. Then a partial differential equation of first order,

$$F(x, y, u, u_x, u_y) = 0$$

with one dependent variable u can be replaced with a difference equation

$$F(x_i, y_j, u_{ij}, (u_{i+1,j} - u_{ij})/h, (u_{i,j+1} - u_{ij})/k) = 0,$$

which is, in fact, a relation connecting the three values u_{ij}, $u_{i+1,j}$, $u_{i,j+1}$ belonging to the three points shown in Fig. 1.

Clearly, if all the values of u on a horizontal (or a vertical) line are given, then repeated use of the difference equation enables us to fill in all values above (or to the right of) the given line of values. Since we might equally well have formed the difference equation with the values of u arranged as shown in Fig. 2, all values on the other side of the line can be obtained in a similar manner, Hence we see that the solution of a first-order equation with two independent variables is determined when we have all the values of the dependent variable on a line. This is equivalent to prescribing an arbitrary function.

More generally, suppose that we have two equations with two dependent variables U and V and three independent variables x, y, z. Suppose that both equations actually contain all the possible partial derivatives of second order. Then the differential equations can be replaced

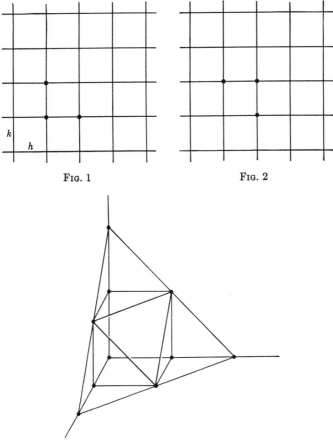

Fig. 1   Fig. 2

Fig. 3

by difference equations involving the values of U and V at ten points arranged as shown in Fig. 3.

Now, if the values of U and V are all given on any two adjacent parallel planes of the system of planes which we have used to make our three-dimensional network, it is clear that we can determine all remaining values of U and V from the difference equations. In such a case we have given the equivalent of four arbitrary functions of two variables.

The generalization to the case of m second-order equations with m dependent and n independent variables is left to the reader. It is important to note that a reduction in the number of required arbitrary functions may occur in equations where not all second-order derivatives are present.

To explain this possibility, we observe that for partial differential equations the idea of *order* is not so precisely defined as for ordinary equations. For example, if we say that

$$(2.3) \qquad u_x = u_y$$

is an equation of first order and

$$(2.4) \qquad u_{xx} = u_{yy}$$

is an equation of second order, what order are we going to assign to

$$(2.5) \qquad u_x = u_{yy}?$$

The general solution of equation 2.3 is

$$u = f(x + y)$$

involving *one* arbitrary function of a single variable, the general solution of equation 2.4 is

$$u = f_1(x + y) + f_2(x - y)$$

involving *two* arbitrary functions of a single variable, but the solution of equation 2.5 may be completely determined by either *one or two* given functions, depending on what is known. For instance, if we require that $u = \cos y$ when $x = 0$, the solution of (2.5) is

$$u = e^{-x} \cos y$$

and is completely determined, whereas, if we require that

$$u = e^{-x}$$

when $y = 0$, the solution is still indeterminate to the extent of one arbitrary function.

3. Solutions: explicit, implicit, numerical

Let us consider the equation of first order

$$(3.1) \qquad y' = f(x, y),$$

where it is assumed that $f(x, y)$ is real, continuous, and single-valued in some region R of the xy plane.

Explicit Solution. If we can find a function $\phi(x)$ with a continuous derivative $\phi'(x)$ in some interval $a < x < b$ in R such that $\phi(x)$ is expressible in finite form in terms of known functions and such that

$$(3.2) \qquad\qquad y = \phi(x), \qquad y' = \phi'(x)$$

together satisfy equation 3.1, we say that equations 3.2 furnishes an *explicit* solution of the differential equation in the interval $a < x < b$.

Example 1. The differential equation

$$y' = 1 + y^2$$

has the explicit solution

$$y = \tan{(x - c)}$$

valid in the interval

$$c - \pi/2 < x < c + \pi/2.$$

Here R may be taken as any finite region of the xy plane.

Example 2. A solution need not be represented everywhere by the same analytical expression. Consider the differential equation

$$y' = +2\sqrt{y},$$

where the region R is the upper half of the xy plane, including the x axis.

An explicit solution of this equation is seen to consist of two parts,

$$y = 0 \qquad \text{for} \quad x < c$$

and

$$y = (x - c)^2 \quad \text{for} \quad x \geqq c.$$

These two pieces, represented by two different analytical expressions, are joined together continuously and with a continuous derivative at $x = c$.

Example 3. It is not always necessary that the function $f(x, y)$ be represented by the same analytical expression throughout the whole of the region R. For example, suppose that

$$y' = x + y \quad \text{if} \quad y > 0,$$

$$y' = x \qquad \text{if} \quad y \leqq 0.$$

The explicit solution is found in this case to consist of three parts joined together continuously, and with continuous slopes:

$$y = (1 - a)e^{x+a} - x - 1 \quad \text{for} \quad x < -a < 0,$$

$$y = \tfrac{1}{2}(x^2 - a^2) \qquad\qquad \text{for} \quad -a < x < a,$$

$$y = (1 + a)e^{x-a} - x - 1 \quad \text{for} \quad a < x.$$

The differential equations applying to bending beams with concentrated loads are familiar examples of this situation.

Implicit Solution. When we can find a relationship

$$(3.3) \qquad\qquad \phi(x, y) = 0$$

where the function ϕ has continuous partial derivatives with respect to x and y in some region R' in R, is expressible in finite form in terms of known functions, and is such that

$$\phi_x + \phi_y y' = 0$$

together with equation 3.3 satisfies equation 3.1, we say that 3.3 provides an *implicit* solution of 3.1.

Example 4. The equation

$$(3x^2 - 8xy + 2y^2)dx - (4x^2 - 4xy + 3y^2)dy = 0$$

has the implicit solution

$$x^3 - 4x^2y + 2xy^2 - y^3 = c.$$

Example 5. The equation

$$\tan y(y + \cot x)dx - (1 + \tan^2 y - x \tan y)dy = 0$$

has the implicit solution

$$\tan y - c \sin xe^{xy} = 0.$$

Numerical Solution. By a *numerical* solution we mean a table of values

$$
\begin{array}{ccc}
x & y & y' \\
x_0 & y_0 & y_0' \\
x_1 & y_1 & y_1' \\
x_2 & y_2 & y_2' \\
\cdot & \cdot & \cdot \\
\cdot & \cdot & \cdot \\
\cdot & \cdot & \cdot \\
x_n & y_n & y_n'
\end{array}
$$

such that $y' = dy/dx$ and $y_i' = f(x_i, y_i)$, $i = 0, 1, 2, \cdots, n$.

Example 6. The differential equation

$$xy' - y = x^2$$

has the numerical solution

x	y	y'	x	y	y'
1.0	3.00	4.0	1.6	5.76	5.2
1.1	3.41	4.2	1.7	6.29	5.4
1.2	3.84	4.4	1.8	6.84	5.6
1.3	4.29	4.6	1.9	7.41	5.8
1.4	4.76	4.8	2.0	8.00	6.0
1.5	5.25	5.0	2.1	8.61	6.2

This table of values is obtained from

$$y = x^2 + 2x,$$

which is a solution of the given differential equation.

We note that the table gives the solution only over a part of the range of values of the independent variable. This is the usual situation since ordinarily a numerical solution represents only a limited part of a solution. We note also that the table does not contain an arbitrary constant and hence represents only one particular solution. For another solution of the same differential equation, say,

$$y = x^2 - x,$$

a new table of values is needed.

Here y is a polynomial of low degree, and it is possible to make a table with values that exactly satisfy the solution

$$y = x^2 + 2x.$$

In this respect the example cited is not at all typical, since in problems for which numerical solutions are needed the solutions are of no such simple form. The values of y corresponding to rational values of x are in general irrational and are incapable of exact representation by terminating decimals. Of necessity then the values given by a numerical solution are not exact but are approximations, the accuracy of which depends on the number of decimal places correctly calculated.

In the example above, the values of x were selected at equal intervals. This is not necessary. It is, however, convenient in most cases to use equal intervals for x, or perhaps for a new independent variable s which may be introduced to facilitate the calculation. The choice of a suitable interval between successive values of x involves the balancing of several different considerations and will be dealt with in later chapters.

4. Need for numerical solution

Whenever an explicit solution of a differential equation can be found, it is usually best to use this explicit solution rather than to resort to numerical methods. Unfortunately in many cases where the solution of a differential equation is needed it proves to be impossible to obtain a solution in elementary form.

Example 1. A simple equation of this type is

$$y' = x^2 + y^2,$$

which has no elementary solution.

Example 2. The equation

$$y'' \pm a(y')^2 + by = 0$$

for vibrations with damping proportional to the square of the velocity cannot be solved in simple form.

Example 3. The equation

$$y'' + ay' + b \sin y = 0$$

for the motion of a pendulum with linear damping is not solvable in terms of elementary functions.

Example 4. A problem of great historical interest, namely, the differential equations defining the motion of three bodies subject to their mutual gravitational forces, cannot be solved.

In problems of this type numerical methods become a necessity due to absence of other methods for getting the requisite information out of the differential equations.

Even in cases where explicit or implicit solutions are known, it is sometimes easier to obtain a numerical solution than attempt to calculate numerical values from the known solution.

Example 5. It is almost as easy to get numerical values from the differential equation in Example 5, Article 3, as from the solution itself.

Example 6. The equation

$$y'' = -xy$$

can be solved in terms of Bessel functions, but, if we want a few values of a particular solution, we can often get them just as quickly directly from the differential equation as to make the necessary transformation and interpolate in a table of Bessel functions.

Example 7. The equation

$$\phi'' = - \sin \phi$$

can be solved by means of elliptic functions, but here again we can often get a few values of a particular solution just as easily from the differential equation as from a table of elliptic integrals.

5. Preliminary investigation of the solution

Before one undertakes the numerical solution of a differential equation, he should glean whatever general information regarding the character of the solution that can be ascertained by an examination of the differential equation itself. The infinite variety of differential equations that may be encountered and the varying amounts of information that it may be possible to secure without excessive labor make a comprehensive treatment of this topic impossible. In the last analysis success in this endeavor depends primarily on the skill and insight of the investigator. We present below a pair of examples which may serve to suggest methods of attack.

Example 1. $y' = x^2 - y^2.$

For differential equation of first order such as this, considerable information can be derived from a chart of the xy plane showing the direction of the tangents to solutions at various points of the plane. Since horizontal and vertical tangents are of particular interest, we look first for the loci on which these occur. In this example horizontal tangents lie on the two lines $y = x$ and $y = -x$, since on these loci $y' = 0$. These lines are shown in Fig. 4. There are no vertical tangents. Next we note that in the regions OAB and OCD the derivative is positive and hence the curve is rising to the right, whereas in the regions OAD and OBC the derivative is negative and the curve is falling to the right. Further information is secured by plotting a few *isoclinal* curves. These are curves (not solutions) on which y' is constant. The curve on which $y' = 1$ is the hyperbola $x^2 - y^2 = 1$. The hyperbolas $x^2 - y^2 = \pm 1$, $x^2 - y^2 = \pm 4$ are sketched in Fig. 4, with short line segments indicating the slope on each. With Fig. 4 before us it is a simple matter to trace roughly the course of any particular solution. For example the curve of the solution through $(-2.5, 0)$ rises rapidly until it reaches the line OD, where y has a maximum, then falls until it reaches OA where y has a minimum, then rises again and recedes toward infinity asymptotically to the line OA. This solution and the solution through $(-1, 0)$ are sketched in Fig. 4. Other particular solutions may be traced in a similar way.

Example 2. $y'' + (y')^2 + 4y^3 = 0.$

Since this equation does not contain x explicitly, there is no loss of generality in limiting the discussion to the solution

$$y = u_a(x)$$

which satisfies the conditions

$$u_a(0) = a, \quad u_a'(0) = 0.$$

Any other solution will be of the form

$$y = u_a(x + c).$$

Since the equation and initial conditions are invariant if x is replaced by $-x$, we see that $u_a(x)$ is an even function. In the yy' plane the points where y' has a maximum or minimum lie on the semicubical parabola $(y')^2 + 4y^3 = 0$, since for these $y'' = 0$, while the points where y has maximum or minimum lie on the y axis, since for these $y' = 0$. The semicubical parabola and the y axis divide the yy' plane into four regions HOG, GOP, TOP, HOT in which the slope of the yy' curves have the signs shown in Fig. 5. The yy' curves are seen to fall into two classes, the class illustrated by AA' in Fig. 5, which intersects the semicubical parabola, and the class illustrated by BB', which is open and does not intersect the semicubical parabola. It appears that there is a critical value a_0 of a such that

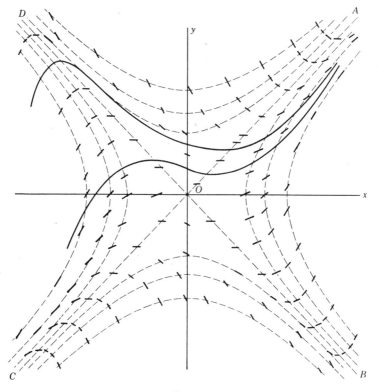

Fig. 4

type AA' occurs if $a < a_0$, whereas type BB' occurs if $a > a_0$. (A more elaborate investigation shows that $a_0 = 0.798 \cdot \cdot \cdot$)

Turning now to the xy plane, we see that for the AA' case where $0 < a < a_0$ the solution is of the form shown in Fig. 6. The slope y' vanishes at $x = 0$ by assumption, and vanishes again for $x = p$, where $y = -b$. We suspect from Fig. 5 that b is greater than a, and this suspicion is confirmed by the equation

$$\tfrac{1}{2} (y')^2 + y^4 = a^4 - \int_0^x (y')^3 \, dx,$$

obtained from the original differential equation by multiplication with y' followed by integration with respect to x. For in the interval $0 < x < p$ the slope y' is

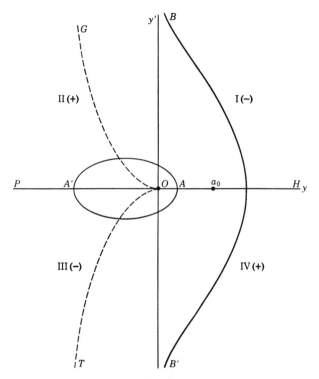

FIG. 5

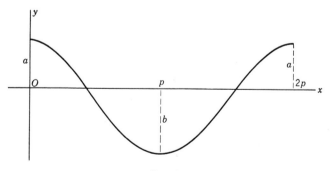

FIG. 6

negative and we have

$$b^4 = a^4 - \int_0^p (y')^3 \, dx,$$

whence $b^4 > a^4$. The solution is seen to be symmetrical with respect to the line $x = p$, and we now see that the whole solution is periodic with period $2p$.

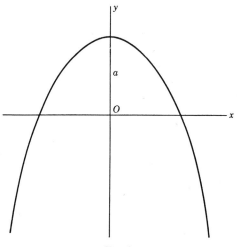

FIG. 7

For the BB' case the solution in the xy plane is always concave down and has the form suggested by Fig. 7.

EXERCISES

Examine the following differential equations for explicit or implicit solutions. Sketch the isoclinal curves and indicate the general trend of the integral curves.

1. $y' = \dfrac{x + y}{x - y}$.

2. $y' = \dfrac{x + 2y}{2x + y}$.

3. $y'^2 = x - y$.

4. $y' = 1 + x^2 y^2$.

5. $y'^2 = x^2 + y^2$.

6. $y' = \sin x + \cos y$.

2

ELEMENTARY NUMERICAL SOLUTION

Not infrequently all that is desired of a differential equation is a solution over a limited range of values of the independent variable with modest accuracy (say two or three significant digits) in the values of the dependent variable. For this purpose the simple procedures outlined in the present chapter are admirably suited. They require a minimum of theory, are easy to grasp, and are simple and rapid in operation. However, when greater accuracy is required, it is better to use some of the more sophisticated methods presented in later chapters.

6. Point-slope formula. Method I

The sketching of integral curves with the aid of slopes as described in Example 1 of Article 5 immediately suggests a simple numerical method of solution. In this method a new point (x_{n+1}, y_{n+1}) is located by means of the slope at a known point (x_n, y_n). The obvious formula is

$$y_{n+1} = y_n + h y_n',$$

where $h = x_{n+1} - x_n$, but an equally simple and more accurate formula is

$$y_{n+1} = y_{n-1} + 2h y_n'.$$

For by Taylor's series we have

$$y_{n+1} = y_n + h y_n' + \frac{h^2}{2} y_n'' + \text{higher powers of } h$$

while

$$y_{n+1} = y_{n-1} + 2h y_n' + \frac{h^3 y_n'''}{3} + \text{higher powers of } h,$$

showing that, if h is sufficiently small, the neglected terms in the second formula are smaller than in the first.

19

In order to get started we need to know y_0 and also y_1'. We assume that the initial point (x_0, y_0) is given, then determine (x_1, y_1) by the first three terms of Taylor's series, next determine y_1' by the differential equation, and from here on proceed by the following sequence of operations:

1. Calculate y_{n+1} by the formula

(6.1) $$y_{n+1} = y_{n-1} + 2hy_n'.$$

2. Calculate y_{n+1}' from the differential equation

$$y_{n+1}' = f(x_{n+1}, y_{n+1}).$$

Repeat the same operations for the next step, etc.

Example. To illustrate the process we select the simple equation

(6.2) $$y' = -2xy^2$$

with the initial point $(0, 1)$. The explicit solution is

$$y = (1 + x^2)^{-1},$$

which we shall later compare with the numerical solution.

In order to start we obtain by differentiation

$$y'' = -2y^2 - 4xyy'.$$

Using the initial values $x_0 = 0$, $y_0 = 1$, we find that

$$y_0 = 1, \quad y_0' = 0, \quad y_0'' = -2,$$

and from these we form Taylor's series to three terms, thus,

$$y = 1 + 0 \cdot h - h^2.$$

A convenient value of h is $h = 0.1$, which in the present example will produce roughly three-decimal-place accuracy in y. This gives $y_1 = 0.990$ at $x = 0.1$, whence, by equation 6.2, $y_1' = -0.196$. Then, by equation 6.1,

$$y_2 = 1.000 + 2(0.1)(-0.196) = 0.961.$$

The values of y_3, y_4, etc. are found by similar steps. The computation can be conveniently arranged as shown in computation 1.

The values of the true solution are given for sake of comparison. It will be noted that, while at any one step formula 6.1 gives about three-place accuracy, the accumulated error finally amounts to about 4 units in the third place after 20 steps. It also appears that an oscillation is starting up, causing the values of y to fluctuate above and below the true values. The cause of this phenomenon will be considered in the following articles.

Method I   Computation 1

Differential equation $y' = -2xy^2$			$(y = 1$ at $x = 0)$	$h = 0.1$
x	y	y'	$(1 + x^2)^{-1}$	Error
0	1	0	1	0
0.1	0.990	−0.196	0.990	0
.2	.961	− .370	.962	−1
.3	.916	− .503	.917	−1
.4	.860	− .592	.862	−2
.5	.798	− .637	.800	−2
.6	.733	− .644	.735	−2
.7	.669	− .627	.671	−2
.8	.608	− .592	.610	−2
.9	.551	− .547	.552	−1
1.0	.499	− .498	.500	−1
1.1	.451	− .447	.452	−1
1.2	.410	− .403	.410	0
1.3	.370	− .356	.372	−2
1.4	.339	− .322	.338	1
1.5	.306	− .282	.308	−2
1.6	.283	− .256	.281	2
1.7	.255	− .221	.257	−2
1.8	.239	− .205	.236	3
1.9	.214	− .175	.217	−3
2.0	.204	− .168	.200	4

EXERCISES

1. Work the example of this article with the initial values $y = 2$, $x = 0$.
2. Work the example with $h = 0.2$. Compare the accuracy attained with that for $h = 0.1$.
3. Work the example with $h = 0.05$. Retain four decimal places. Carry the computation to $x = 1$.
4. Solve the equation

$$y' = x + y$$

numerically. Start with $x = 1$, $y = 0$, and carry to $x = 2$ with $h = 0.1$. Compare results with values of the explicit solution.
5. Solve the equation

$$y' = x^2 - y^2.$$

Start with $x = -1$, $y = 0$, and carry to $x = 1$ with $h = 0.25$.

7. Error of Method I

It is evident that the computational routine described in the preceding article provides no check on the accuracy either of the numerical calculations or of the process itself. Consequently, special care must be taken to avoid numerical mistakes, since such mistakes are liable to escape detection. Leaving aside numerical mistakes on the part of the operator, there are essentially three sources of error in the process itself.   These are:

(a) *Round-off error*, due to chopping off all numbers after a specified number of decimal places.   The analysis of this source of error is complicated and discussion of it is postponed to Appendix A.   Let us therefore ignore round-off error for the present by assuming that all numbers are carried to enough places so that round-off will not affect the places that we ultimately desire to retain.

(b) *Truncation error*, defined here as being the error committed in making the step from x_n to x_{n+1} by use of the approximate formula

$$y_{n+1} - y_{n-1} = 2hy_n'.$$

It can be shown that, if y has a continuous third derivative y''' in the interval $x_{n-1} \leqq x \leqq x_{n+1}$, the truncation error for this formula is

$$h^3 y'''(s)/3,$$

where s is somewhere in the interval $x_{n-1} < s < x_{n+1}$.

(c) *Inherited error*, defined as the error produced in y_{n+1} by the errors occurring in previous steps.

If we ignore round-off error and suppose that e_n is the error of y_n due to b and c, we have in general $y_n - z_n = e_n$, where y_n is the computed and z_n the true value of y at x_n.   Then

$$y_{n+1} - y_{n-1} = 2hf(x_n, y_n),$$

$$z_{n+1} - z_{n-1} = 2hf(x_n, z_n) + T_n,$$

where $T_n = h^3 y'''(s)/3$, $x_{n-1} < s < x_{n+1}$.   Hence by subtraction

(7.1) $e_{n+1} - e_{n-1} = 2h[f(x_n, y_n) - f(x_n, z_n)] - T_n.$

Using the mean value theorem, we have

$$f(x_n, y_n) - f(x_n, z_n) = f_y(x_n, y)(y_n - z_n)$$

where y lies between y_n and z_n.   With the symbol g_n defined by

$$g_n = f_y(x_n, y),$$

we may now write equation 7.1 in the form

(7.2) $e_{n+1} - 2hg_n e_n - e_{n-1} = -T_n.$

This recurrence relation establishes the law of propagation of error throughout the computation, provided the values of T_n and g_n are known for each n.   In general these quantities are variables, but in many instances they vary rather slowly, so that a fair idea of the behavior of the error may be secured if we treat T_n and g_n as constants T and g, respectively, at least over a limited range of steps. With this understanding we may readily show that, if g is positive, we have

$$(7.3) \qquad e_n = A(q)^n + B(-q)^{-n} + T/2hg,$$

where $q = hg + \sqrt{1 + h^2 g^2} > 1$, whereas, if g is negative,

$$(7.4) \qquad e_n = A(-q)^n + B(q)^{-n} + T/2hg,$$

where $q = -hg + \sqrt{1 + h^2 g^2} > 1$. The constants A and B are to be determined by the values of e_0 and e_1. From equation 7.3 we observe that, since $q > 1$, the first term on the right soon dominates and

$$e_n - Aq^n \to 0.$$

Ultimately, therefore, the error has one sign and increases exponentially. From equation 7.4 in like manner it appears that, if g is negative,

$$e_n - A(-q)^n \to 0,$$

and hence the magnitude of the error increases exponentially, but the sign alternates from step to step.

In the example of the preceding article the function $f(x, y)$ is $-2xy^2$ so that $g = -4xy$ and is negative for positive values of x and y. Hence we should expect the error eventually to oscillate, and in fact it does.

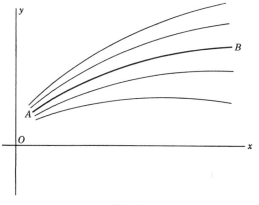

FIG. 8

Another observation may be made here. If the particular solution AB being computed lies in a field of integral curves that diverge as x increases, as shown in Fig. 8, slight errors tend to be magnified as the computation proceeds, to a greater extent than in the case illustrated by Fig. 9. This observation is general, not limited to any particular method of solution.

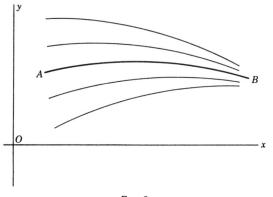

Fig. 9

EXERCISES

1. Show by substitution that the expressions given by equations 7.3 and 7.4 actually satisfy equation 7.2 if T_n and g_n are constant.

2. On the assumption that the first two values of y are correct so that $e_0 = e_1 = 0$, determine A and B in equations 7.3 and 7.4.

3. Using the result of exercise 2, describe the trend of the error if T is positive; if T is negative.

4. Apply the present theory to the errors in the exercises of Article 6. Estimate bounds for the error in each exercise. In the cases where explicit solutions can be found compare the theoretical error with the actual error.

8. Trapezoidal formula. Method II

We have noted that Method I contains in itself no check on the accuracy either of the numerical calculations or of the process itself. In the preceding article we made an investigation of the error in general, but for any particular problem this requires some additional labor. A definite improvement in accuracy, a check on the accuracy of the process, and a partial check on the arithmetical work are provided by the use of an additional formula in conjunction with Method I. We shall call the new process Method II.

Briefly stated, Method II consists of Method I with the added feature that each line is checked and corrected by means of the trapezoidal formula

$$(8.1) \qquad y_{n+1} = y_n + \frac{h}{2}(y_{n+1}' + y_n').$$

The truncation error of this formula is

$$-h^3 y'''(t)/12,$$

where t lies in the interval $x_n < t < x_{n+1}$. The routine (except for the starting line, where three terms of Taylor's series are initially used) is as follows:
We suppose that the computation has been completed through the line x_n. Then:

Step 1. Predict $\bar{y}_{n+1}$ by

(8.2) $$\bar{y}_{n+1} = y_{n-1} + 2hy_n'.$$

Step 2. Calculate $\bar{y}_{n+1}'$ by the differential equation

$$\bar{y}_{n+1}' = f(x_{n+1}, \bar{y}_{n+1}).$$

Step 3. Correct y_{n+1} by the trapezoidal formula 8.1, using the value $\bar{y}_{n+1}'$ just obtained for y_{n+1}'.

Step 4. Correct y_{n+1}' by the differential equation.

Step 5. Repeat steps 3 and 4, if necessary, until no further change occurs.

Method II					Computation 2
Differential equation $y' = -2xy^2$			($y = 1$ at $x = 0$)		$h = 0.1$
x	y	y'	C_0	$(1 + x^2)^{-1}$	Error
0	1	0	0	1	0
0.1	0.990	−0.196	0	0.990	0
.2	.962	− .370	1	.962	0
.3	.918	− .506	2	.917	1
.4	.863	− .596	2	.862	1
.5	.801	− .642	2	.800	1
.6	.737	− .650	1	.735	1
.7	.672	− .632	1	.671	1
.8	.611	− .597	1	.610	1
.9	.554	− .552	1	.552	2
1.0	.501	− .502	0	.500	1
1.1	.453	− .451	−1	.452	1
1.2	.410	− .403	−1	.410	0
1.3	.372	− .360	0	.372	0
1.4	.338	− .320	0	.338	0
1.5	.308	− .285	0	.308	0
1.6	.281	− .253	0	.281	0
1.7	.257	− .225	0	.257	0
1.8	.236	− .201	0	.236	0
1.9	.217	− .179	0	.217	0
2.0	.200	− .160	0	.200	0

Example. The process is illustrated by the same example that was used for Method I.

The values of y shown here are the corrected values. The predicted values were originally written with pencil, and the corrected values were then inserted after erasure of such digits as required changing. The column headed C_0 preserves a record of the correction made in y, that is,

$$C_0 = corrected\ value - predicted\ value.$$

In article 10 we discuss in some detail the use to which the column of values of C_0 is put. The values of the explicit solution are also tabulated as well as the error.

EXERCISES

Work the exercises of Article 6 by Method II.

9. The convergence factor

In Article 8 we assumed without critical examination that the process of repeating steps 3 and 4 was a convergent process, and moreover tacitly assumed that the convergence was rapid enough to secure the desired accuracy very quickly.

Now the correction to y_{n+1} obtained by the first application of equation 8.1 is C_0. Then the correction to y_{n+1}' as found from the differential equation is C_0', where

$$C_0' = gC_0,$$

in which as usual g denotes a value of f_y. The second correction C_1 obtained by applying equation 8.1 again is

$$C_1 = \frac{h}{2} C_0' = \frac{hg}{2} C_0,$$

and, in general, if C_m denotes the mth correction to y_{n+1}, we have

$$C_m = (hg/2)^m C_0.$$

This relationship shows that the process is convergent if $\left|\dfrac{hg}{2}\right| < 1$. Accordingly we define a quantity θ by the equation

$$\theta = hg/2,$$

and call θ the **convergence factor.**

Ordinarily we desire the convergence to be so rapid that one application of steps 3 and 4 will suffice. Hence, the step interval h must be chosen small enough that the correction

$$C_1 = \theta C_0$$

is negligible to the number of places retained. An examination of computation 2 reveals that this condition is satisfied throughout for three-decimal-place accuracy in this particular problem.

10. Check column. Practical estimation of truncation error.

It is recommended that the first correction C_0 be recorded for each step of computation 2. Any sudden fluctuation in the recorded values suggests an arithmetical mistake in either that line or the line above, and the work should be checked. If in the completed computation the values of C_0 vary smoothly, we are reasonably confident that no serious accidental mistake has been made. Obviously, however, this check may fail to reveal systematic errors and errors in the first two lines of the computation.

The column of C_0 is also used for the even more important purpose of estimating the approximate magnitude of the truncation error at each step. We recall that C_0 is the difference between the value of y_{n+1} given by equation 8.2, which has a truncation error

$$T' = h^3 y'''(s)/3, \qquad x_{n-1} < s < x_{n+1},$$

and the value given by equation 8.1, which has a truncation error

$$T = - h^3 y'''(t)/12, \qquad x_n < t < x_{n+1}.$$

Then, if we neglect other sources of error,

$$- C_0 = T - T' = - \frac{h^3 y'''(t)}{12} - \frac{h^3 y'''(s)}{3}$$

$$= - \frac{5h^3}{12} y'''(r), \qquad x_{n-1} < r < x_{n+1},$$

if y''' is continuous. Now

$$T + \frac{C_0}{5} = \frac{h^3}{12} [y'''(r) - y'''(t)],$$

and, if we assume the existence of a bounded fourth derivative, $y^{(4)}$, where $\left| y^{(4)} \right| < M$, we have

$$T + \frac{C_0}{5} = \frac{h^3}{12} [y'''(r) - y'''(t)] = \frac{h^3}{12} y^{(4)}(u)(r - t)$$

where u lies between r and t. Then

$$\left| T + \frac{C_0}{5} \right| < \frac{h^3}{12} M |r - t| < \frac{h^4 M}{6},$$

since $|r - t| < 2h$. Hence,

(10.1) $$T = -C_0/5,$$

with an error of the order of h^4. This formula 10.1 serves the very practical purpose of furnishing an estimate of the truncation error at each step. Formula 10.1 can have a significant error only when the fourth derivative is large, i.e., when C_0 itself is changing very rapidly. Hence we have ample warning if it is not safe to use 10.1.

In computation 2 the quantities discussed in Articles 9 and 10, namely, C_0, g, θ, and T, vary rather slowly and remain within reasonable bounds. In contrast with computation 2, computation 3 exhibits a rapid increase in all of these quantities, together with the expected loss of accuracy. This example employs the same differential equation as before,

$$y' = -2xy^2,$$

but with the initial value $y = -1$ at $x = 0$.

Method II   Computation 3

| Differential equation $y' = -2xy^2$ | | | | $(y = -1$ at $x = 0)$ | | | | $h = 0.1$ |
x	y	y'	C_0	T	g	θ	$(x^2 - 1)^{-1}$	Error
0.0	$-1.$	0					$-1.$	0
0.1	-1.010	-0.204			0.40	0.02	-1.010	0
0.2	-1.042	-0.434	-0.001		0.84	.04	-1.042	0
0.3	-1.100	-0.726	$-.003$		1.32	.07	-1.099	-1
0.4	-1.193	-1.139	$-.006$	0.001	1.90	.10	-1.190	-3
0.5	-1.338	-1.790	$-.010$	.002	2.68	.13	-1.333	-5
0.6	-1.572	-2.965	$-.021$	.004	3.78	.19	-1.563	-9
0.7	-1.981	-5.494	$-.050$	.010	5.55	.27	-1.961	-20
0.8	-2.826	-12.778	$-.155$	.031	9.04	.45	-2.778	-48
0.9	-5.317	-50.887	$-.780$	.156	19.14	.96	-5.263	-54
1.0						∞		

The explicit solution is $y = (x^2 - 1)^{-1}$, and its values are tabulated together with the difference between the computed and the correct solutions. Though the error of the computed solution is rather glaring, it is obvious from the size of C_0, T, and g that we have ample warning of the error to be expected.

In computation 3 we carried the work along for several steps beyond the point where we had any reason to expect three place accuracy. This was done just to show what happens. In a practical problem where three-decimal-place accuracy is needed we should shorten the interval h as soon as C_0 exceeds 0.0025. Also in computation 3 we did not in any step use formula 8.1 a second time, even after θ reached a size where such repetition was called for. In fact, for $x = 0.9$ where $\theta = 0.96$ it is evident that very many repetitions would be required.

To emphasize the importance of some kind of check we work the same example by Method I, as shown in computation 4. Here, as far as the computer can tell

Method I   Computation 4

Differential equation $y' = -2xy^2$ $(y = -1$ at $x = 0)$ $h = 0.1$

x	y	y'
0.0	-1	0
0.1	-1.010	$-\ 0.204$
.2	-1.041	$-\ .433$
.3	-1.097	$-\ .722$
.4	-1.185	$-\ 1.124$
.5	-1.322	$-\ 1.748$
.6	-1.535	$-\ 2.827$
.7	-1.887	$-\ 4.985$
.8	-2.532	-10.258
.9	-3.939	-27.929
1.0	-8.118	

from internal evidence, everything is proceeding merrily, and he has no suspicion that at $x = 1.0$ his error is actually infinite!

11. Error of Method II

Just as in the case of Method I we shall evade the difficult problem of round-off error and consider only the cumulative effects of truncation and inherited errors. The value finally recorded for y in Method II is calculated by equation 8.1, and we accordingly have

$$y_{n+1} - y_n = \frac{h}{2}(y_{n+1}' + y_n'),$$

$$z_{n+1} - z_n = \frac{h}{2}(z_{n+1}' + z_n') + T_n,$$

where, as before, y is the computed and z the true value of y. Then setting $e_n = y_n - z_n$ and subtracting, we get approximately

$$e_{n+1} - e_n = \frac{h}{2}g_{n+1}e_{n+1} + \frac{h}{2}g_ne_n - T_n,$$

where $g_n = f_y$ at x_n. This equation connecting e_{n+1} with e_n gives the law of growth of the error. We get a pretty good idea of the character of e_n by examining the case where T_n and g_n are constant with respect to n. For then we have the simple equation

$$e_{n+1} = \left(\frac{1 + hg/2}{1 - hg/2}\right)e_n - \frac{T}{1 - hg/2},$$

and we readily verify that this relationship is satisfied by

(11.1) $$e_n = Aq^n + T/hg,$$

where A is a constant, and

$$q = \frac{1 + hg/2}{1 - hg/2}.$$

Since $hg/2 = \theta$, the convergence factor, we must have $|\theta| < 1$, and hence q must be positive. It is clear that, if g is positive, q is greater than 1, and q^n increases exponentially as n increases, whereas, if g is negative, q is less than 1 and q^n decreases exponentially as n increases.

The situation is clearly exhibited by comparing computations 2 and 3, for in the former $g = -4xy$ and is negative since x and y are both positive, but in the latter g is positive since y is negative. In computation 2 the error remains bounded while in 3 the error increases rapidly. Both phenomena are in accord with what equation 11.1 leads us to expect.

The process of solution described in Method II employs formulas to predict and to check which have error terms of degree 3 in h. In order to secure greater accuracy per step, it is often desirable to use more accurate formulas. Methods exactly like Method II in spirit but based on formulas with error terms of degree 5 and degree 7 are presented in Articles 28 and 30, respectively. The reader who is interested in these methods may proceed directly to those articles.

EXERCISES

Examine the error of Method II in connection with each of the differential equations in the exercises of Article 6.

3

ANALYTICAL FOUNDATIONS

Two elementary methods for the numerical solution of differential equations were presented in Chapter 2. A large number of methods, similar in spirit but differing in various details of procedure, in attainable accuracy, and in complexity, have been developed by different writers and in almost bewildering variety. Since the theoretical foundation on which a large class of these methods rests is the same for all methods of the class, we shall devote this chapter to foundations and leave details to later chapters.

12. Formal solution by Taylor's series

In the differential equation

$$(12.1) \qquad w' = f(z, w)$$

where the prime denotes differentiation with respect to z we suppose that $f(z, w)$ is an analytic function of the two complex variables z and w in the region R defined by the inequalities

$$\left| z - z_0 \right| \leqq A, \qquad \left| w - w_0 \right| \leqq B.$$

Then in R the partial derivatives of $f(z, w)$ exist, and by successive differentiations we obtain from formula 12.1 the equations

$$
\begin{aligned}
w' &= f, \\
(12.2) \qquad w'' &= f_z + w'f_w, \\
w''' &= f_{zz} + 2w'f_{zw} + (w')^2 f_{ww} + w''f_w,
\end{aligned}
$$

$$\cdots \cdots \cdots \cdots \cdots \cdots \cdots \cdots \cdots .$$

Derivatives of w with respect to z are denoted by primes, and partial derivatives of f are expressed by means of subscripts. We assume that $w = w_0$ when $z = z_0$. By substituting these values in the first of equations 12.2, we obtain w_0' (i.e., the value of w' at $z = z_0$), then, putting z_0, w_0, w_0' on the right in the second equation, we find w_0''.

Proceeding step by step in this manner, we obtain the sequence

$$w_0, \ w_0', \ w_0'', \ w_0''', \ \cdots \cdot$$

Once these are determined, it is a simple matter to set up the formal Taylor's series for w,

$$(12.3) \qquad w = w_0 + w_0'(z - z_0) + \frac{w_0''}{2!}(z - z_0)^2 + \cdots \cdot$$

It remains to determine a lower bound for the radius of convergence of the series, and next to show that in the region where the series converges it defines a solution of the differential equation.

13. Radius of convergence

The equation $\left| z - z_0 \right| = A$ defines a circle C_1 in the z plane with center at z_0 and radius A. Similarly the equation $\left| w - w_0 \right| = B$ defines a circle C_2 in the w plane. If t_1 is a variable on C_1, t_2 a variable on C_2 and, if z is inside C_1 and w inside C_2, we can express $f(z, w)$ by Cauchy's integral formula as follows:

$$f(z, w) = \frac{1}{(2\pi i)^2} \int_{C_1} \frac{dt_1}{t_1 - z} \int_{C_2} \frac{f(t_1, t_2) dt_2}{t_2 - w}.$$

Differentiating this equation m times with respect to z, and n times with respect to w, we get

$$(13.1) \qquad \frac{\partial^{m+n} f}{\partial z^m \, \partial w^n} = \frac{m! n!}{(2\pi i)^2} \int_{C_1} \frac{dt_1}{(t_1 - z)^{m+1}} \int_{C_2} \frac{f(t_1, t_2) dt}{(t_2 - w)^{n+1}}.$$

If we assume that M is the maximum of the absolute value of $f(t_1, t_2)$ for t_1 on C_1 and t_2 on C_2, we can derive from equation 13.1 the inequality

$$(13.2) \qquad \left| \frac{\partial^{m+n} f}{\partial z^m \, \partial w^n} \right|_0 \leqq \frac{m! n! M}{A^m B^n},$$

where the left-hand member is evaluated at (z_0, w_0).

Let us also consider the differential equation

$$(13.3) \qquad \frac{dW}{dZ} = \frac{M}{\left(1 - \dfrac{Z - z_0}{A}\right)\left(1 - \dfrac{W - w_0}{B}\right)} = F(Z, W).$$

Successive differentiations show that

$$\frac{\partial^{m+n} F}{\partial Z^m \, \partial W^n} = \frac{m! n! M}{A^m B^n},$$

at $Z = z_0$, $W = w_0$, and this together with inequality 13.2 gives

$$(13.4) \qquad \left| \frac{\partial^{m+n} f}{\partial z^m \, \partial w^n} \right|_0 \leqq \left(\frac{\partial^{m+n} F}{\partial Z^m \, \partial W^n} \right)_0,$$

where we note that the right member is real and positive. In particular for $m = n = 0$

$$|f(z_0, w_0)| \leqq F(z_0, w_0),$$

whence

$$(13.5) \qquad |w_0'| \leqq W_0',$$

where we use primes on w to denote differentiation with respect to z and primes on W to denote differentiation with respect to Z. From the equations

$$w'' = f_z + f_w w'$$

and

$$W'' = F_Z + F_W W'$$

it is easy to show with the aid of equations 13.4 and 13.5 that W_0'' is real and positive and that

$$|w_0''| \leqq W_0''.$$

By successive differentiations of the equations 12.1 and 13.3 and comparison of the results, we prove in a similar manner that

$$(13.6) \qquad |w_0^{(k)}| \leqq W_0^{(k)}$$

for all positive integral values of k. If ρ is the radius of convergence of equation 12.3 in the z plane and P the radius of convergence of the corresponding W series in the Z plane, we see from equation 13.6 that

$$\rho \geqq P.$$

Now the explicit solution of 13.3 with the given initial values is

$$W = w_0 + B \left[1 - \sqrt{1 + \frac{2AM}{B} \ln \left(1 - \frac{Z - z_0}{A} \right)} \right].$$

From this explicit formula for W we find that the singular point in the Z plane nearest to $z = z_0$ is located by solving the equation

$$1 + \frac{2AM}{B} \ln \left(1 - \frac{Z - z_0}{A} \right) = 0,$$

and from this follows

$$P = |Z - z_0| = A(1 - e^{-B/2AM}).$$

Hence finally we have the result that

$$\rho \geqq A(1 - e^{-B/2AM}).$$

Within its circle of convergence, the series 12.3 defines a single-valued analytic function of z, which takes the value w_0 when $z = z_0$ and satisfies the differential equation

$$(13.7) \qquad w' = f(z, w).$$

We can verify this last statement by expanding both members of equation 13.7 in a series of powers of $(z - z_0)$, and using equations 12.2 for $z = z_0$, $w = w_0$,

14. Analytic continuation. Method III

The Taylor's series 12.3 provides a theoretical solution of the differential equation for z in the circle of convergence. This series also supplies an approximate numerical solution in a more restricted region R' defined by $|z - z_0| < \rho'$, where ρ' is considerably less than ρ. For we can explicitly calculate as many terms of the series 12.3 as we are willing to pay for in terms of labor, and then take $|z - z_0|$ so small that the remainder is negligible to the degree of accuracy required. For values of z thus limited we have a numerical solution.

It often happens that the circle of convergence does not cover the entire set of values of z for which a solution is wanted, and even more frequently we find that the numerical solution falls far short of covering the desired values of z. In such cases the familiar process of analytic continuation enables us by successive steps to cover an open region in which the solution is single-valued and analytic. In most practical problems the region for which a solution is wanted is a segment of the real axis, but the method is obviously not restricted to reals, nor (in the case of multiple-valued solutions) to a single sheet of the Riemann surface.

The analytic continuation is effected by the following steps. By the method of Article 12 we calculate the Taylor's series 12.3. For a value of z, $z = z_1$, where the series converges with sufficient rapidity for the desired accuracy, we compute from the series the corresponding w_1. Using z_1 and w_1 as initial values, we obtain a new Taylor's series

$$(14.1) \qquad w = w_1 + w_1'(z - z_1) + \frac{w_1''}{2!}(z - z_1)^2 + \cdots.$$

From this a new value w_2 of w is found for $z = z_2$, z_2 being chosen where equation 14.1 converges. With z_2, w_2, as initial values another

Taylor's series is computed, and this chain of steps is carried on until either the region where we want a solution is covered or the process is halted by singularities.

Example 1.   To facilitate the explanation let us use the differential equation

$$w' = 1 - w^2,$$

for which we can obtain the explicit solution and can locate the singular points. The explicit solution with initial values $(0, 0)$ is

$$w = \tanh z,$$

and this solution has poles at $z = \pm \left(\dfrac{2n + 1}{2} \right) \pi i.$

For this problem the analytic continuation along the axis of reals could be made by steps as indicated in Fig. 10, where z_0, z_1, z_2, $\cdots$ are the points for which

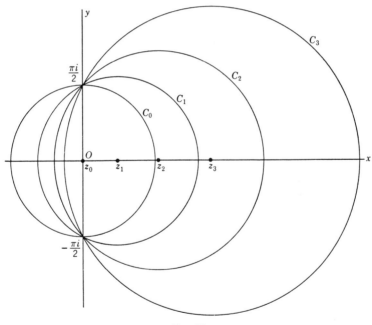

Fig. 10

the successive Taylor's series are formed and are the centers of the successive circles of convergence C_0, C_1, C_2, $\cdots$. Figure 10 was drawn on the assumption that each new point z_n is taken a distance from the preceding z_{n-1} equal to half the radius of C_{n-1}. Although such large steps are theoretically possible, they are often unsuited to actual computation, because adequate accuracy would require a large number of terms of the Taylor's series.

Example 2. Let us take the same problem as in Example 1, but assume that we do not know anything about the explicit solution and must rely on numerical calculation alone.

Differentiation of the differential equation gives

$$w' = 1 - w^2,$$

$$w'' = -2ww',$$

(14.2) $$w''' = -2(w')^2 - 2ww'',$$

$$w^{(4)} = -6w'w'' - 2ww'''$$

$$w^{(5)} = -6(w'')^2 - 8w'w''' - 2ww^{(4)}.$$

The series at the ith step will be

$$w = w_i + w_i'(z - z_i) + \frac{w_i''}{2!}(z - z_i)^2 + \cdots .$$

In this example it is convenient to make the step length $z_{i+1} - z_i$ constant and equal to 0.3.   Then

(14.3) $$w_{i+1} = w_i + (0.3)w_i' + (0.045)w_i'' + (0.0045)w_i'''$$
$$+ (0.000\ 3375)w_i^{(4)} + (0.0000\ 2025)w_i^{(5)} + \cdots .$$

The computational routine consists of the following steps:

1. Using the initial values, compute w', w'', $\cdots$ from 14.2.
2. Insert the values from step 1 into 14.3 and calculate w_1.
3. With w_1 known, calculate the derivatives from 14.2.
4. Put these in 14.3 to obtain w_2.

The steps are repeated with advancing subscripts as far as may be necessary or until the process hits a snag. In this particular problem we may infer from Fig. 10 that no trouble will be encountered since we are moving farther from the singular points with each step.

Method III   Computation 5

Differential equation $w' = 1 - w^2$			$(w = 0, z = 0)$			$h = 0.3$
Coef.	1	0.300 000	0.045 000	0.004 500	0.000 337	0.000 020
z	w	w'	w''	w'''	$w^{(4)}$	$w^{(5)}$	E

z	w	w'	w''	w'''	$w^{(4)}$	$w^{(5)}$	E
0	0	1	0	−2	0	16	
0.3	0.29132	0.91513	−0.5332	−1.364	3.72	6.1	0.00001
.6	.53710	.71152	− .7643	− .192	3.47	−6.1	.00005
.9	.71635	.48684	− .6975	.525	1.29	−6.8	.00005
1.2	.83368	.30498					.00003

No explanations are required beyond the observation that the entry in the w column is computed from the preceding line by means of equation 14.3 and the entries in the w', w'', $\cdots$, $w^{(5)}$ columns are obtained in sequence from equations 14.2. Since in this illustrative example we know the explicit solution, it is of interest to tabulate the errors of the computed values. These are shown in the column headed E.

Greater accuracy can be secured with the same value of h by the use of more terms in the Taylor's series. But equations 14.2 become progressively more complex, and the labor grows more excessive as the number of terms increases. Greater accuracy can be secured with the same number of terms if we shorten the interval h. But again the price must be paid, this time in a greater number of steps to cover the same range. The computer has to decide on the number of terms and length of interval best suited to the particular differential equation and to the range and accuracy demanded for the solution. It is often advisable to change h in the course of the computation.

Actually the method of analytic continuation is rarely used for an extended interval because the labor required almost always exceeds that of other methods to be explained later. However, it is a common practice to employ one or two steps of this procedure in order to get other methods started.

EXERCISES

1. Solve $y' = xy + 1$, $y = 0$, at $x = 0$. Use $h = 0.3$, and carry to $x = 1.2$.
2. Solve $y' = 1 + y/(x + 1)$, $y = 0$, at $x = 0$. Use $h = 0.1$, and carry to $x = 1$.
3. Using the remainder term for the truncated Taylor's series employed in exercises 1 and 2, estimate the error in the result found for exercises 1 and 2.
4. About how many terms of the Taylor's series should be used in exercises 1 and 2 to secure four-figure accuracy?

15. Successive substitutions

The method of analytic continuation requires $f(z, w)$ to be analytic in z and w or at least to have continuous partial derivatives up to an order sufficient to form the needed partial sum of Taylor's series. The method of *successive substitutions*, generally known as Picard's method [152, 153, 154, 7], imposes much milder restrictions on the differential equation.

We seek an approximate solution of the equation

$$y' = f(x, y)$$

with initial values $x = a$, $y = b$. A sequence of functions $y_0(x)$, $y_1(x)$, $y_2(x)$, $\cdots$ is obtained satisfying the recurrence equations

$$y_{n+1}' = f(x, y_n)$$

and such that $y_n(a) = b$. These may be expressed in the equivalent form

(15.1) $$y_{n+1} = b + \int_a^x f(s, y_n(s))ds.$$

It will be shown in the next article that with a few limitations on $f(x, y)$ the sequence $y_0, y_1, y_2, \cdots$ will converge to a solution of the differential equation.

Example 1. We apply the method to the equation

$$y' = x - y$$

with initial values $x = 0$, $y = 1$.

Let us choose $y_0(x) = 1$. Then equation 15.1 gives

$$y_1 = 1 + \int_0^x (s - 1)ds$$

$$= 1 - x + x^2/2.$$

Likewise,

$$y_2 = 1 + \int_0^x \left(s - 1 - \frac{s^2}{2} + s \right) ds$$

$$= 1 - x + x^2 - x^3/6.$$

Similarly we obtain

$$y_3 = 1 - x + x^2 - \frac{x^3}{3} + \frac{x^4}{24}$$

$$y_4 = 1 - x + x^2 - \frac{x^3}{3} + \frac{x^4}{12} - \frac{x^5}{120}$$

and, in general,

$$y_n = 1 - x + \frac{2x^2}{2!} - \frac{2x^3}{3!} + \cdots + (-1)^n \frac{2x^n}{n!} - (-1)^n \frac{x^{n+1}}{(n+1)!}.$$

In this simple case we readily find a finite expression for the limit of y_n, namely,

$$y = 2e^{-x} - (1 - x),$$

which is the required solution.

This example is too simple to give a real idea of the process. A more realistic picture is furnished by

Example 2. Solve the equation

$$y' = x^2 - y^2$$

with initial values $x = -1$, $y = 0$.

Let $y_0 = 0$. Then

$$y_1 = \int_{-1}^x s^2 \, ds = \frac{x^3}{3} + \frac{1}{3},$$

$$y_2 = \int_{-1}^x \left[s^2 - \frac{1}{9} (s^6 + 2s^3 + 1) \right] ds,$$

$$y_2 = \frac{1}{126}(33 - 14x + 42x^3 - 7x^4 - 2x^7),$$

$$y_3 = \int_{-1}^{x}\left[s^2 - \frac{1}{15\,876}(33 - 14s + 42s^3 - 7s^4 - 2s^7)^2\right]ds,$$

$$y_3 = \frac{1}{2\,619\,540}(724\,780 - 179\,685x + 76230x^2 + 862\,400x^3$$
$$- 114\,345x^4 + 54054x^5 - 5390x^6 - 41580x^7$$
$$+ 14850x^8 - 1925x^9 + 2520x^{11} - 385x^{12} - 44x^{15}).$$

It is apparent that the labor of obtaining successive approximations increases rapidly. Evidently y_4 will be a polynomial of degree 31, y_5 of degree 63, etc.

The third example illustrates an additional difficulty.

Example 3. Consider the differential equation

$$y' = \sin x + \cos y$$

with $y = 0$ when $x = 0$.

Here we take $y_0 = 0$ and have $y_1' = \sin x + 1$, whence

$$y_1 = 1 + x - \cos x.$$

Next we have to solve

$$y_2' = \sin x + \cos (1 + x - \cos x).$$

At this stage, however, progress is thwarted because the expression on the right cannot be integrated by elementary methods. In fact, Example 3 illustrates the situation usually encountered in successive substitutions—sooner or later a point is reached where the integrations cannot be explicitly carried out.

The method of successive substitutions as here presented is seldom used for the practical solution of differential equations because of the difficulties shown by Examples 2 and 3. However, with suitable modifications this method furnishes the theoretical justification of the purely numerical methods given in the next chapter. Hence we shall examine its convergence in some detail.

16. Convergence of successive substitutions

Let

(16.1) $y' = f(x, y).$

where the function $f(x, y)$ is continuous in x and y in a region R defined by $|x - x_0| < A$, $|y - y_0| < B$. Assume that $|f(x, y_0)|$ is less than M, and that $f(x, y)$ satisfies the Lipschitz condition

(16.2) $|f(x, y_i) - f(x, y_j)| < L|y_i - y_j|$

for (x, y_i) and (x, y_j) in R. The quantities A, B, M, L are understood to be positive constants.

In successive substitutions the values of y_1, y_2, y_3, $\cdots$ are given by the sequence of equations

$$y_1 = y_0 + \int_{x_0}^{x} f(x, y_0)dx,$$

$$y_2 = y_0 + \int_{x_0}^{x} f(x, y_1)dx,$$

and, in general,

(16.3)
$$y_{n+1} = y_0 + \int_{x_0}^{x} f(x, y_n)dx.$$

We take the difference of two successive equations of type 16.3 and obtain the relation

$$y_{n+1} - y_n = \int_{x_0}^{x} [f(x, y_n) - f(x, y_{n-1})]dx.$$

From this it follows that

$$\left| y_{n+1} - y_n \right| \leqq \int_{x_0}^{x} \left| f(x, y_n) - f(x, y_{n-1}) \right| dx,$$

and inequality 16.2 applied to the right-hand member gives

(16.4)
$$\left| y_{n+1} - y_n \right| \leqq \int_{x_0}^{x} L \left| y_n - y_{n-1} \right| dx.$$

For $n = 0$ inequality 16.4 becomes

$$\left| y_1 - y_0 \right| \leqq \int_{x_0}^{x} \left| f(x, y_0) \right| dx,$$

and, since $\left| f(x, y_0) \right|$ is less than M, we have

$$\left| y_1 - y_0 \right| \leqq \int_{x_0}^{x} M \, dx,$$

which gives

(16.5)
$$\left| y_1 - y_0 \right| \leqq M(x - x_0), \quad \text{for} \quad x > x_0.$$

Setting $n = 1$ in 16.4 and using inequality 16.5, we get

$$\left| y_2 - y_1 \right| \leqq ML \frac{(x - x_0)^2}{2!}.$$

By successive substitutions into 16.4 we arrive at the general inequality

(16.6)
$$\left| y_n - y_{n-1} \right| \leqq ML^{n-1} \frac{(x - x_0)^n}{n!}.$$

If for brevity we now write $x - x_0 = h$ and make use of the fact that the absolute value of a sum does not exceed the sum of the absolute values, we find that

$$(16.7) \quad |y_n - y_0| \leq Mh + MLh^2/2! + ML^2h^3/3!$$
$$+ \cdots + ML^{n-1}h^n/n!.$$

Since the right-hand member of equation 16.7 is less than $(M/L)(e^{Lh} - 1)$, it follows that

$$|y_n - y_0| \leq (M/L)(e^{Lh} - 1).$$

The value of y_n will lie within the prescribed limits $|y - y_0| < B$ provided

$$B \geq (M/L)(e^{Lh} - 1).$$

Hence we choose h so that

$$(16.8) \qquad h \leq \frac{1}{L} \ln \left(1 + \frac{LB}{M} \right).$$

With this limitation on h we are assured that y_n will not get out of the region in which 16.2 holds, no matter how large n may be taken. This being the case, we may allow n to become infinite.

If x is in the interval

$$x_0 \leq x \leq x_0 + h,$$

we have from inequality 16.6 that $y_n - y_{n-1}$ is less in absolute value than

$$ML^{n-1}h^n/n!,$$

which is the nth term of a convergent series of positive constants. It follows that the series

$$y_0 + (y_1 - y_0) + (y_2 - y_1) + (y_3 - y_2) + \cdots$$

is absolutely convergent, and therefore its partial sums $y_0, y_1, y_2, \cdots$ converge to a continuous function $y(x)$ in the interval x_0 to $x_0 + h$.

Similarly by differentiating equation 16.3 we get

$$(16.9) \qquad y_{n+1}' = f(x, y_n),$$

and this equation, together with 16.2, shows that

$$(16.10) \qquad |y_{n+1}' - y_n'| \leq L|y_n - y_{n-1}|.$$

Since the series whose general term is $y_n - y_{n-1}$ was just seen to be absolutely convergent, the same is true of the series

$$(y_1' - y_0') + (y_2' - y_1') + (y_3' - y_2') + \cdots.$$

Hence we can conclude that the sequence y_1', y_2', y_3', $\cdots$ converges uniformly to a continuous function which is the derivative of y. That is, the limit of the derivatives y_i' is actually the derivative of the limit of the y_i. Finally, from 16.3, we see that

$$y_{n+1}' = f(x, y_n)$$

for all values of n. As n becomes infinite y_{n+1}' approaches y' and y_n approaches $y(x)$, both uniformly, so that in the limit

$$y' = f(x, y).$$

Hence the limit function $y(x)$ is a solution of equation 16.1.

17. Numerical integration

We saw in Article 15 that the method of successive substitutions frequently leads to insuperable difficulties of integration, and even when explicit integrations are possible the analytical work rapidly becomes excessively complicated. We can to a large extent evade both these difficulties by using numerical integration in place of formal integration. Accordingly, the remainder of this chapter will be devoted to the derivation and use of formulas for numerical integration.

For a more detailed discussion of the material of this article as well as Articles 19, 20, and 21 the reader may consult any one of a number of texts: e.g., Boole [15], Kowalewski [101], Milne [133], Milne-Thompson [137], von Sanden [175], Scarborough [180], Steffensen [195], or Whittaker and Robinson [219].

Suppose that $z = P_n(x)$ is a polynomial of degree n in the variable x. Then the derivative $z' = P_n'(x)$ is a polynomial of degree $n - 1$ which is completely determined if its n coefficients are given. These coefficients in turn are completely determined if the values of $P_n'(x)$ namely, z_0', z_1', $\cdots$, z_{n-1}', are known at n distinct points, x_0, x_1, $\cdots$, x_{n-1}. Then the quantity

$$z_k - z_j = \int_{x_j}^{x_k} P_n'(s)ds$$

can also be expressed in terms of the values z_0', z_1', $\cdots$, z_{n-1}'. One way by which we could express $z_k - z_j$ in terms of z_0', z_1', $\cdots$, z_{n-1}' would be to represent $P_n'(x)$ in terms of the z_i' $(i = 0, \cdots, n - 1)$ by Lagrange's interpolation formula and to integrate the resulting polynomial between the given limits. However, the same result can be reached with less trouble by the following reasoning.

We assume that the equation

$$(17.1) \qquad z_k - z_j = A_0 z_0' + A_1 z_1' + \cdots + A_{n-1} z_{n-1}'$$

is valid provided z is a polynomial of the form

$$(17.2) \qquad z = a_0 + a_1 x + a_2 x^2 + \cdots + a_n x^n,$$

no matter what values of the coefficients a_i are chosen. In particular then equation 17.1 holds if $z = x$, if $z = x^2$, $\cdots$, if $z = x^n$. Putting these n choices of z into equation 17.1 in succession, we obtain n equations that are just sufficient to determine the n unknown A's.

Example. If $n = 5$, obtain the value of $z_1 - z_0$ in terms of z_0', z_1', z_2', z_3', z_4', the values of x being equally spaced with interval h.

Here we set

$$(17.3) \qquad z_1 - z_0 = A_0 z_0' + A_1 z_1' + A_2 z_2' + A_3 z_3' + A_4 z_4'$$

with $x_0 = 0$, $x_1 = h$, $x_2 = 2h$, $x_3 = 3h$, $x_4 = 4h$. If now we let $z = x$, x^2, x^3, x^4, x^5 in turn we have

$$h = A_0 + A_1 + A_2 + A_3 + A_4,$$

$$h^2 = 2h(A_1 + 2A_2 + 3A_3 + 4A_4),$$

$$h^3 = 3h^2(A_1 + 4A_2 + 9A_3 + 16A_4),$$

$$h^4 = 4h^3(A_1 + 8A_2 + 27A_3 + 64A_4),$$

$$h^5 = 5h^4(A_1 + 16A_2 + 81A_3 + 256A_4).$$

When these equations are solved and the values of the A's are put back into equation 17.3, the final formula is

$$z_1 - z_0 = \frac{h}{720} (251 z_0' + 646 z_1' - 264 z_2' + 106 z_3' - 19 z_4').$$

In a similar manner we derive formulas for $z_2 - z_0$, $z_3 - z_0$, $z_4 - z_0$. The complete set is

$$z_1 - z_0 = \frac{h}{720} (251 z_0' + 646 z_1' - 264 z_2' + 106 z_3' - 19 z_4'),$$

$$z_2 - z_0 = \frac{h}{90} (29 z_0' + 124 z_1' + 24 z_2' + 4 z_3' - z_4'),$$

$$(17.4)$$

$$z_3 - z_0 = \frac{h}{80} (27 z_0' + 102 z_1' + 72 z_2' + 42 z_3' - 3 z_4'),$$

$$z_4 - z_0 = \frac{4h}{90} (7 z_0' + 32 z_1' + 12 z_2' + 32 z_3' + 7 z_4').$$

EXERCISES

Derive the last three formulas above.

The four formulas 17.4 just derived are true only if z is any polynomial of degree not exceeding 5; yet in actual usage they are applied to functions that are not polynomials at all. The justification of this usage rests on the fact that we can usually take h small enough to make these formulas as nearly exact as we wish, even when z is not a polynomial. In fact, it can be shown that, if $y(x)$ has a continuous derivative of order 7 in the interval $z_0 \le x \le x_4$, we have in place of formula 17.4 the four similar formulas:

$$(17.5) \quad y_1 - y_0 = \frac{h}{720} (251y_0' + 646y_1' - 264y_2'$$

$$+ 106y_3' - 19y_4') + \frac{27h^6 y^{(6)}(s)}{1440},$$

$$(17.6) \quad y_2 - y_0 = \frac{h}{90} (29y_0' + 124y_1' + 24y_2' + 4y_3' - y_4')$$

$$+ \frac{16h^6 y^{(6)}(s)}{1440},$$

$$(17.7) \quad y_3 - y_0 = \frac{h}{80} (27y_0' + 102y_1' + 72y_2' + 42y_3' - 3y_4')$$

$$+ \frac{27h^6 y^{(6)}(s)}{1440},$$

$$(17.8) \quad y_4 - y_0 = \frac{4h}{90} (7y_0' + 32y_1' + 12y_2' + 32y_3' + 7y_4')$$

$$- \frac{8h^7 y^{(7)}(s)}{945}.$$

In each remainder term the quantity s is some value between x_0 and x_4.

This particular set of formulas makes use of five consecutive values of $y'(x)$. In like manner other sets may be obtained using 2, 3, 4, 6, 7, etc., consecutive values of $y'(x)$. Several such sets are tabulated in Article 19.

18. Successive substitutions. Method IV

Before we go on to derive a more extensive set of formulas for numerical integration let us see how those obtained in Article 17 may be employed in the solution of a differential equation.

Example. Solve the equation

$$y' = \sin x + \cos y$$

with $y = 0$ when $x = 0$.

This is Example 3 of Article 15 in which the explicit integrations could not be carried out. Before starting to solve the problem numerically we should select a value of h which will guarantee the desired accuracy. However, since the question of accuracy will be treated in later chapters, let us arbitrarily take $h = 0.2$ and carry the computation to five decimal places without justifying this decision at

Method IV   Computation 6

| Differential equation $y' = \sin x + \cos y$ | | | $(y = 0, x = 0)$ | | |
x	y	y'	$\sin x$	$\cos y$	Approximation
0	0	1	0	1	
0.2	0.2	1.17874	0.19867	0.98007	
.4	.4	1.31048	.38942	.92106	$y_0(x)$
.6	.6	1.38998	.56464	.82534	
.8	.8	1.41407	.71736	.69671	
0	0	1			
0.2	0.21860	1.17487		.97620	
.4	.46836	1.28173		.89231	$y_1(x)$
.6	.73931	1.30357		.73893	
.8	1.02065	1.24017		.52281	
0	0	1			
0.2	0.21839	1.17492		.97625	
.4	.46538	1.28307		.89365	$y_2(x)$
.6	.72538	1.31289		.74825	
.8	.98108	1.27348		.55612	
0	0	1			
0.2	0.21840	1.17492			
.4	.46547	1.28303		.89361	$y_3(x)$
.6	.72636	1.31224		.74760	
.8	.98596	1.26942		.55206	
0	0	1			
0.2	0.21840	1.17492			
.4	.46547	1.28303			$y_4(x)$
.6	.72631	1.31227		.74763	
.8	.98552	1.26979		.55243	
0	0	1			
0.2	0.21840	1.17492			
.4	.46547	1.28303			$y_5(x)$
.6	.72631	1.31227			
.8	.98555	1.26976			

present. The fact that $y' = 1$ when $x = 0$ suggests that we may take $y_0(x) = x$. The completed calculation shown above is effected by the following steps.

1. Having selected $y_0(x) = x$, we calculate $y'(x)$ for $x = 0$, 0.2, 0.4, 0.6, 0.8, by substitution into the differential equation. These values appear in the first group of five lines under y'.

2. Using formulas 17.5 to 17.8, we obtain the values of $y(x)$ and record them in the second group of five lines under y.

3. The values of y' for the new values of y are obtained from the differential equation as before and recorded under y' in the second group of five lines.

4. Step 2 is applied to these, and this closed ring of operations is repeated until no further change takes place.

The process has carried us as far as $x = 0.8$. If still further values are needed, we could use the last line of this computation as the first line of a new computation which would carry us on to $x = 1.6$, and repetitions of this procedure would take us as far as we wished to go.

Actually this method of continuing the computation is highly inefficient and is not recommended. In the next chapter we develop methods that make more effective use of the information already obtained.

On the other hand, Method IV affords an effective procedure for securing several initial values which most methods require for starting. The following modification, called Method IVa, is usually more rapidly convergent than Method IV and provides one of the best methods for starting. The underlying idea is to compute values symmetrically on both sides of the initial point. The formulas for the case of five points are:

$$(18.1) \quad y_{-2} - y_0 = -\frac{h}{90}(29y_{-2}' + 124y_{-1}' + 24y_0' + 4y_1' - y_2')$$
$$+ \frac{41h^6 y^{(6)}}{3600},$$

$$(18.2) \quad y_{-1} - y_0 = -\frac{h}{720}(-19y_{-2}' + 346y_{-1}' + 456y_0' - 74y_1'$$
$$+ 11y_2') + \frac{11h^6 y^{(6)}}{1440},$$

$$(18.3) \quad y_1 - y_0 = \frac{h}{720}(11y_{-2}' - 74y_{-1}' + 456y_0' + 346y_1'$$
$$- 19y_2') + \frac{11h^6 y^{(6)}}{1440},$$

$$(18.4) \quad y_2 - y_0 = \frac{h}{90}(-y_{-2}' + 4y_{-1}' + 24y_0' + 124y_1' + 29y_2')$$
$$+ \frac{41h^6 y^{(6)}}{3600}.$$

To illustrate the process we apply it to the differential equation $y' = x^2 - y^2$ with $y = 0$ when $x = -1$.

Method IVa   Computation 7

Differential equation $y' = x^2 - y^2$ $(y = 0, x = -1)$ $h = 0.1$

x	y	y'	
-1.2	-0.2	1.40	
-1.1	-0.1	1.20	
-1.0	0	1.00	1st approximation
-0.9	0.1	.80	
$-\ .8$	0.2	.60	
-1.2	-0.240	1.3824	
-1.1	-0.110	1.1979	
-1.0	0	1.0000	2d approximation
-0.9	0.090	.8019	
$-\ .8$	0.160	.6144	
-1.2	-0.23914	1.38281	
-1.1	-0.10995	1.19791	
-1.0	0	1.00000	3d approximation
$-\ .9$	0.09005	.80189	
$-\ .8$	0.16074	.61416	
-1.2	-0.23916	1.38280	
-1.1	-0.10995	1.19791	
-1.0	0	1.00000	4th approximation
$-\ .9$	0.09005	.80189	
$-\ .8$	0.16073	.61417	

EXERCISES

Using $h = 0.2$ and keeping four decimal places, apply Method IV to Example 2 of Article 15. How do the numerical results compare with the approximate analytical expression obtained in Article 15?

19. Formulas in terms of ordinates

Sets of formulas for numerical integration analogous to equations 17.5 through 17.8 are derived in the same manner and are assembled in Table I for convenience and reference.

Formulas for seven points, eight points, etc., become increasingly complicated and are seldom used in spite of the fact that, other things being equal, the accuracy increases as the number of points increases.

In numerical solution of differential equations much use is made of a set of symmetric formulas using an odd number of values of y' and characterized by the fact that the expression for $y_m - y_0$ does not contain y_m'. For our purposes the two most important of these are formulas 19.15 and 19.16. Finally we include Weddle's rule, formula 19.17, which is convenient for checking. It is the author's custom to apply Weddle's rule to each six lines of a computation as an independent check on the integrations.

Table I.   Quadrature Formulas in Terms of Ordinates

I. Three-point formulas

(19.1) $y_1 - y_0 = \dfrac{h}{12}(5y_0' + 8y_1' - y_2') + \dfrac{h^4 y^{(4)}(s)}{24}.$

(19.2) $y_2 - y_0 = \dfrac{h}{3}(y_0' + 4y_1' + y_2') - \dfrac{h^5 y^{(5)}(s)}{90}.$

II. Four-point formulas

(19.3) $y_1 - y_0 = \dfrac{h}{24}(9y_0' + 19y_1' - 5y_2' + y_3') - \dfrac{19h^5 y^{(5)}(s)}{720}.$

(19.4) $y_2 - y_0 = \dfrac{h}{3}(y_0' + 4y_1' + y_2' + 0) - \dfrac{h^5 y^{(5)}(s)}{90}.$

(19.5) $y_3 - y_0 = \dfrac{3h}{8}(y_0' + 3y_1' + 3y_2' + y_3') - \dfrac{3h^5 y^{(5)}(s)}{80}.$

III. Five-point formulas

(19.6) $y_1 - y_0 = \dfrac{h}{720}(251y_0' + 646y_1' - 264y_2' + 106y_3' - 19y_4')$
$$+ \frac{27h^6 y^{(6)}(s)}{1440}.$$

(19.7) $y_2 - y_0 = \dfrac{h}{90}(29y_0' + 124y_1' + 24y_2' + 4y_3' - y_4') + \dfrac{16h^6 y^{(6)}(s)}{1440}.$

(19.8) $y_3 - y_0 = \dfrac{h}{80}(27y_0' + 102y_1' + 72y_2' + 42y_3' - 3y_4') + \dfrac{27h^6 y^{(6)}(s)}{1440}.$

(19.9) $y_4 - y_0 = \dfrac{4h}{90}(7y_0' + 32y_1' + 12y_2' + 32y_3' + 7y_4') - \dfrac{8h^7 y^{(7)}(s)}{945}.$

IV. Six-point formulas

(19.10) $y_1 - y_0 = \dfrac{h}{1440}(475y_0' + 1427y_1' - 798y_2' + 482y_3'$
$$- 173y_4' + 27y_5') - \frac{863h^7 y^{(7)}(s)}{60{,}480}.$$

Table I.   (Continued)

(19.11) $y_2 - y_0 = \dfrac{h}{90} (28y_0' + 129y_1' + 14y_2' + 14y_3' - 6y_4' + y_5')$

$$- \frac{37h^7 y^{(7)}(s)}{3780}.$$

(19.12) $y_3 - y_0 = \dfrac{3h}{160} (17y_0' + 73y_1' + 38y_2' + 38y_3' - 7y_4' + y_5')$

$$- \frac{29h^7 y^{(7)}(s)}{2240}.$$

(19.13) $y_4 - y_0 = \dfrac{4h}{90} (7y_0' + 32y_1' + 12y_2' + 32y_3' + 7y_4' + 0)$

$$- \frac{8h^7 y^{(7)}(s)}{945}.$$

(19.14) $y_5 - y_0 = \dfrac{5h}{288} (19y_0' + 75y_1' + 50y_2' + 50y_3' + 75y_4' + 19y_5')$

$$- \frac{275h^7 y^{(7)}(s)}{12,096}.$$

V. Special formulas

(19.15) $y_4 - y_0 = \dfrac{4h}{3} (2y_1' - y_2' + 2y_3') + \dfrac{28h^5 y^{(5)}}{90}.$

(19.16) $y_6 - y_0 = \dfrac{3h}{10} (11y_1' - 14y_2' + 26y_3' - 14y_4' + 11y_5') + \dfrac{41h^7 y^{(7)}}{140}.$

(19.17) $y_6 - y_0 = \dfrac{3h}{10} (y_0' + 5y_1' + y_2' + 6y_3' + y_4' + 5y_5' + y_6')$

$$- \frac{h^7 y^{(7)}}{140} - \frac{9h^9 y^{(9)}}{1400}.$$

20. Formulas in terms of backward differences

The backward difference operator ∇ is defined by the equation

$$\nabla y_n = y_n - y_{n-1},$$

in which $y_0, y_1, \cdots, y_n$ denote values of the function y corresponding to equally spaced values of the independent variable x. Repeated applications of the difference operator are indicated by exponents, thus

$$\nabla^2 y_n = \nabla y_n - \nabla y_{n-1} = y_n - 2y_{n-1} + y_{n-2},$$

$$\nabla^3 y_n = \nabla^2 y_n - \nabla^2 y_{n-1} = y_n - 3y_{n-1} + 3y_{n-2} - y_{n-3},$$

and, in general,

(20.1) $\nabla^m y_n = \nabla^{m-1} y_n - \nabla^{m-1} y_{n-1} = \displaystyle\sum_{k=0}^{m} (-1)^k \binom{m}{k} y_{n-k}.$

Table II. Quadrature Formulas in Terms of Backward Differences

	y_n'	$\dfrac{\nabla y_n'}{2}$	$\dfrac{\nabla^2 y_n'}{12}$	$\dfrac{\nabla^3 y_n'}{24}$	$\dfrac{\nabla^4 y_n'}{720}$	$\dfrac{\nabla^5 y_n'}{1440}$	$\dfrac{\nabla^6 y_n'}{60480}$	$\dfrac{\nabla^7 y_n'}{120960}$	$\dfrac{\nabla^8 y_n'}{3628800}$	$\dfrac{\nabla^9 y_n'}{7257600}$
(20.3)	$y_n - y_{n+1} = h\,(-1$	-1	-5	-9	-251	-475	$-19\,087$	$-36\,799$	$-1\,070\,017$	$-2\,082\,753$ $-\cdots)$
(20.4)	$y_n - y_n = h\,(0$	0	0	0	0	0	0	0	0	0 $)$
(20.5)	$y_n - y_{n-1} = h\,(1$	-1	-1	-1	-19	-27	-863	$-1\,375$	$-33\,953$	$-57\,281$ $-\cdots)$
(20.6)	$y_n - y_{n-2} = h\,(2$	-4	$+4$	0	-8	-16	-592	$-1\,024$	$-26\,656$	$-46\,656$ $-\cdots)$
(20.7)	$y_n - y_{n-3} = h\,(3$	-9	$+27$	-9	-27	-27	-783	$-1\,215$	$-29\,889$	$-51\,138$ $-\cdots)$
(20.8)	$y_n - y_{n-4} = h\,(4$	-16	$+80$	-64	$+224$	0	-512	$-1\,024$	$-27\,392$	$-48\,128$ $-\cdots)$
(20.9)	$y_n - y_{n-5} = h\,(5$	-25	$+175$	-225	$+2125$	-475	$-1\,375$	$-1\,375$	$-30\,625$	$-51\,138$ $-\cdots)$
(20.10)	$y_n - y_{n-6} = h\,(6$	-36	$+324$	-576	$+8856$	-4752	$+17\,712$	0	$-23\,328$	$-46\,656$ $-\cdots)$
(20.11)	$y_n - y_{n-7} = h\,(7$	-49	$+539$	-1225	$+26117$	-22491	$+216\,433$	$-36\,799$	$-57\,281$	$-57\,281$ $-\cdots)$
(20.12)	$y_n - y_{n-8} = h\,(8$	-64	$+832$	-2304	$+62848$	-74752	$+1\,160\,192$	$-471\,040$	$+1\,012\,736$	0 $)$

The ordinates y can be expressed in terms of differences by the formula

$$(20.2) \qquad y_{n-m} = \sum_{k=0}^{m} (-1)^k \binom{m}{k} \nabla^k y_n.$$

Any formula expressed in terms of finite number of equally spaced ordinates can be rewritten in terms of differences by use of equation 20.2, while conversely any formula in terms of differences can be written also in terms of ordinates with the aid of equation 20.1. Thus, in particular, all the formulas of Article 19 could equally well have been given in terms of differences. A selected set of formulas using differences and especially suited to solution of differential equations appears in Table II. The numbers in the body of this table are the coefficients of the terms at the head of corresponding columns. The dummy formula 20.4 is included for sake of symmetry only.

EXERCISES

1. Express the last set of formulas in Article 19 in terms of differences.

2. Express formula 20.8 in terms of ordinates, neglecting differences of the sixth and higher orders.

21. Formulas in terms of central differences

It will be noted from formula 20.1 that all even-ordered differences are symmetrical with respect to the mid-ordinate. It is often convenient to refer the even differences to the mid-ordinate rather than the end ordinate as in the case of backward differences, and this is accomplished by use of the symbol δ^2 where

$$\delta^2 y_n = \nabla^2 y_{n+1}.$$

Then

$$\delta^2 y_n = y_{n+1} - 2y_n + y_{n-1},$$

$$\delta^4 y_n = y_{n+2} - 4y_{n+1} + 6y_n - 4y_{n-1} + y_{n-2},$$

and, in general,

$$\delta^{2m} y_n = \sum_{k=0}^{2m} (-1)^k \binom{2m}{k} y_{n+m}.$$

Certain symmetric formulas are most conveniently given in terms of central differences, and, accordingly, a list of the most useful of these is provided in Table III. The subscript notation is modified so as to emphasize the symmetry.

Table III. Quadrature Formulas in Terms of Central Differences

(21.1) $\quad y_1 - y_{-1} = h \left(2y_0' + \frac{1}{3} \delta^2 y_0' - \frac{1}{90} \delta^4 y_0' + \frac{5}{3780} \delta^6 y_0' \right.$

$$\left. - \frac{23}{113\,400} \delta^8 y_0' + \frac{263}{7484\,400} \delta^{10} y_0' - \cdots \right),$$

(21.2) $\quad y_2 - y_{-2} = h \left(4y_0' + \frac{8}{3} \delta^2 y_0' + \frac{28}{90} \delta^4 y_0' - \frac{32}{3780} \delta^6 y_0' \right.$

$$\left. + \frac{13}{14175} \delta^8 y_0' - \frac{62}{467\,775} \delta^{10} y_0' + \cdots \right),$$

(21.3) $\quad y_3 - y_{-3} = h \left(6y_0' + \frac{27}{3} \delta^2 y_0' + \frac{297}{90} \delta^4 y_0' + \frac{1107}{3780} \delta^6 y_0' \right.$

$$\left. - \frac{9}{1400} \delta^8 y_0' + \frac{19}{30800} \delta^{10} y_0' - \cdots \right),$$

(21.4) $\quad y_4 - y_{-4} = h \left(8y_0' + \frac{64}{3} \delta^2 y_0' + \frac{1376}{90} \delta^4 y_0' + \frac{14720}{3780} \delta^6 y_0' \right.$

$$\left. + \frac{3956}{14175} \delta^8 y_0' - \frac{2368}{467\,775} \delta^{10} y_0' + \cdots \right),$$

(21.5) $\quad y_5 - y_{-5} = h \left(10y_0' + \frac{125}{3} \delta^2 y_0' + \frac{4375}{90} \delta^5 y_0' + \frac{86125}{3780} \delta^6 y_0' \right.$

$$\left. + \frac{20225}{4536} \delta^8 y_0' + \frac{80335}{299\,376} \delta^{10} y_0' - \cdots \right).$$

A more complete set of coefficients for numerical differentiation and integration is given by Salzer [169, 170, 171, 172]. The error is treated in most of the texts previously mentioned and also by Milne [134]. For a discussion of central difference formulas and operators see Sheppard [188] and Bickley [9].

4

METHODS BASED ON NUMERICAL
INTEGRATION

Most of the principal methods for the numerical solution of ordinary differential equations are based in one way or another on formulas for numerical integration. It is helpful to classify these methods under three headings.

1. *Forward integration*, comprising those methods that obtain y_{n+1} by means of any one of several quadrature formulas that do not involve y_{n+1}'.

2. *Successive integrations*, comprising methods in which a trial value of either y_{n+1} or y_{n+1}' is estimated in some manner (usually from a table of differences) and subsequently improved by successive applications of a quadrature formula that does involve y_{n+1}'.

3. *Forward and Successive integrations*, a title that includes those methods where a trial value of y_{n+1} is obtained by forward integration as in type 1 and subsequently improved by successive integrations as in type 2.

Each one of the above types may be further subdivided according to the kind of quadrature formula employed, the number of terms used, and the way in which the work is organized. It is, however, quite unnecessary (in fact it is impossible) to include all these different cases, many of which differ only in trivial respects. We shall be content to describe and illustrate two or three of the most significant types under each of the three headings just mentioned.

22. Forward integration. Method V

The procedure described in Article 6 and designated as Method I can be extended by the use of a more elaborate formula in place of 6.1. One such formula is 20.3, which we rewrite in the form

$$(22.1) \quad y_{n+1} = y_n + h\left(y_n' + \frac{1}{2}\nabla y_n' + \frac{5}{12}\nabla^2 y_n' \right.$$
$$\left. + \frac{3}{8}\nabla^3 y_n' + \frac{251}{720}\nabla^4 y_n' + \cdots \right).$$

53

This formula requires a table of differences of y', which at the start of the computation is not available and hence must be obtained by other means. Methods III, IV, and IVa, already discussed, can be used for starting, as well as other methods described later on. In any case we suppose that several lines of a computation for equally spaced values of x have already been obtained and the differences of y' have been computed, so that the computation appears as shown below.

x	y	y'	$\nabla y'$	$\nabla^2 y'$	$\nabla^3 y'$	$\nabla^4 y'$
x_{n-3}	y_{n-3}	y_{n-3}'		$\nabla^2 y_{n-2}'$		$\nabla^4 y_{n-1}'$
			$\nabla y_{n-2}'$		$\nabla^3 y_{n-1}'$	
x_{n-2}	y_{n-2}	y_{n-2}'		$\nabla^2 y_{n-1}'$		$\nabla^4 y_n'$
			$\nabla y_{n-1}'$		$\nabla^3 y_n'$	
x_{n-1}	y_{n-1}	y_{n-1}'		$\nabla^2 y_n'$		
			$\nabla y_n'$			
x_n	y_n	y_n'				

Actually it saves space and causes no particular confusion if the tabulation above is compressed into the following horizontal form:

x	y	y'	$\nabla y'$	$\nabla^2 y'$	$\nabla^3 y'$	$\nabla^4 y'$
x_{n-4}	y_{n-4}	y_{n-4}'				
x_{n-3}	y_{n-3}	y_{n-3}'	$\nabla y_{n-3}'$			
x_{n-2}	y_{n-2}	y_{n-2}'	$\nabla y_{n-2}'$	$\nabla^2 y_{n-2}'$		
x_{n-1}	y_{n-1}	y_{n-1}'	$\nabla y_{n-1}'$	$\nabla^2 y_{n-1}'$	$\nabla^3 y_{n-1}'$	
x_n	y_n	y_n'	$\nabla y_n'$	$\nabla^2 y_n'$	$\nabla^3 y_n'$	$\nabla^4 y_n'$

In this illustration the differences are carried to the fourth order. In practice, the number actually required will vary, depending on how rapidly the differences of higher order become negligible. However, it is not customary to retain more than about nine at most. If more are needed to secure accuracy, it is usually better to shorten the interval h. On the other hand, if second differences are negligible, it is evidence that a larger interval can safely be used. The beginner will do well to keep the number of significant differences rather small.

The essence of Method V is contained in these steps:

Step 1. Use formula 22.1 to compute y_{n+1}.

Step 2. From the given differential equation $y' = f(x, y)$ calculate y_{n+1}'.

Step 3. Complete the difference table for the $(n + 1)$th line.

Using the $(n + 1)$th line in place of the nth line, repeat the three steps. The computation proceeds by repetition of these operations.

Example. Apply Method V to the equation $y' = x^2 - y^2$ with initial values $y = 0$ at $x = -1$.

In computation 7 we have already started the solution of this equation.   Using the values obtained there, we set up the difference table and then proceed with Method V as far as $x = 0$.

Method V   Computation 8

| Differential equation $y' = x^2 - y^2$ $(y = 0, x = -1)$ | | | | | | | $h = 0.1$ |
x	y	y'	$\nabla y'$	$\nabla^2 y'$	$\nabla^3 y'$	$\nabla^4 y'$	$\nabla^5 y'$
-1.2	-0.23916	1.38280					
-1.1	-0.10995	1.19791	-18489				
-1.0	0	1	-19791	-1302			
$-.9$	0.09005	0.80189	-19811	-20	1282		
$-.8$	$.16073$	$.61417$	-18772	1039	1059	-223	
$-.7$	$.21350$	$.44442$	-16975	1797	758	-301	-78
$-.6$	$.25036$	$.29732$	-14710	2265	468	-290	11
$-.5$	$.27376$	$.17505$	-12227	2483	218	-250	40
$-.4$	$.28619$	$.07809$	-9696	2531	48	-170	80
$-.3$	$.29019$	$.00579$	-7230	2466	-65	-113	57
$-.2$	$.28814$	$-.04302$	-4881	2349	-117	-52	61
-0.1	$.28233$	$-.06971$	-2669	2212	-137	-20	32
0	$.27490$						

The process just described is due to Adams [4] and is generally known as *Adams' method*.   For additional literature see Collatz and Zurmühl [39], Nyström [145], Sauer and Posch [179], Tollmien [211], and Whittaker and Robinson [219].

23. Variants of Method V*

Other formulas suitable for forward integration with differences can be constructed from the list 20.3 to 20.12.   For instance, if we subtract 20.3 from 20.5, we have

$$y_{n+1} - y_{n-1} = h(2y_n' + \tfrac{1}{3}\nabla^2 y_n' + \tfrac{1}{3}\nabla^3 y_n' + \tfrac{29}{90}\nabla^4 y_n' + \tfrac{28}{90}\nabla^5 y_n' + \cdots),$$

and this may be put in the form

$$(23.1) \quad y_{n+1} = y_{n-1} + 2hy_n' + \frac{h}{3}(\nabla^2 y_n' + \nabla^3 y_n' + \nabla^4 y_n' + \nabla^5 y_n')$$
$$- \frac{h}{90}\nabla^4 y_n' - \frac{2h}{90}\nabla^5 y_n' + \cdots.$$

Formula 23.1 is easier to calculate than formula 22.1, especially if the terms $\dfrac{h}{90}\nabla^4 y_n'$, $\dfrac{2h}{90}\nabla^5 y_n'$, etc. are negligible.   Since 23.1 is used in exactly the same way as was 22.1 in Computation 8, it is unnecessary to give an additional numerical example.

*This is unstable.   Replace by methods of Appendix E.

If we wish to avoid computing differences in Method V, we may employ any one of several quadrature formulas expressed in terms of ordinates. Two particularly useful ones are equations 19.15 and 19.16. These two are chosen from the list of Newton-Cotes open-type formulas. Others in this list with more terms are more accurate, but the greater number of terms and more complicated coefficients reduce their usefulness. The two that we have cited are the ones most commonly employed.

As an illustration we apply formula 19.15 to the solution of the equation

$$y' = (1 + y^2)^{-1}$$

with initial values $y = 25$, $x = 0$.

Here we choose $h = 10$ and select Method IVa for the start. The third approximation of Method IVa and eight steps of Method Va are shown.

Method Va   Computation 9

Differential equation $y' = (1 + y^2)^{-1}$ $h = 10$

x	y	y'	
-20	24.9680 1026	0.0016 01534	
-10	24.9840 1536	.0015 99486	Method IVa
0	25	.0015 97444	3d approximation
10	25.0159 6426	.0015 95409	
20	25.0319 0820	.0015 93381	
30	25.0478 3191	.0015 91359	
40	25.0637 3540	.0015 89343	
50	25.0796 1878	.0015 87334	Method Va
60	25.0954 8211	.0015 85331	
70	25.1113 2543	.0015 83334	
80	25.1271 4880	.0015 81344	
90	25.1429 5233	.0015 79360	
100	25.1587 3603	.0015 77382	

EXERCISES

1. Carry computation 8 five lines farther.
2. Carry computation 9 five lines farther.
3. Use formula 23.1 to obtain a solution of

$$y' = -xy$$

with $y = 1$ and $x = 0$ from $x = 0$ to $x = 1.2$. Let $h = 0.1$, and use Taylor's series to start.

4. Use formula 19.15 to solve the equation

$$y' = (x^2 + y^2)^{-1/2}$$

with $y = 1$ at $x = 0$. Take $h = 0.2$. Use Method IVa to start.

24. Discussion of Method V

The error in any particular case can be analyzed in the same way as in Article 7, but no general formula can be given because the error depends on the particular quadrature formula employed. As a typical example we shall investigate Method Va, where the formula is

$$y_{n+1} = y_{n-3} + \frac{4h}{3}(2y_n' - y_{n-1}' + 2y_{n-2}')$$

with truncation error

$$T_n = \tfrac{28}{90}h^5 y^{(5)}(s), \qquad x_{n-3} < s < x_{n+1}.$$

If z denotes the true value of y and $y_n - z_n = e_n$, we have

$$z_{n+1} = z_{n-3} + \frac{4h}{3}(2z_n' - z_{n-1}' + 2z_{n-2}') + T_n.$$

The mean value theorem gives us

$$y_n' - z_n' = g_n(y_n - z_n) = g_n e_n,$$

where g_n denotes a value of f_y.

By subtracting the equation for y_{n+1} and z_{n+1} above and using the mean-value theorem we get the difference equation of fourth order for e_n:

$$e_{n+1} - e_{n-3} - \frac{4h}{3}(2g_n e_n - g_{n-1}e_{n-1} + 2g_{n-2}e_{n-2}) = -T_n$$

The quantities g_n and T_n, of course, vary as n varies, but we can secure considerable information about the behavior of the error e_n if we solve the above difference equation on the assumption that g_n and T_n are constants equal to g and T, respectively. The difference equation is then

$$e_{n+1} - 2qe_n + qe_{n-1} - 2qe_{n-2} - e_{n-3} = -T, \quad \text{where} \quad q = 4hg/3$$

and the solution is

$$e_n = c_1 r_1^{n} + c_2 r_2^{n} + c_3 r_3^{n} + c_4 r_4^{n} + T/4hg.$$

Here the c_i's are constants and the four quantities r_i are the four roots of the algebraic equation

$$r^4 - 2qr^3 + qr^2 - 2qr - 1 = 0.$$

When q is small, these roots may be represented by a series of powers of q and, to terms of the second degree, we find that

$$r_1 = (1 + \tfrac{3}{4}q + \tfrac{9}{32}q^2 + \cdots),$$

$$r_2 = (-1 + \tfrac{5}{4}q - \tfrac{25}{32}q^2 + \cdots),$$

$$r_3 = (\tfrac{8}{32}q^2 + \cdots) + i(1 + \tfrac{1}{4}q + \tfrac{1}{32}q^2 + \cdots),$$

$$r_4 = (\tfrac{8}{32}q^2 + \cdots) - i(1 + \tfrac{1}{4}q + \tfrac{1}{32}q^2 + \cdots).$$

For q sufficiently small it is apparent that the largest root in absolute value is r_1 if q is positive and is r_2 if q is negative.

In computation 9 for example, we have $h = 10$, $g = -0.000\,1276$, and $q = -0.0017$ approximately over the range of values in the computation. The roots are $r_1 = 0.9987$, $r_2 = -1.0021$, $r_3 = 0.9996i$, $r_4 = -0.9996i$, and

$$T = \tfrac{28}{90}h^5 y^{(5)} = 7.08 \times 10^{-13}.$$

With these data we see that the initial error is very small, that it grows rather slowly as n increases, and eventually (since the largest root is real and negative) alternates in sign from step to step. It is understood that this analysis of the error does not account for round-off error or for arithmetical mistakes.

It is now apparent that computation 9 could have been carried to several more decimal places with the same h, or a larger h could have been safely used to secure eight-decimal-place accuracy.

It may be said of Method V in general that of all methods of comparable accuracy it is the simplest and easiest to apply. Its principal defect is the lack of a suitable current check on the accuracy of the process itself and of the arithmetical work. This showed up in computation 9, where we had little indication of the accuracy attained, and only learned after the investigation just concluded that we might have made a better choice of h.

25. Successive integrations. Method VI

The particular virtue of the formulas for numerical integration used in Method V is that y_{n+1} can be calculated without the use of y_{n+1}'. Formulas that do use y_{n+1}' are, other things being equal, more accurate than those that do not. One way to take advantage of this increased accuracy is to estimate the value of y_{n+1}', say, from the table of differences, integrate to get y_{n+1}, calculate y_{n+1}' from the differential equation, integrate again, and continue until no change occurs. This is essentially the method of successive substitutions of Article 15 applied to the step from x_n to x_{n+1}, and with a quadrature formula replacing formal integration.

Any one of the quadrature formulas previously listed which give y_{n+1} in terms of y_{n+1}', y_n', y_{n-1}', $\cdots$ etc. may be used. To take a concrete case we choose equation 20.6, advance the subscripts by unity, observe that $2y_{n+1}' - 2\nabla y_{n+1}' = 2y_n'$, and then we have

$$(25.1) \quad y_{n+1} - y_{n-1} = h[2y_n' + \tfrac{1}{3}\nabla^2 y_{n+1}' - \tfrac{1}{90}(\nabla^4 y_{n+1}' + \nabla^5 y_{n+1}')].$$

Example. Solve the problem in computation 8 by Method VI, using equation 25.1.

To save time we take the value of y for $x = -0.9$ and the entire line for $x = -0.8$ directly from computation 8. It will be recalled that these values were originally obtained from computation 7. These are shown in lines 1 and 2 of computation 10.

To proceed we assume that $\nabla^4 y'$ for the next line also will be -223. With this guess we fill in line 3 as far as $\nabla^2 y'$. Note that y_{n+1}' and $\nabla y_{n+1}'$ do not appear in 25.1. Now by means of 25.1 we calculate y for $x = -0.7$ and record the result in line 4. Then from the differential equation we get y' for line 4 and fill in the corrected differences. Using these corrected differences, we recalculate y by 25.1, enter the value in line 5, and fill in the rest of line 5 just as was done for line 4.

Recalculation of y by 25.1 gives no change, and hence line 5 is accepted as correct.

For the next step we take $\nabla^5 y'$ in line 6, then obtain y by 25.1, and record y in line 7. Line 7 is completed and y again obtained by 25.1 and recorded in line 8.

Method VI   Computation 10

Line	x	y	y'	$\nabla y'$	$\nabla^2 y'$	$\nabla^3 y'$	$\nabla^4 y'$	$\nabla^5 y'$	$\nabla^6 y'$
									$h = 0.1$

Differential equation $y' = x^2 - y^2$ (See computation 8)

Line	x	y	y'	$\nabla y'$	$\nabla^2 y'$	$\nabla^3 y'$	$\nabla^4 y'$	$\nabla^5 y'$	$\nabla^6 y'$
1	-0.9	0.09005							
2	$-.8$	.16073	0.61417	-18772	1039	1059	-223		
3					(1875	836	-223	0	)
4	$(- .7$	.21351	.44441	-16976	1796	757	-302	-79	)
5	$- .7$	.21349	.44442	-16975	1797	758	-301	-78	
6					(2176	379	-379	-78	)
7	$(- .6$	.25034	.29733	-14709	2266	469	-289	12	)
8	$- .6$	.25037	.29731	-14711	2264	467	-291	10	88
9					(2450	186	-281	10	)
10	$(- .5$	.27377	.17505	-12226	2485	221	-246	45	)
11	$- .5$	.27378	.17504	-12227	2484	220	-247	44	34
12					(2501	17	-203	44	)
13	$(- .4$	.28621	.07808	$- 9696$	2531	47	-173	74	)
14	$- .4$	.28622	.07808	$- 9696$	2531	47	-173	74	30
15					(2479	-52	$- 99$	74	)
16	$(- .3$	.29022	.00577	$- 7231$	2465	-66	-113	60	)
17	$- .3$	.29022	.00577	$- 7231$	2465	-66	-113	60	-14

Recalculation does not change y, and line 8 is therefore assumed to be correct. The entire computation is carried out by repetitions of these steps.

Actually the work is done with pencil, and corrections are made by erasures, so that in the end the tentative lines such as 3, 4, 6, 7, 9, 10 do not appear, and only the finally corrected lines 5, 8, 11, · · · remain.

In comparison with Method V it is clear that Method VI provides much greater security against arithmetical mistakes, and, since the quadrature formulas available for Method VI are generally more accurate than those for Method V, we expect greater accuracy in the solution by Method VI. On the other hand, the numerical labor for Method VI is considerably greater than for Method V.

Method VI is substantially that devised by F. R. Moulton to integrate the differential equations of exterior ballistics and may be called *Moulton's method*, although this name seems not to be in general use. See e.g. Jackson [94], Moulton [142, 143, 144], Bennett, Milne and Bateman [7]. Cf. also Bickley [8].

26. Variants of Method VI

Formula 20.5 may be used for the integration in Method VI. It has the advantage of covering one step only and hence is less subject to dangerous oscillation than the formula of Article 25, where y_2, y_4, y_6, etc. form one chain of values and y_1, y_3, y_5 form another. On the other hand, formula 20.5 is not so convenient for numerical calculation.

In situations where the term $\dfrac{5h}{3780} \delta^6 y'$ is negligible and the fourth difference small we may use the central difference formula 21.1 to advantage. For this purpose we have to estimate $\delta^4 y'$ *two steps ahead* and $\delta^2 y'$ *one* step ahead. If $\dfrac{h}{90} \delta^4 y'$ is also negligible, this method is very convenient.

The computation may be simplified by the use of

$$\tfrac{1}{3}\delta^2 y_{n-1}' = \tfrac{1}{3}(y_n' - 2y_{n-1}' + y_{n-2}')$$

to avoid writing the first difference, and the work is set up as follows:

$$x \quad y \quad y' \quad \tfrac{1}{3}\delta^2 y'.$$

As an illustration we carry computation 10 a few steps farther, using this new procedure.

The entries in computation 10 for $x = -0.5$, -0.4, -0.3 provide the following tabulation:

x	y	y'	$\delta^2 y'/3$
-0.5	0.27378	0.17504	844
$-.4$	.28622	.07808	822
$-.3$	.29022	.00577	

On the basis of the recorded values of $\delta^2 y'/3$ we estimate the entry for $x = -0.3$, say, 800. With this the formula

$$y_{n+1} = y_{n-1} + 2hy'_n + \frac{h}{3}\delta^2 y_n$$

gives $y = 0.28817$ at $x = -0.2$. Then y' is calculated, and a corrected value of $\delta^2 y'/3$ replaces the first estimate. Recalculation gives $y = 0.28816$ but no other change. The computation as far as $x = +0.3$ is shown in number 11. Only the corrected values appear.

Method VIa   Computation 11

| Differential equation $y' = x^2 - y^2$ | | (See computation 10) | | $h = 0.1$ |
x	y	y'	$\delta^2 y'/3$	Diff.
-0.5	0.27378	0.17504	844	
$-.4$	.28622	.07808	822	-22
$-.3$	.29022	.00577	783	-39
$-.2$	.28816	$-.04304$	738	-45
$-.1$	.28235	$-.06972$	694	-44
$.0$	.27491	$-.07558$	656	-38
$.1$	.26789	$-.06177$	623	-33
$.2$	.26318	$-.02926$	592	-31
$.3$	.26263	$+.02103$		

EXERCISES

1. Carry computation 10 five steps farther by the same method.
2. Carry computation 11 five steps farther by Method VIa.
3. Carry computation 9 five steps farther by the use of equation 20.5.

27. Use of differences at the start

It has been noted that special devices are required for getting a computation under way, and Methods III or IV have been suggested for this purpose. However, if differences are to be used in the body of the computation, it is advisable to use differences at the start also. This can be done in many different ways. We shall illustrate one of these ways by means of an example, which should be regarded as a suggestive illustration rather than a rigid model.

The process about to be described uses formulas adapted from equations 20.3 and 20.5. The first two are

(27.1) $y_n = y_{n-1} + h(y_n' - \frac{1}{2}\nabla y_n' - \frac{1}{12}\nabla^2 y_n' - \frac{1}{24}\nabla^3 y_n' - \cdots)$

from 20.5 and

(27.2) $y_{n+1} = y_n + h(y_n' + \frac{1}{2}\nabla y_n' + \frac{5}{12}\nabla^2 y_n' + \frac{3}{8}\nabla^3 y_n' + \cdots)$

from 20.3. If in equation 27.1 we reverse the direction of integration but preserve the present subscript notation for the differences, it will be found that

$$(27.3) \quad y_{n-1} = y_n - h(y_{n-1}' + \tfrac{1}{2}\nabla y_n' - \tfrac{1}{12}\nabla^2 y_{n+1}'$$
$$+ \tfrac{1}{24}\nabla^3 y_{n+2}' - \cdots)$$

This formula contains the underscored values in the difference diagram shown below:

$$
\begin{array}{ccccccc}
\underline{y_{n-1}} & \underline{y_{n-1}'} & & & & & \\
& & \underline{\nabla y_n'} & & & & \\
\underline{y_n} & y_n' & & \nabla^2 y_{n+1}' & & & \\
& & \nabla y_{n+1}' & & \nabla^3 y_{n+2}' & & \\
y_{n+1} & y_{n+1}' & & \nabla^2 y_{n+2}' & & & \\
& & \nabla y_{n+2}' & & & & \\
y_{n+2} & y_{n+2}' & & & & &
\end{array}
$$

Similarly, upon reversal, equation 27.2 becomes

$$(27.4) \quad y_n = y_{n+1} - h(y_{n+1}' - \tfrac{1}{2}\nabla y_{n+2}' + \tfrac{5}{12}\nabla^2 y_{n+3}'$$
$$- \tfrac{3}{8}\nabla^3 y_{n+4}' + \cdots).$$

The general idea of the method is to build up values of y (and from these the values of y' and its differences) on both sides of the initial values (x_0, y_0). For instance, at some stage of the work we are going to have a difference table like this:

$$
\begin{array}{ccccccc}
x_{-2} & y_{-2} & y_{-2}' & & & & \\
& & & \nabla y_{-1}' & & & \\
x_{-1} & y_{-1} & y_{-1}' & & \nabla^2 y_0' & & \\
& & & \nabla y_0' & & \nabla^3 y_1' & \\
x_0 & y_0 & y_0' & & \nabla^2 y_1' & & \nabla^4 y_2' \\
& & & \nabla y_1' & & \nabla^3 y_2' & \\
x_1 & y_1 & y_1' & & \nabla^2 y_2' & & \\
& & & \nabla y_2' & & & \\
x_2 & y_2 & y_2' & & & &
\end{array}
$$

At this stage we can use 27.1 with $n = 1$ to improve y_1, then with $n = 2$ to improve y_2, then use 27.2 with $n = 2$ to get a trial value of y_3, from which another line of differences is found. Also we use equation 27.3 with $n = 0$ to improve y_{-1}, with $n = -1$ to improve y_{-2}, and then use equation 27.4 with $n = -3$ to estimate y_{-3} and thus extend the difference table.

Example. Use differences to start the solution of

$$y' = x^2 - y^2, \quad (x = 0, \quad y = 1),$$

with $h = 0.1$.

Starting Method, Using Differences   Computation 12

Differential equation $y' = x^2 - y^2$ ($y = 1$ at $x = 0$) $h = 0.1$

x	y	y'	$\nabla y'$	$\nabla^2 y'$	$\nabla^3 y'$	$\nabla^4 y'$	$\nabla^5 y'$
−0.1	1.1	−1.2					
			2				
.0	1.0	−1.0		0			
			2				
.1	.9	− .8					

. . . .

− .2	1.24	−1.50					
			28				
− .1	1.11	−1.22		− 6			
			22		2		
.0	1.00	−1.00		− 4		0	
			18		2		
.1	.91	− .82		− 2			
			16				
.2	.84	− .66					

. . . .

− .3	1.415	−1.912					
			399				
− .2	1.246	−1.513		− 110			
			289		45		
− .1	1.111	−1.224		− 65		− 20	
			224		25		10
.0	1.000	−1.000		− 40		− 10	
			184		15		4
.1	.909	− .816		− 25		− 6	
			159		9		
.2	.835	− .657		− 16			
			143				
.3	.777	− .514					

. . . .

− .3	1.4171	−1.9182					
			4032				
− .2	1.2470	−1.5150		− 1121			
			2911		449		
− .1	1.1108	−1.2239		− 672		− 186	
			2239		263		85
.0	1.0000	−1.0000		− 409		− 101	
			1830		162		50
.1	.9094	− .8170		− 247		− 51	
			1583		111		
.2	.8359	− .6587		− 136			
			1447				
.3	.7772	− .5140					

. . . .

− .3	1.41751	−1.91933					
			40427				
− .2	1.24702	−1.51506		−11302			
			29125		4558		
− .1	1.11077	−1.22381		− 6744		−1901	
			22381		2657		885
.0	1.00000	−1.00000		− 4087		−1016	
			18294		1641		418
.1	.90943	− .81706		− 2446		− 598	
			15848		1043		
.2	.83581	− .65858		− 1403			
			14445				
.3	.77726	− .51413					

To get our first toe-hold on the problem we note that $y_0' = -1$, and hence estimate y_{-1} as 1.1, y_1 as 0.9.

With these values stage one of computation 12 is effected. Stage two shows the result of one improvement on y_{-1} and y_1 with estimates for y_{-2} and y_2. At each stage in this particular example we have made *one* improvement in existing values and added *two* new values (except for the last stage). Other examples may require more improvements before new values are added. As a matter of fact, the computer should follow his judgment in selecting the precise order of operations. Five stages of the process are shown in computation 12.

28. Forward and successive integrations.   Method VII

This method is just the same in spirit as Method II described in Chapter 2, the only difference being that we are now prepared to use higher-powered quadrature formulas. A trial value of y_{n+1} is obtained by one of the quadrature formulas not containing y_{n+1}', say, equation 19.15 or 19.16, just as was done in Method V. With this approximate value of y_{n+1} we can then compute an approximate value for y_{n+1}'. From here on, we improve the values of y_{n+1} and y_{n+1}' by the use of a suitable quadrature formula containing y_{n+1}', just as in Method VI. It is advantageous to select a pair of quadrature formulas that are symmetrical about a mid-point and that involve an odd number of ordinates. Such pairs, expressed in terms of ordinates are: (A) the 3-point pair, 19.15 to predict and 19.2 to recalculate, (B) the 5-point pair; 19.16 to predict and 19.9 to recalculate. The pairs for 7, 9, 11, etc. points are excellent in theory but in practice are rarely used. The reason for selecting the formulas A or B is that they are more accurate than any other quadrature formulas employing the same number of equally spaced ordinates. (Of course, Gauss's formulas cannot be used for step-by-step integration because of irrational abscissas). Whether or not pairs A and B cited above are used in terms of ordinates or are expressed otherwise is a matter of computational convenience and has no bearing on the basic merits of the formulas. Actually it is here recommended that we use the central difference forms where the pair A becomes

$$(28.1) \qquad y_{n+1} = y_{n-3} + 4hy_{n-1}' + \frac{8h}{3} \delta^2 y_{n-1}' + \frac{28}{90} h^5 y^{(5)}$$

from equation 21.2 and

$$(28.2) \qquad y_{n+1} = y_{n-1} + 2hy_n' + \frac{h}{3} \delta^2 y_n' - \frac{1}{90} h^5 y^{(5)}$$

from equation 21.1.

As an illustration we take the same problem that was started in computation 7 and carried forward several more steps in computations 8 and 10. To start, we take initially four values from computation 7, and compute $\delta^2 y'/3$. Then equation 28.1 gives $y = 0.21357$ as a trial value of y at $x = -0.7$. With this we obtain y' and $\delta^2 y'/3$ and then use equation 28.2 to secure an improved value of y. This turns out to be 0.21348. The difference between this and the trial value is -0.00009, and this difference we record in the column headed C_0. Here the quantity C_0 plays precisely the same role as in computation 2. The corrections are made by erasures, so that only corrected values remain. After the entries in column C_0 have been started, we can often avoid recomputation by guessing ahead for the next correction and applying the correction to y before calculating y' and $\delta^2 y'/3$.

Method VII Computation 13

| Differential equation $y' = x^2 - y^2$ | | | $(y = 0, x = -1)$ | $h = 0.1$ |
x	y	y'	$\delta^2 y'/3$	C_0
-1.1	-0.10995	1.19791		
-1.0	$.0$	1.00000	$- 7$	(From
$- .9$	$.09005$	$.80189$	346	computation 7)
$- .8$	$.16073$	$.61417$	599	
$- .7$	$.21348$	$.44443$	754	-9
$- .6$	$.25037$	$.29731$	829	-9
$- .5$	$.27377$	$.17505$	843	-8
$- .4$	$.28622$	$.07808$	822	-7 Method VII
$- .3$	$.29021$	$.00578$	782	-3
$- .2$	$.28816$	$- .04304$	738	-2
-0.1	$.28234$	$- .06972$	694	0
-0	$.27491$	$- .07558$		0

It is of interest to compare the results of computations 8, 10, 11, and 13 on the same problem. The differences between the results in computations 11 and 13 are probably mainly due to vagaries in round-off.

An alternative procedure for Method VII is to omit the tabulation of $\delta^2 y'/3$ and to compute the predicted and corrected values of y directly from equations 19.15 and 19.2, respectively. As long as all goes well, there is little choice between the two procedures, but in case of more than one recalculation there is a slight advantage in having the second difference tabulated.

It may be noted that 19.2 is Simpson's rule.

The treatment of the convergence factor as given in Article 9 applies to Method VII with just one modification. Using the notation of Article 9, we find that in the present case the convergence factor θ is defined by $\theta = hg/3$. In order for the work to proceed without tedious recalculation it is necessary to choose the step interval h small enough to make the second correction $C_1 = \theta C_0$ negligible to the required number of decimal places.

The truncation error of 28.1 is $\frac{28}{90}h^5 y^{(5)}$, whereas that of 28.2 is

$$T_n = -\frac{1}{90}h^5 y^{(5)}.$$

Neglecting the variation in $y^{(5)}$ over one step, we see that approximately

$$- C_0 = +\frac{28}{90}h^5 y^{(5)}.$$

whence

$$T_n = - C_0/29.$$

Hence, the value of C_0 recorded for each step in the right-hand column provides a rough and ready way of estimating the truncation error at that step. In computation 13 it appears that the maximum truncation error to be feared is about 0.000003 and is completely masked by the round-off error. The column C_0 not only serves as a control for the truncation error but also reveals arithmetical mistakes. Whenever the values in the column of C_0 make abrupt changes, it is wise to check the last line or two of the computation.

Method VII is frequently referred to as *Milne's method.** In the original publication (Milne [128]) the formulas used were 19.15 and 19.2 instead of the central difference formulas used in the text. For a particularly readable account of this method see Morris and Brown [141]. See also references [112], [114], and [115].

29. Analysis of the error

Articles 7, 11, and 24 gave a somewhat perfunctory treatment of the error for Methods I, II, and V, respectively. In this article we shall attempt a more complete and critical examination of the error associated with any method based on Simpson's rule, in particular

*See Appendix D.

Method VIa as exemplified by computation 11 and Method VII as exemplified by computation 13. We shall again ignore the cumulative effect of round-off error by assuming that all calculations are carried to enough decimal places so that round-off error is negligible in comparison with the inherited and truncation errors.

Suppose then that the differential equation

$$y' = f(x, y)$$

has been integrated numerically over a segment of the x axis of length L with N steps of length h, where $h = L/N$. Let M be a positive constant such that

(29.1) $$\left| y^{(5)}(x) \right| < M$$

throughout the range of integration. Since the truncation error is given by

(29.2) $$T_n = - h^5 y^{(5)}/90$$

we can readily estimate $y^{(5)}$ and hence find a suitable value for M directly from the C_0 column in the case where Method VII is used. For Method VIa the same information can be secured from the fourth differences of y', and these are easily found from the column headed $\delta^2 y'/3$.

Let G be a positive constant such that

(29.3) $$\left| f_y \right| < G$$

throughout the range of the integration. In practice the value of G is estimated by inserting the tabulated values of x and y in

$$\left| f_y(x,y) \right|$$

and thus G also is obtainable directly from the computation. It is assumed that the step interval h is chosen small enough to satisfy the inequality

$$hG/3 < 1,$$

a choice that will guarantee the convergence of the successive integrations at each step.

Applying the same methods as in Articles 7, 11, and 24, we derive from Simpson's rule the linear difference equation

(29.4) $$e_{n+1} = \alpha e_n + \beta e_{n-1} + \gamma,$$

where $e_n = y_n - z_n$. Here y_n is the calculated, and z_n the true value of y. The coefficients α, β, γ are defined by

$$\alpha = \frac{4hg_n}{3 - hg_{n+1}}, \qquad \beta = \frac{3 + hg_{n-1}}{3 - hg_{n+1}}, \qquad \gamma = \frac{3T_n}{3 - hg_{n+1}},$$

in which $g_n = f_y$, evaluated for $x = x_n$, y between y_n and z_n. Equation 29.4 with its variable coefficients is not readily solved, but we shall show that

(29.5) $$E_{n+1} = 4KE_n + (1 + 2K)E_{n-1} + R,$$

with

$$K = \frac{hG}{3 - hG}, \qquad R = \frac{h^5 M}{30(3 - hG)},$$

is a dominating difference equation for 29.4. For from 29.1, 29.2, and 29.3 there follow the inequalities

$$|\alpha| < 4K, \qquad |\beta| < (1 + 2K), \qquad |\gamma| < R,$$

and now a comparison of 29.4 with 29.5 shows that, if $|e_n| \leq E_n$ and $|e_{n-1}| \leq E_{n-1}$, then $|e_{n+1}| < E_{n+1}$. In particular, if we suppose that the two initial values of y, namely y_0 and y_1, are correct and, consequently, $e_0 = e_1 = E_0 = E_1 = 0$, we have $|e_n| < E_n$ for $n = 2, 3, \cdots, N$.

Equation 29.5 is a linear difference equation of the second order with constant coefficients. A solution of 29.5 satisfying the conditions $E_0 = E_1 = 0$ is

(29.6) $$E_n = \frac{R}{6K} \left[\frac{(1 - r_2)r_1{}^n + (r_1 - 1)r_2{}^n}{r_1 - r_2} - 1 \right],$$

in which r_1 and r_2 are the two roots of the quadratic equation

$$r^2 - 4Kr - 1 - 2K = 0.$$

The roots r_1 and r_2 satisfy the inequalities

$$-1 < r_2 < 0, \qquad 1 < r_1 < 1 + 4K.$$

Hence, the term $r_2{}^n$ in equation 29.6 decreases exponentially while $r_1{}^n$ increases exponentially as n increases, and for n sufficiently large the significant part of 29.6 is

$$E_n = \frac{R}{6K} \left[\frac{(1 - r_2)}{(r_1 - r_2)} r_1{}^n - 1 \right],$$

whence

$$E_n < \frac{R}{6K} (r_1{}^n - 1),$$

since $1 - r_2 < r_1 - r_2$. Moreover,

$$E_n < \frac{R}{6K} [(1 + 4K)^n - 1],$$

since $r_1 < 1 + 4K$. With the values for R and K inserted this becomes

$$(29.7) \qquad E_n < \frac{h^4 M}{180G} \left[\left(\frac{1 + hG}{1 - hG/3} \right)^n - 1 \right].$$

Though a slightly sharper inequality might have been obtained this appears to be adequate for a practical estimation of the error after n steps.

We can make the error of any individual step as small as we please by reducing the length of the step interval h, but this increases the number of steps and thus increases the n in equation 29.7. To examine the resultant effect of these two opposing factors in the error we replace h in 29.7 by L/N and consider the value of the right-hand member when n attains its maximum value, namely N. The inequalities

$$\left(1 + \frac{LG}{N} \right)^N < e^{LG}$$

and

$$\left(1 - \frac{LG}{3N} \right)^{-N} < e^{LG} \quad \text{if} \quad N > \frac{LG}{2}$$

enable us to put 29.7 in the final form

$$(29.8) \qquad E_N < \frac{L^4 M}{180 N^4 G} (e^{2LG} - 1)$$

provided $N > LG/2$. Formula 29.8 gives a bound for the maximum accumulated error of a numerical integration over a range of length L with N equal steps. The two other constants involved, M and G, are easily estimated from the computation itself. Thus the practical evaluation of the right member of equation 29.8 does not require any lengthy independent calculation.

A recent paper by Richter [252] gives a detailed treatment of the error in the Milne Method.

30. Method VII, five-point formulas

The use of the three-point formulas for Method VII has been explained in some detail in the two preceding articles. The five-point formulas 19.16 and 19.9 are used in exactly the same way, subject to a few rather obvious differences. For example, in the five-point case the truncation error is estimated from the entries in the C_0 column by the approximate formula

$$T_n = C_0/35.$$

The convergence factor is

$$\theta = 28hf_y/90.$$

The treatment of the error is somewhat more involved because we need the roots of an algebraic equation of fourth degree, but the method of Article 24 can be applied to express the roots in power series in a parameter. The details of the analysis are left to the reader.

The five-point formulas are applied to the same example as in computation 13, but this time with the interval $h = 0.2$. The necessary starting values are taken from number 13, and computation 14 then is carried forward to $x = 1.2$. From the size of the entries in the column headed C_0 we may expect the fifth decimal place of y to be in error by at least one unit.

Method VII (Five-point Formulas)   Computation 14

Differential equation $y' = x^2 - y^2$ ($y = 0$, $x = -1$) $h = 0.2$

x	y	y'	C_0	
-1.0	0	1.00000		
$-.8$	0.16073	.61417		
$-.6$	.25037	.29731		From
$-.4$	.28622	.07808		Computation 13
$-.2$	.28816	$-.04304$		
$.0$	.27491	$-.07558$		
$.2$	.26317	$-.02926$	-52	
$.4$	.26794	$+.08821$	-46	
$.6$	.30265	$+.26840$	-30	Five-point
$.8$	.37854	$+.49671$	-16	Formulas
1.0	.50284	$+.74715$	0	
1.2	.67641	$+.98247$	33	

EXERCISES

1. Work out for the five-point method an inequality for the error analogous to equation 29.8.

2. Use Method VII with three-point formulas to calculate e^x from the equation

$$y' = y$$

with $y = 1$ where $x = 0$. Choose $h = 0.1$. How many places in the solution are reliable? Carry to $x = 2$.

3. Use Method VII with three-point formulas to solve the equation

$$y' = \frac{\pi}{180} \sqrt{1 - y^2}$$

where $y = 0$ when $x = 0$. From this obtain a table of sin x with entries at intervals of 1 degree. How many decimal places in the solution are reliable? Carry to 20 degrees. Compare results with a table of natural sines.

4. Use Method VII with three-point formulas to solve the equation

$$y' = 12(x^2 + y^2)^{-1}$$

with $y = 10$ at $x = 0$. Carry to $x = 5$. Choose h to give three-place accuracy. Use four places in the computation.

5. Examine the convergence factor in each of the exercises 2, 3, 4. How many repetitions are required to secure convergence with the accuracy allowed by the truncation error?

6. What effect does reversing the direction of integration have on the convergence factor? On the truncation error? On the accumulated error?

There is a wealth of literature relating to the material of this chapter. Methods differing in varying degrees from those of the text are proposed by Bukovics [20], Chadaja [22], Collatz [34], Duncan [50], Mikeladze [120, 123], and many others. Levy and Baggot [106] summarize in brief form a large number of methods. A rather complete account of methods based on numerical quadrature is given by Massera [116]. von Mises [138], Chadaja [23], and Duncan [49] treat the error. See also references [33], [108], [186], [214], [215], [235], [238], and, in particular, Collatz's book [36].

5

METHODS OF RUNGE-KUTTA
METHODS BASED ON HIGHER
DERIVATIVES

The methods of Chapter 4 are based on formulas for numerical integration. In this chapter we lump together two sets of methods which rest upon different bases. The first of these is celebrated, especially on the continent of Europe, and is usually designated by the names of the principal authors, Runge and Kutta. The second is not well known; it is not even mentioned in many treatments of numerical integration. It is based on the use of additional derivatives obtained by differentiation of the given differential equation.

31. Kutta's fourth-order method. Method VIII

The fundamental idea involved in this method is to obtain an expression for y_{n+1} which coincides up to terms of a certain order in h with the development of y_{n+1} in a power series in h, without, however, using in the actual computation the derivatives of $f(x, y)$ as was done in the method of Taylor's series. It differs from the processes using quadrature formulas in that the step from y_n to y_{n+1} is made without the use of values of y preceding y_n, so that no special methods are required to start the computation. Kutta showed that approximations of the second, third, and fourth orders, respectively, require two, three, and four substitutions into the differential equation. He derived several sets of formulas for these cases, the most noted of which is

(31.1) $\qquad y_{n+1} = y_n + (k_1 + 2k_2 + 2k_3 + k_4)/6,$

where

$$k_1 = hf(x_n, y_n),$$
$$k_2 = hf(x_n + \tfrac{1}{2}h, y_n + \tfrac{1}{2}k_1),$$
$$k_3 = hf(x_n + \tfrac{1}{2}h, y_n + \tfrac{1}{2}k_2),$$
$$k_4 = hf(x_n + h, y_n + k_3).$$

This gives y_{n+1} with an accuracy of the fourth order in h and requires four substitutions into the differential equation for each value of y obtained.

Example. The arrangement of the computation is illustrated by applying it to the differential equation $y' = -xy$. The first step is to compute k_1, k_2, k_3, k_4, in succession by substitutions in the equation $y' = -xy$. Then y_{n+1} is found from equation 31.1, after which a new set of k's is obtained, y_{n+2} is found from 31.1, and so on. The computation appears as follows, beginning with $x = 0.5$.

Method VIII   Computation 15

Differential equation $y' = -xy$ ($x = 0.5$, $y = 0.88249690$) $h = 0.1$

x	y	y'	k_1	k_2	k_3	k_4
0.5	0.8824 9690	−0.4412 4845	−0.0441 2485	−0.0473 2390	−0.0472 3592	−0.0501 1566
.6	.8352 7021	− .5011 6213	− .0501 1621	− .0526 6379	− .0525 8099	− .0547 8825
.7	.7827 0454	− .5478 9318	− .0547 8932	− .0566 4824	− .0565 7853	− .0580 9008
.8	.7261 4905	− .5809 1924	− .0580 9192	− .0592 5376	− .0592 0438	− .0600 2502
.9	.6669 7685	− .6002 7916	− .0600 2792	− .0605 1147	− .0604 8851	− .0606 4883
1.0	.6065 3073					

The original paper on this method, that of Runge [167], appeared in 1895, and was followed with a paper by Heun [89] in 1900 and one by Kutta [103] in 1901. Subsequent writers, notably Nyström [145], have made additional contributions to the method. See also Zurmühl, [225], [226].

32. Fifth-order methods

To secure an accuracy of the mth order in h by the Runge-Kutta method when m is greater than 4, more than m substitutions are required. Kutta and Nyström have obtained sets of formulas giving y_{n+1} with an accuracy of the fifth order in h and making use of six substitutions. The complexity of the formulas increases rapidly as the order increases.

For example, one of Kutta's formulas as corrected by E. J. Nyström is

$$y_{n+1} = y_n + \tfrac{1}{192}(23k_1 + 125k_3 - 81k_5 + 125k_6)$$

where

$$k_1 = hf(x_n, y_n),$$

$$k_2 = hf\left(x_n + \frac{h}{3}, y_n + \frac{k_1}{3}\right),$$

$$k_3 = hf\left(x_n + \frac{2h}{5}, y_n + \frac{6k_2 + 4k_1}{25}\right),$$

$$k_4 = hf\left(x_n + h,\, y_n + \frac{15k_3 - 12k_2 + k_1}{4}\right),$$

$$k_5 = hf\left(x_n + \frac{2h}{3},\, y + \frac{8k_4 - 50k_3 + 90k_2 + 6k_1}{81}\right),$$

$$k_6 = hf\left(x_n + \frac{4h}{5},\, y_n + \frac{8k_4 + 10k_3 + 36k_2 + 6k_1}{75}\right).$$

Clearly the pursuit of greater accuracy by the derivation of formulas of higher order is a losing game. The formulas rapidly assume formidable complexity, and the large number of substitutions per step demands an exorbitant amount of toil in the case of any but the very simplest differential equations.

33. Comparison of Method VIII with Method VII

In contrast to step-by-step procedures based on formulas for numerical quadrature the Runge-Kutta method (as it is usually called) enjoys two conspicuous advantages:

1. No special devices are required for starting the computation.
2. The length of the step can be modified at any time in the course of the computation without additional labor.

On the other hand, it is open to two major objections:

1. The process does not contain in itself any simple means for estimating the error or for detecting arithmetical mistakes. It is true that Bieberbach [11] has found an expression that provides an upper bound for the error at a given step of the Runge-Kutta process (or, more accurately, the Kutta process). However, this estimate depends on quantities that do not appear directly in the computation, and therefore requires some additional separate calculation.

2. Each step requires four substitutions into the differential equation. For complicated equations this may demand an excessive amount of labor per step.

The example used in computation 15 displayed the Runge-Kutta method in an exceptionally favorable light. We shall now select other examples to bring out the essential weakness of this method. Take, for instance, the equation $y' = 5y/(1 + x)$ with the simple polynomial solution $y = (1 + x)^5$. The comparison of Methods VII and VIII is in computation 16.

For this simple differential equation a comparison of the times required by the two methods is not very significant. However, since in this example Kutta's method requires 40 substitutions while the second method requires 15 or so (depending on exactly how the start is made), the difference would obviously be significant for more difficult equations.

Comparison of Methods VII and VIII   Computation 16

Differential equation $y' = \dfrac{5y}{1 + x}$ $(y = 1, \quad x = 0)$ $h = 0.1$

x	Method VIII	Error	Method VII	Error	$(1 + x)^5$
0	1.0000	......	1.0000	......	1.0000
0.1	1.6103	0.0002	1.6105	......	1.6105
.2	2.4878	.0005	2.4883	......	2.4883
.3	3.7119	.0010	3.7129	......	3.7129
.4	5.3765	.0017	5.3782	......	5.3782
.5	7.5911	.0027	7.5937	0.0001	7.5938
.6	10.4819	.0039	10.4857	.0001	10.4858
.7	14.1931	.0055	14.1985	.0001	14.1986
.8	18.8882	.0075	18.8956	.0001	18.8957
.9	24.7509	.0101	24.7609	.0001	24.7610
1.0	31.9867	.0133	31.9999	.0001	32.0000

It is of some interest to make the comparison for the equations $y' = 4y/(1 + x)$ and $y' = 2y/(1 + x)$ with $y = 1$ at $x = 0$, and $h = 0.1$.

For the first of these Kutta's method is in error at $x = 1$ by 0.00242, and for the second by 0.0000205. In both of these cases the second method gives *exact* values.

The foregoing examples serve to show that even for very innocent-looking differential equations the Runge-Kutta method may give very bad results. Since the accuracy is in any case difficult to ascertain, the possibility of such errors occurring is a serious criticism of the Runge-Kutta method.

34. Methods based on higher derivatives

Numerical integration can be made more accurate by the use of more terms but only at the expense of simplicity and ease in starting. The search for greater accuracy without increase in the number of steps or in the number of lines used per step leads to the use of higher derivatives.

A number of pairs of formulas analogous in a general way to the predictor and corrector formulas of Method VII but containing derivatives of order higher than the first will be derived and illustrated in the following articles.

35. Method IX. Formulas with first and second derivatives

Taylor's series applied to y_1, y_1', and y_1'' give the equations

$$y_{n+1} = y_n + hy_n' + \frac{h^2 y_n''}{2!} + \frac{h^3 y_n'''}{3!} + \frac{h^4 y_n^{(4)}}{4!} + \frac{h^5 y_n^{(5)}}{5!} + \cdots,$$

$$hy_{n+1}' = hy_n' + h^2 y_n'' + \frac{h^3 y_n'''}{2!} + \frac{h^4 y_n^{(4)}}{3!} + \frac{h^5 y_n^{(5)}}{4!} + \cdots,$$

$$h^2 y_{n+1}'' = h^2 y_n'' + h^3 y_n''' + \frac{h^4 y_n^{(4)}}{2!} + \frac{h^5 y_n^{(5)}}{3!} + \cdots,$$

We multiply the second of these equations by $-\frac{1}{2}$, the third by $+\frac{1}{12}$, and add. The coefficients of h^3 and h^4 prove to be zero in the result, and after rearrangement we get

$$(35.1) \quad y_{n+1} - y_n = \frac{h}{2}(y_{n+1}' + y_n') - \frac{h^2}{12}(y_{n+1}'' - y_n'') + R_1.$$

It may be shown that

$$(35.2) \quad R_1 = h^5 y^{(5)}(s)/720, \qquad x_n < s < x_{n+1}.$$

By the use of Taylor's series in a similar manner the reader can verify that

$$(35.3) \quad y_{n+1} - y_{n-2} = 3(y_n - y_{n-1}) + h^2(y_n'' - y_{n-1}'') + R_2,$$

where

$$(35.4) \quad R_2 = 60h^5 y^{(5)}(s)/720, \qquad x_{n-2} < s < x_{n+1}.$$

Given the differential equation

$$(35.5) \qquad y' = f(x, y),$$

we obtain from it by differentiation

$$(35.6) \qquad y'' = f_x(x, y) + f_y(x, y)y'.$$

The routine for the numerical solution (except for the first three lines) proceeds according to the following steps. We have the values in the tabulation below:

$$
\begin{array}{cccc}
x_{n-2} & y_{n-2} & y_{n-2}' & y_{n-2}'' \\
x_{n-1} & y_{n-1} & y_{n-1}' & y_{n-1}'' \\
x_n & y_n & y_n' & y_n'' \\
x_{n+1} & & &
\end{array}
$$

Then a trial value of y_{n+1} is obtained from equation 35.3; y_{n+1}' and y_{n+1}'' are calculated from formulas 35.5 and 35.6, respectively. Next a corrected value for y_{n+1} is furnished by equation 35.1, y_{n+1}' and y_{n+1}'' are corrected, and, if necessary, y_{n+1} is again computed by 35.1. If the h has been properly chosen, further recomputation should not be needed.

At the start of the computation we can use 35.1 after making an estimate of y_1 from

$$(35.7) \qquad y_1 = y_0 + h y_0' + \tfrac{1}{2} h^2 y_0''.$$

Example. Solve the differential equation $y' = x^2 - y^2$ with $x = 0$, $y = 1$, and $h = 0.1$.

The successive approximations required to get the values of y for $x = 0.1$, 0.2, 0.3 are shown in detail in the first part of computation 17. The numbers in paren-

Method IX   Computation 17

Differential equation $y' = x^2 - y^2$ ($x = 0$, $y = 1$) $h = 0.1$

x	y	y'	y''	
0	1.000	-1.000	2.000	
(0.1	.910	$-$ .818	1.688)	(35.7)
(.1	.909 36	$-$ .816 94	1.6858)	(35.1)
(.1	.909 415	$-$ .817 036	1.6860)	(35.1)
.1	.909 410	$-$ .817 026	1.6860	(35.1)
(.2	.835 68	$-$ .662 8	1.3720)	*
(.2	.835 680	$-$ .658 36	1.5004)	(y_2 copied from line above)
(.2	.835 796	$-$ .658 555	1.5008)	(35.1)
(.2	.835 785	$-$ .658 536	1.5008)	(35.1)
.2	.835 786	$-$ .658 538	1.5008	(35.1)
(.3	.777 276	$-$ .514 158	1.3994)	(35.3)
(.3	.777 236	$-$ .514 096	1.3991)	(35.1)
.3	.777 239	$-$ .514 100	1.3992	(35.1)

x	y	y'	y''	C_0
0	1.000 000	$-1.000\ 000$	2.0000	
0.1	.909 410	$-$ .817 026	1.6860	
.2	.835 786	$-$ .658 538	1.5008	
.3	.777 239	$-$ .514 100	1.3992	-40
.4	.732 728	$-$ .376 890	1.3523	-26
.5	.701 769	$-$ .242 480	1.3403	-16
.6	.684 230	$-$ .108 169	1.3480	-13

* Assume $y_2'' - y_1'' = y_1'' - y_0''$ to get y_2''; use Simpson's rule to get y_2', equation 35.1 to get y_2.

theses at the right indicate the formula used for the calculation of y in that line. The second part shows the completed and corrected calculation from $x = 0$ to $x = 0.6$.

36. Variations of Method IX

Following the procedure of Article 35, we may obtain the pair of formulas,

$$(36.1) \quad y_{n+1} - y_{n-2} = 3(y_n - y_{n-1}) + \frac{h^3}{2}(y_n''' + y_{n-1}''')$$
$$+ \frac{420h^7y^{(7)}}{100,800},$$

which we use as a predictor, and

$$(36.2) \quad y_n - y_{n-1} = \frac{h}{2}(y_n' + y_{n-1}') - \frac{h^2}{10}(y_n'' - y_{n-1}'')$$
$$+ \frac{h^3}{120}(y_n''' + y_{n-1}''') - \frac{h^7y^{(7)}}{100,800}$$

which we use for the final determination of y_n. The remainder term in equation 36.2 shows that this formula provides a high degree of accuracy. Accuracy is, of course, only obtained at a price, and must be paid for by the additional labor required by the calculation of the second and third derivatives. As a general rule, this method should be applied only where the form of the differential equation is such that the additional derivatives can be computed without excessive labor. In Chapter 6 formula 36.2 will be applied to linear differential equations of second order.

Another pair of formulas of the same sort is

$$(36.3) \quad y_{n+2} - y_{n-2} = 2(y_{n+1} - y_{n-1}) - 2h\delta^2 y_n'$$
$$+ 2h^2(y_{n+1}'' - y_{n-1}'') + \frac{105h^7y^{(7)}}{4725}$$

$$(36.4) \quad y_{n+1} - y_{n-1} = 2hy_n' + \frac{7h}{15}\delta^2 y'$$
$$- \frac{h^2}{15}(y_{n+1}'' - y_{n-1}'') + \frac{h^7y^{(7)}}{4725}.$$

This pair is of seventh order, and uses one less derivative than equations 36.1 and 36.2 but requires more lines of the computation.*

*This is unstable. Replace by methods of Appendix E.

It is evident that by comparison of the remainder terms the accuracy of the calculation can be controlled by the aid of the C_0 column just as was done in Method VII.

Formulas 35.1 and 36.2 are special cases of a general formula due to Obrechkoff [148]. The methods of the text are patterned after that of Milne [132]. Another method based on Obrechkoff's formula is given by Beck [6], and Hartree [84] has a still different procedure that uses an additional derivative.

EXERCISES

1. Carry computation 17 five steps farther by means of equations 36.3 and 36.4.

2. Take initially needed values from computation 17, compute y''', and then obtain the solution of the problem in computation 17 with $h = 0.2$, with the aid of equations 36.1 and 36.2.

6

SYSTEMS OF EQUATIONS
HIGHER-ORDER EQUATIONS

The methods of numerical solution of differential equations given in the preceding chapters have been developed and illustrated for the case of a single differential equation of the first order. The extension to systems of first-order equations and to equations of higher order involves relatively little that is new. There are, however, certain special methods that merit consideration. These will be considered later in this chapter after the extension of the general methods has been made.

37. A system of first-order equations

Any one of the methods already described can be at once applied to a system of simultaneous differential equations of the first order,

$$y_1' = f_1(x, y_1, \cdots, y_n),$$
$$y_2' = f_2(x, y_1, \cdots, y_n),$$

(37.1)

$$\cdots \cdots \cdots \cdots \cdots$$

$$y_n' = f_n(x, y_1, \cdots, y_n).$$

The values of y_i at the $(n + 1)$th step are obtained from y_i' and the preceding values of y_i and y_i' by the same method as in the case of a single equation. The calculation of the y_i' from equations 37.1 is the only part of the process where the presence of several equations instead of one makes any difference in the method of solution, and this difference consists simply in the substitution of the several values $y_1, y_2, \cdots, y_n$ instead of the single value y.

As an illustration we present a few steps in the numerical solution of the system of three equations.

(37.2) $$x' = yz, \qquad y' = -xz, \qquad z' = -\tfrac{1}{2}xy,$$

in which primes denote differentiations with respect to the independent variable u and where

(37.3) $$x = 0, \quad y = 1, \quad z = 1 \quad \text{when} \quad u = 0.$$

We shall employ Method III to get starting values and Method VII as given in Article 28 for the body of the calculation. Differentiation of equations 37.2 gives

$$x' = yz,$$

$$x'' = y'z + yz',$$

$$x^{(3)} = y''z + 2y'z' + yz'',$$

$$x^{(4)} = y^{(3)}z + 3y''z' + 3y'z'' + yz^{(3)},$$

with similar equations for y and z. From equations 37.3 these provide coefficients for Taylor's series,

$$x = u - u^3/4 + \cdots,$$

$$y = 1 - u^2/2 + u^4/8 - \cdots,$$

$$z = 1 - u^2/4 + 3u^4/32 - \cdots.$$

Setting $u = -0.1, 0, 0.1, 0.2$ in succession, we find the three starting values and a trial value for $u = 0.2$. From these we calculate x', y', z', by 37.2 and proceed with the computation, using Method VII.

Method VII   Computation 18

Differential equations $x' = yz$, $y' = -xz$, $z' = -\tfrac{1}{2}xy$

u	x	x'	$\tfrac{1}{3}\delta^2 x'$	C_0	y	y'	$\tfrac{1}{3}\delta^2 y'$	C_0
					($x = 0$,	$y = 1$,	$z = 1$ at $u = 0$)	$h = 0.1$
−0.1	−0.09975	0.99253			0.99501	0.09950		
0	0	1	−498		1	0	0	
0.1	0.09975	0.99253	−484		0.99501	−0.09950	98	
.2	.19802	.97055	−446	2	.98020	− .19607	187	0
.3	.29341	.93518	−385	2	.95598	− .28703	261	−1
.4	.38467	.88826	−313	2	.92306	− .37017	314	−1
.5	.47075	.83194	−230	1	.88226	− .44390	343	−3
.6	.55083	.76871	−153	1	.83462	− .50733	352	−2
.7	.62434	.70090	− 77	−1	.78115	− .56020	340	−1
.8	.69093	.63077	− 16	0	.72292	− .60286	314	−2
.9	.75048	.56016	+ 35	−1	.66089	− .63609	277	−1
1.0	.80300	.49059		−1	.59598	− .66100		−1

u	z	z'	$\tfrac{1}{3}\delta^2 z'$	C_0
−0.1	0.99751	0.04963		
0	1	0	0	
0.1	0.99751	−0.04963	74	
.2	.99015	− .09705	141	0
.3	.97824	− .14025	197	−1
.4	.96230	− .17754	239	−1
.5	.94297	− .20766	264	−2
.6	.92103	− .22987	274	−2
.7	.89727	− .24385	270	−1
.8	.87253	− .24974	255	−1
.9	.84758	− .24799	232	−1
1.0	.82316	− .23929		−1

The system of equations 37.2 and 37.3 has as its solution the elliptic functions of Jacobi,

$$x = \text{sn } (u, k), \qquad y = \text{cn } (u, k), \qquad z = \text{dn } (u, k),$$

for the particular case where $k^2 = \frac{1}{2}$. The values obtained in computation 18 check with those in published tables except in a few cases where the fifth digit differs by one unit.

38. Equations of second or higher order

A single differential equation of order higher than the first can always be replaced by an equivalent system of first-order equations so that the numerical solution of higher-order equations is merely a special case of systems of first-order equations. However, we may call attention to a slight simplification that occurs in the use of those numerical methods which predict and subsequently correct. In considering an equation of second order

$$y'' = f(x, y, y')$$

suppose that the computation has been completed through $x = x_n$, giving the lines

$$. \; . \; . \; . \; . \; . \; . \; . \; . \; . \; . \; . \; . \; .$$

$$x_{n-1} \quad y_{n-1} \quad y_{n-1}' \quad y_{n-1}''$$

$$x_n \quad y_n \quad y_n' \quad y_n''$$

Then we use our predictor formula in the usual manner to obtain y_{n+1}' from the values of y'', but obviously it is unnecessary to use the predictor for y_{n+1} itself. If we denote the predictor formula by A and the formula used for correction by B, the steps are:

1. Obtain y_{n+1}' from y_n'', y_{n-1}'', etc. by A.
2. Obtain y_{n+1} from y_{n+1}', y_n', etc. by B.
3. Obtain y_{n+1}'' from the differential equation.
4. Recompute y_{n+1}' from y_{n+1}'', y_n'', etc., by B.

If necessary

5. Recompute y_{n+1} from y_{n+1}', y_n', etc. by B.
6. Repeat 3, 4, 5 until no change occurs.

As an example we apply Method VII to the second-order equation

$$60 - H = 77.7H'' + 19.42(H')^2$$

with $H = 0$, $H' = 0$, when $t = 0$. This specific equation arose in the analysis of a proposed sewage ejector for a sewage disposal plant. Here H denotes the distance in feet from initial level to instantaneous level of the fluid in the ejection chamber and t the time in seconds from the beginning of the down stroke. The solution, carried to hundredths of a foot at intervals of half a second, appears in computation 19.

Method VII   Computation 19

Differential equation 60 $- H = 77.7H'' + 19.42(H')^2$			$h = 0.5$
t	H	H'	H''
0	0	0	0.77
0.5	0.10	0.38	.73
1.0	.37	.72	.64
1.5	.81	1.01	.51
2.0	1.37	1.23	.38
2.5	2.03	1.39	.26
3.0	2.75	1.50	.18
3.5	3.52	1.57	.11
4.0	4.31	1.61	.07

39. Special equations of second order. Method X*

Differential equations of the type

$$y_1'' = f_1(t, y_1, \cdots, y_n),$$

$$y_2'' = f_2(t, y_1, \cdots, y_n),$$

$$\cdots \cdots \cdots \cdots \cdots$$

$$y_n'' = f_n(t, y_1, \cdots, y_n),$$

in which first-order derivatives do not appear are found so frequently in applied problems, particularly those arising from the laws of motion, that special methods have been devised for their solution.   The essential idea of these methods is to go directly from the second derivatives y_i'' to the functions y_i by a single formula of double integration instead of using one integration to get y_i' and another integration for y_i.

A suitable formula for our purposes may be obtained by two-fold integration of Stirling's interpolation formula.   We have in general

$$y = y_0 + (x - x_0)y_0' + \int_{x_0}^{x} \left[\int_{x_0}^{z} y''(t)dt \right] dz,$$

whence

$$y_1 = y_0 + hy_0' + \int_{x_0}^{x_1} \left[\int_{x_0}^{z} y''(t)dt \right] dz,$$

$$y_{-1} = y_0 - hy_0' + \int_{x_0}^{x_{-1}} \left[\int_{x_0}^{z} y''(t)dt \right] dz,$$

and finally

$$(39.1) \quad \delta^2 y_0 = \int_{x_0}^{x_1} \left[\int_{x_0}^{z} y''(t)dt \right] dz + \int_{x_0}^{x_{-1}} \left[\int_{x_0}^{z} y''(t)dt \right] dz.$$

The importance of equation 39.1 lies in the fact that it does not contain the first derivative y'.

*Replace this Article by Appendix F.

The second derivative y'' may be expressed in terms of Stirling's central difference formula in the form

$$(39.2) \quad y'' = y_0'' + \mu\delta y_0'' s + \frac{\delta^2 y_0'' s^2}{2!} + \frac{\mu\delta^3 y_0'' s(s^2 - 1)}{3!}$$
$$+ \frac{\delta^4 y_0'' s^2(s^2 - 1)}{4!} + \frac{\mu\delta^5 y_0'' s(s^2 - 1)(s^2 - 4)}{5!}$$
$$+ \frac{\delta^6 y_0'' s^2(s^2 - 1)(s^2 - 4)}{6!} + \cdots ,$$

where $s = (x - x_0)/h$. When we apply 39.1 to this formula, we find that the terms of odd degree drop out and that the final result is

$$(39.3) \quad \delta^2 y_0 = h^2 \left(y_0'' + \frac{1}{12} \delta^2 y_0'' - \frac{1}{240} \delta^4 y_0'' + \frac{31}{60,480} \delta^6 y_0'' \right.$$
$$\left. - \frac{289}{3,628,800} \delta^8 y_0'' + \cdots \right).$$

In symbolic form this may be written

$$\delta^2 y_0 = h^2 \left[\frac{\delta/2}{\sinh^{-1} (\delta/2)} \right]^2 y_0''.$$

One method of applying equation 39.3 to the numerical solution of differential equations in the form

$$y'' = f(x, y)$$

is analogous to Method VI for equations of first order. Here the unknown terms on the right in 39.3 at a given stage are estimated by extrapolation in the table of differences of y'', then $\delta^2 y$ is calculated, and the value of y is obtained by summation. After this, y'' is calculated, and another line of differences is added. The cycle is to be repeated until no change occurs.

As an illustration we apply the method to the differential equation

$$y'' = -xy$$

with the initial values $y = 1$, $y' = 0$ when $x = 0$. Starting values are found from the Taylor's series

$$y = 1 - \frac{1}{3!} x^3 + \frac{1 \cdot 4}{6!} x^6 - \frac{1 \cdot 4 \cdot 7}{9!} x^9 + \cdots .$$

Method X Computation 20

Differential equation $y'' = -xy$ ($y = 1$, $y' = 0$, at $x = 0$) $h = 0.1$

x	y	δy	$\delta^2 y$	y''	$\delta y''$	$\delta^2 y''$	$\delta^3 y''$	$\delta^4 y''$	$\delta^5 y''$	$\delta^6 y''$	$\delta^7 y''$
-0.2	1.0013 33689			0.200 2667							
		-116 7017			-1002 500						
-.1	1.0001 66672		100 0345	0.100 0167		2333					
		-16 6672			-1000 167		-1999				
.0	1.0000 00000		11	0		334		3997			
		-16 6661			-999 833		1998		-4		
.1	.9998 33339		-999 656	-0.099 9833		2332		3993		-37	
		-116 6317			-997 501		5991		-41		-13
.2	.9986 67022		-1996 657	-.199 7334		8323		3952		-50	
		-316 2974			-989 178		9943		-91		-36
.3	.9955 04048		-2985 006	-.298 6512		18266		3861		-86	
		-614 7980			-970 912		13804		-177		-20
.4	.9893 56068		-3954 767	-.395 7424		32070		3684		-106	
		-1010 2747			-938 842		17488		-283		(-20)
.5	.9792 53321		-4892 150	-.489 6266		49558		3401		(-126)	
		-1499 4897			-889 284		20889		(-409)		(-20)
.6	.9642 58424		-5779 692	-.578 5550		70447		(2992)		(-146)	
		-2077 4589			-818 837		(23881)		(-555)		(-20)
.7	.9434 83835			-.660 4387		(94328)		(2437)		(-166)	

x	y	δy	$\delta^2 y$	y''	$\delta y''$	$\delta^2 y''$	$\delta^3 y''$	$\delta^4 y''$	$\delta^5 y''$	$\delta^6 y''$	$\delta^7 y''$
.5	—	—	—	—	—	—	—	—	-415		
.6	—	—	—					2986			
.7	—	—	-6596 537		-724 515		23875				
.8	.9161 12709	-2737 1126		-.732 8902		94322					

Computation 20 displays the corrected and completed work through $x = 0.7$. In order to proceed we extrapolate in the difference table, assuming constant seventh differences, and thus supply the numbers in parenthesis. To the line $x = 0.7$ we now apply formula 39.3 and obtain $\delta^2 y = -0.0065\ 96537$, $\delta y = -0.0273\ 71126$, and $y = 0.9161\ 12709$ at $x = 0.8$. Then the differential equation gives $y'' = -0.7328902$, and we correct the corresponding diagonal of differences. The corrected values are shown at the bottom of computation 20. For this diagonal of differences the correction turned out to be 6 in the seventh decimal place, or 600 if carried to nine places. This correction of 600 is to be multiplied by $h^2/12 = 1/1200$ and hence is negligible to nine places. The fourth and sixth differences in the line for $x = 0.7$ are still not exactly known, but, since they are multiplied by $1/24000$ and $1/195\ 000$, respectively, only unreasonably large errors in their estimated values could change the computed value of y.

This method appears to be due to Störmer [199, 200, 201] and is essentially that used by Cowell and Crommelin [41] in computing the orbit of Halley's comet. See also Nyström [145].

40. Method of summation*

In Method X we used formula 39.3 to calculate the second difference of y and then found y itself by summation from the second difference. This procedure may be varied by applying a double summation directly to formula 39.3 so as to obtain

$$(40.1) \quad y_0 = h^2 \left(\sum^2 y_0'' + \frac{1}{12} y_0'' - \frac{1}{240} \delta^2 y_0'' \right.$$

$$\left. + \frac{31}{60480} \delta^4 y_0'' - \cdots \right).$$

The expression $\sum^2 y_0''$ is derived from y_0'' by double summation and contains two arbitrary constants which have to be determined by the initial conditions.

For sake of comparison we apply equation 40.1 to the solution of the problem given in computation 20. Computation 21 shows the calculation through $x = 0.7$. As before we extrapolate in the difference table on the assumption that seventh differences are constant and thus obtain the differences shown in parentheses. Applying 40.1 to the line $x = 0.8$, we obtain $y = 0.9161\ 12709$. Now that y is known, the differential equation gives $y'' = -0.732\ 8901$, and we correct the differences of y''. Again it is clear that the correction does not affect y and hence a fortiori does not affect y''.

It is of some interest to compare computations 20 and 21 with respect to the effect of errors arising from the assumption of constant seventh differences. Let the errors of estimation of seventh differences be in order e_1, e_2, e_3, $\cdots$. These

*See Appendix F.

Method Xa

Computation 21

Differential equation $y'' = -xy$ ($y = 1$, $y' = 0$, at $x = 0$) $h = 0.1$

x	y	$\Sigma^2 y''$	$\Sigma y''$	y''	$\delta y''$	$\delta^2 y''$	$\delta^3 y''$	$\delta^4 y''$	$\delta^5 y''$	$\delta^6 y''$	$\delta^7 y''$
0	1.0000 00000	100.000 0000		0		334		3997			
			−0.008 3334		−999 833		1998		− 4		
0.1	.9998 33339	99.991 6666		−0.099 9833		2 332		3993		− 37	
			− .108 3167		−997 501		5991		− 41		− 13
.2	.9986 67021	99.883 3499		− .199 7334		8 323		3952		− 50	
			− .308 0501		−989 178		9943		− 91		− 36
.3	.9955 04048	99.575 2998		− .298 6512		18 266		13804		− 86	
			− .606 7013		−970 912		13804		−177		− 20
.4	.9893 56068	98.968 5985		− .395 7424		32 070		17488		−106	
			−1.002 4437		−938 842		17488		−283		(− 20)
.5	.9792 53321	97.966 1548		− .489 6266		49 558		20889		(−126)	
			−1.492 0703		−889 284		20889		(−409)		(− 20)
.6	.9642 58424	96.474 0845		− .578 5550		70 447		(2992)		(−146)	
			−2.070 6253		−818 837		(23881)		(−555)		(− 20)
.7	.9434 83835	94.403 4592		− .660 4387		(94 328)		(2437)		(−166)	
			−2.731 0640		(−724 509)		(26318)		(−721)		
.8	.9161 12709	91.672 3952		(− .732 8896)		(120 646)		(1716)			
.8				− .732 8901							

errors are propagated through the difference table in the manner shown in the diagram below.

x	y''	$\delta y''$	$\delta^2 y''$	$\delta^3 y''$	$\delta^4 y''$	$\delta^5 y''$	$\delta^6 y''$	$\delta^7 y''$
0.4 —	—	—		—			—	
		—	—		—			e_1
.5 —	—		—		—		e_1	
		—		—		e_1		e_2
.6 —	—			e_1		$e_1 + e_2$		
		—	e_1		$2e_1 + e_2$			e_3
.7 —	e_1		$3e_1 + e_2$		$e_1 + e_2 + e_3$			
	e_1	$4e_1 + e_2$		$3e_1 + 2e_2 + e_3$				
.8 e_1	$5e_1 + e_2$	$6e_1 + 2e_2 + e_3$						

In computation 20 the formula 39.3 is applied to the line $x = 0.7$, and the error due to faulty estimate is evidently

$$(40.2) \quad E_1 = h^2 \left[\frac{1}{12} e_1 - \frac{1}{240} (3e_1 + e_2) + \frac{31}{60480} (e_1 + e_2 + e_3) + \cdots \right].$$

whereas in computation 21 formula 40.1 is applied to the line $x = 0.8$, and the error is

$$(40.3) \quad E_2 = h^2 \left[\frac{1}{12} e_1 - \frac{1}{240} (5e_1 + e_2) + \frac{31}{60480} (6e_1 + 2e_2 + e_3) - \cdots \right].$$

If $|e_i| < e$ for $i = 1, 2, 3$, we find that

$$|E_1| < 0.076h^2 e,$$
$$|E_2| < 0.070h^2 e.$$

This analysis must of course be modified if we use some other difference column than the seventh as the basis for extrapolation.

The Method Xa was proposed by Cowell and Crommelin [41], and its merits are clearly presented by Herrick [86].

41. Method XI[*]

Equations of the second order in which first derivatives do not occur can be neatly solved by a method quite analogous to Method VII. The formula

$$(41.1) \quad y_{n+1} - 2y_{n-1} + y_{n-3} = 4h^2 \left(y_{n-1}'' + \frac{1}{3} \delta^2 y_{n-1}'' \right) + \frac{16}{240} h^6 y^{(6)}$$

is used to integrate ahead, and

$$(41.2) \quad y_{n+1} - 2y_n + y_{n-1} = h^2 \left(y_n'' + \frac{1}{12} \delta^2 y_n'' \right) - \frac{h^6 y^{(6)}}{240}$$

[*]See Appendix F.

supplies the correction. The error of equation 41.1 is seen to be about one-seventeenth the difference between the predicted and the corrected value, and hence this difference supplies a current check on the accuracy attained, just as in Article 28. The actual calculation is compactly arranged in tabular form as follows:

x	y	y''	$\delta^2 y''$
x_{n-3}	y_{n-3}	y_{n-3}''	$\delta^2 y_{n-3}''$
x_{n-2}	y_{n-2}	y_{n-2}''	$\delta^2 y_{n-2}''$
x_{n-1}	y_{n-1}	y_{n-1}''	$\delta^2 y_{n-1}''$
x_n	y_n	y_n''	
x_{n+1}			

To proceed we apply equation 41.1 to the line $x = x_{n-1}$, obtaining a trial value of y_{n+1}. The y_{n+1}'' is calculated from the differential equation, and the second difference from the equation $\delta^2 y_n'' = y_{n+1}'' - 2y_n'' + y_{n-1}''$. We now apply equation 41.2 to the line $x = x_n$ and correct the values of y_{n+1}, y_{n+1}'', and $\delta^2 y_n''$. If h has been properly chosen, further correction should be unnecessary.

To illustrate the process we take a problem arising in a study of an electron moving in a field due to two charged parallel wires, one with negative and one with positive charge. If the motion is in a plane perpendicular to the wires the equations may be reduced to

(41.3)
$$\frac{d^2x}{dt^2} = \frac{x+1}{(x+1)^2+y^2} - \frac{x-1}{(x-1)^2+y^2},$$
$$\frac{d^2y}{dt^2} = \frac{y}{(x+1)^2+y^2} - \frac{y}{(x-1)^2+y^2}.$$

We take the particular solution where $x = 0$, $y = 2$, $x' = 1$, $y' = 0$, at $t = 0$.

Method XI					Computation 22	
See equations 41.3					$h = 0.4$	
t	x	x''	$\delta^2 x''$	y	y''	$\delta^2 y''$
0	0	0.400		2.000	0	
0.4	0.432	.369	−0.064	1.996	−0.132	0.019
.8	.921	.274	− .042	1.972	− .246	.062
1.2	1.454	.136	.010	1.909	− .298	.074
1.6	2.009	.008	.048	1.799	− .276	.043
2.0	2.566	− .072		1.645	− .212	

This method is essentially that of Milne [130].

42. Linear equations of second order. Method XII

The general linear differential equation of second order

$$\frac{d^2u}{dt^2} + P(t)\frac{du}{dt} + q(t)u = r(t)$$

in which $P(t)$, dP/dt, $q(t)$, and $r(t)$ are assumed continuous, can be transformed into an equation of the type

(42.1) $$d^2y/dx^2 + g(x)y = f(x)$$

by the change of variables

$$u = e^{-\frac{1}{2}\int P\,dt}y, \qquad t = x$$

or by the change of variables

$$u = y, \qquad x = \phi(t), \quad \text{where} \quad d\phi/dt = Ce^{-\int P(t)\,dt}.$$

One method of solving equation 42.1 numerically replaces the second-order differential equation by a suitable chosen difference equation.

To derive an appropriate difference equation we differentiate Stirling's central difference interpolation formula twice, set the argument equal to the central value, and obtain (using Sheppard's central difference operator δ)

$$y'' = \frac{1}{h^2}\left(2\sinh^{-1}\frac{\delta}{2}\right)^2 y$$

(42.2) $$= \frac{1}{h^2}\left(\delta^2 y - \frac{\delta^4 y}{12} + \frac{\delta^6 y}{90} - \cdots\right),$$

in which h denotes the length of the interval between equally spaced values of x. Next we operate on equations 42.1 and 42.2 with the operator δ^2, which gives

(42.3) $$\delta^2 y'' + \delta^2[g(x)y] = \delta^2 f(x)$$

and

(42.4) $$\delta^2 y'' = \frac{1}{h^2}\left(\delta^4 y - \frac{\delta^6 y}{12} + \cdots\right).$$

From equations 42.1, 42.2, 42.3, and 42.4 we eliminate the three quantities y'', $\delta^2 y''$, and $\delta^4 y$.

The resulting equation is

$$(42.5) \quad \delta^2 \left[1 + \frac{h^2 g(x)}{12} \right] y + h^2 g(x) y$$

$$= h^2 f(x) + \frac{h^2 \delta^2 f(x)}{12} - \frac{\delta^6 y}{240} + \cdots.$$

The substitutions

$$\left[1 + \frac{h^2 g(x)}{12} \right] y = z, \qquad \frac{h^2 g(x)}{1 + \frac{h^2 g(x)}{12}} = G(x)$$

and

$$h^2 f(x) + \frac{h^2 \delta^2 f(x)}{12} = F(x)$$

carry this equation into

$$(42.6) \qquad \delta^2 z + G(x) z = F(x) - \frac{\delta^6 y}{240} + \cdots.$$

Now, if the interval is small enough so that the term $\delta^6 y / 240$ is negligible, equation 42.6 reduces to a simple linear difference equation of the second order.

The procedure in solving 42.6 numerically is obvious. First we select a suitable value for h, then tabulate the values of $G(x)$ and $F(x)$ according to the scheme shown below:

x	$F(x)$	$G(x)$	z	Δz	$\Delta^2 z$
x_0	F_0	G_0	z_0		
x_1	F_1	G_1	z_1	Δz_0	
x_2	F_2	G_2	z_2	Δz_1	$\Delta^2 z_0 = \delta^2 z_1$
x_3	F_3	G_3			
x_4	F_4	G_4			

The first two values z_0 and z_1 are assumed to be known, and from them we get $\Delta z_0 = z_1 - z_0$. Then from 42.6 we have

$$\Delta^2 z_0 = \delta^2 z_1 = -G_1 z_1 + F_1,$$

and we complete the line for x_2 with the formulas

$$\Delta z_1 = \Delta z_0 + \Delta^2 z_0,$$

$$z_2 = z_1 + \Delta z_1.$$

The integration proceeds by repetitions of this process. When it is completed, the values of y, or at least as many of them as are desired, are obtained from

$$y = \frac{z}{1 + \dfrac{h^2 g(x)}{12}}.$$

If the general solution of 42.1 is required, it may be obtained by solving

$$\delta^2 u + G(x)u = 0, \qquad u_0 = 0, \qquad u_1 = h,$$

$$\delta^2 v + G(x)v = 0, \qquad v_0 = v_1 = 1,$$

and

$$\delta^2 w + G(x)w = F(x), \qquad w_0 = w_1 = 0.$$

The general solution of 42.6 is then

$$z = C_1 u + C_2 v + w,$$

and from z we readily obtain y.

Consider the differential equation

$$y'' + \frac{3 + 4x}{16x^2} y = 0 \quad y = 1, \quad y' = 0, \quad \text{at} \quad x = 1.$$

Since $f(x) = 0$, we need calculate only $G(x)$, which for $h = 0.1$ proves to be

$$G(x) = \frac{12(3 + 4x)}{19200x^2 + (3 + 4x)}.$$

The values of $G(x)$ are calculated from this expression and tabulated as shown below. We find by a few terms of Taylor's series that $y_1 = 0.997911$, so that

Method XII   Computation 23

Differential equation $y'' = -(3 + 4x)y/16x^2$, $y = 1$ $y' = 0$ at $x = 1$

$h = 0.1$

x	$G(x)$	z	Δz	$\Delta^2 z$	y
1.0	0.004 3734	1.000 364			1.000 000
1.1	.003 8211	.998 228	−2136		
1.2	.003 3845	.992 278	−5950	−3814	
1.3	.003 0318	.982 970	−9308	−3358	
1.4	.002 7417	.970 682	−12288	−2980	
1.5	.002 4995	.955 733	−14949	−2661	
1.6	.002 2945	.938 395	−17338	−2389	
1.7	.002 1190	.918 904	−19491	−2153	
1.8	.001 9673	.897 466	−21438	−1947	
1.9	.001 8349	.874 262	−23204	−1766	
2.0	.001 7185	.849 454	−24808	−1604	.849 333

$z_0 = 1.000\,364$, $z_1 = 0.998\,228$. Everything needed is now known, and the calculation is completed to $x = 2.0$ as in computation 23.

The equation chosen above can be solved explicitly, and the particular solution desired proves to be

$$y = x^{\frac{1}{4}} [\cos\,(x^{\frac{1}{2}} - 1) - 0.5 \sin\,(x^{\frac{1}{2}} - 1)].$$

The substitution of $x = 2$ in this expression gives $y = 0.849\,3293$ which differs only by about four units in the sixth decimal place from the value found by numerical integration.

43. Linear equations of second order.   Method IX

A more powerful method than that shown in Article 42 is provided by Method IX, using formulas 36.1 and 36.2 together with the same two formulas applied to y', y'', y''', $y^{(4)}$ instead of to y, y', y'', y'''. Let us designate these two, respectively, as formulas 36.1' and 36.2'. After the computation is under way through $x = x_n$, we predict y_{n+1}' by 36.1' and y_{n+1} by 36.1, then from the differential equation (and its derivatives) we get y_{n+1}'', y_{n+1}''', $y^{(4)}{}_{n+1}$. We now use 36.2' to correct y_{n+1}' and 36.2 to correct y_{n+1}.

As an example we apply this method to Bessel's equations of order zero

$$xy'' + y' + xy = 0$$

with $y = 1$, $y' = 0$ when $x = 0$. The required initial values were obtained from series after which Method IX was used. The completed work is shown in computation 24. The values of y and y' to ten decimal places as given in existing tables are supplied for comparison.

44. Linear equations.   Method XIII

One of the most obvious ways to solve a linear differential equation numerically is to replace the derivatives in the equation by their formal expressions in terms of differences, truncated to a suitable order, and then to solve the resulting difference equation. In this crude form, however, the method presents the following dilemma: If the formulas for the derivatives are truncated to the same order as the differential equation, the resulting difference equation does not give good accuracy except for very small intervals; if more terms are retained the difference equation is of too high an order to permit easy numerical solution.

To see how these difficulties are circumvented we consider for example a linear differential equation of second order

$$(44.1) \qquad [P_0(x)D^2 + P_1(x)D + P_2(x)]y = f(x),$$

Method IX — Computation 24

Differential equation $xy'' + y' + xy = 0$. $y = 1$, $y' = 0$, at $x = 0$ — $h = 0.1$

x	y	y'	y''	y'''	$y^{(4)}$	c	c'
0.0	1 —	0 —	−0.5 —	0 —	0.375 —	—	—
.1	0.99750 15621	−0.04993 75260	− .49812 63017	0.03744 79394	.37343 86390	—	—
.2	.99002 49723	− .09950 08326 (7)	− .49252 08093	.07458 40641	.36876 81738	0	−1
.3	.97762 62466	− .14831 88162 (6)	− .48323 01926	.11109 92782	.36102 95182	0	−4
.4	.96039 82267	− .19602 65779 (81)	− .47033 17820	.14668 99212	.35029 02625	0	−2
.5	*.93846 98073 (2)	− .24226 84576 (9)	− .45393 28921	.18106 04114	.33664 42541	−1	−3
.6	.91200 48636 (5)	− .28670 09880 (1)	− .43416 98836	.21392 58273	.32021 07071	−1	−1
.7	.88120 08887	− .32899 57415 (6)	− .41120 69723	.24501 43928	.30113 31217	0	−1
.8	.84628 73528 (7)	− .36884 20461 (2)	− .38523 47952	.27406 98431	.27957 79988	−1	−1
.9	.80752 37982 (0)	− .40594 95461 (2)	− .35646 87470	.30085 36525	.25573 33411	−2	−1
1.0	.76519 76866 (0)	− .44005 05858 (9)	− .32514 71008	.32514 71008	.22980 69700	−6	−1

True Values

x	y	y'
—	—	—
—	—	—
0.2	0.99002 49722	−0.09950 08326
.3	0.97762 62465	− .14831 88163
.4	0.96039 82267	− .19602 65780
.5	0.93846 98072	− .24226 84577
.6	0.91200 48635	− .28670 09881
.7	0.88120 08886	− .32899 57415
.8	0.84628 73528	− .36884 20461
.9	0.80752 37981	− .40594 95461
1.0	0.76519 76866	− .44005 05857

* In the column for y and y' appear the corrected values. The digits of the predicted values when different from the corrected are shown in parentheses.

where, as is customary, $D = d/dx$.   In terms of central differences in Sheppard's notation the differential operator D is given formally by

$$(44.2) \quad D = \frac{1}{h}\left(\mu\delta - \frac{\mu\delta^3}{3!} + \frac{2^2\mu\delta^5}{5!} - \frac{2^2 \cdot 3^2\mu\delta^7}{7!} \right.$$
$$\left. + \frac{2^2 \cdot 3^2 \cdot 4^2\mu\delta^9}{9!} - \cdots \right).$$

From this, together with the symbolic identity

$$\mu^2 = 1 + \delta^2/4,$$

we can obtain formulas for $D^2, D^3, \cdots$.   In particular,

$$(44.3) \quad D^2 = \frac{1}{h^2}\left(\delta^2 - \frac{2\delta^4}{4!} + \frac{2 \cdot 2^2\delta^6}{6!} - \frac{2 \cdot 2^2 \cdot 3^2\delta^8}{8!} \right.$$
$$\left. + \frac{2 \cdot 2^2 \cdot 3^2 \cdot 4^2\delta^{10}}{10!} - \cdots \right).$$

In equation 44.1 we substitute for D and D^2 the values given by equations 44.2 and 44.3, and in the resulting equation collect together in a group on the right all differences of order higher than the second, so that we now have the equation

$$(44.4) \qquad\qquad Ly = h^2f(x) + Ky,$$

where

$$Ly = P_0(x)\delta^2 y + hP_1(x)\mu\delta y + h^2P_2(x)y$$

and

$$Ky = \left[hP_1(x)\left(\mu\delta^3/6 - \frac{\mu\delta^5}{30} + \cdots\right) + P_0(x)\left(\frac{\delta^4}{12} - \frac{\delta^6}{90} + \cdots\right)\right]y.$$

To get a first approximation z_0 for the solution of equation 44.4 we ignore Ky and solve the simple second-order difference equation

$$Lz_0 = h^2f(x).$$

From the values of z_0 thus obtained we can now compute the differences and hence the values of Kz_0.   The first correction z_1 is the solution of the simple difference equation

$$Lz_1 = Kz_0,$$

the second correction satisfies

$$Lz_2 = Kz_1$$

and so on, the desired value of y being given finally by

$$y = z_0 + z_1 + z_2 + \cdots,$$

provided the process converges.

Several comments are pertinent. First, the interval h should be taken small enough so that the desired accuracy can be obtained with not more than two or three approximations. Otherwise the method loses some of its advantages in comparison with other methods.

Second, the central differences will require values beyond the range of the computation. Normally the easiest way to get these is to solve the difference equation for additional values at both ends of the range. When because of singularities or lack of data or for other reasons this may be difficult or impossible, one may have recourse to extrapolation to secure the needed differences.

Third, it is much more convenient to start the solution of a difference equation with given values of $y(x_0)$, $y(x_1)$ than with given values of $y(x_0)$, $y'(x_0)$, etc. If possible then the problem to be solved by Method XIII should be formulated so that initially we have given values of y, rather than of y and its derivatives. For example, if we seek the general solution of 44.1, we may find w to satisfy 44.1 and the conditions $w(x_0) = 0$, $w(x_1) = 0$, and find u and v to satisfy the homogeneous equation derived from 44.1 with $u(x_0) = 1$, $u(x_1) = 1$, $v(x_0) = 0$, $v(x_1) = 1$. The general solution is then

$$y = Au + Bv + w.$$

Fourth, if the initial values $y(x_0)$ and $y(x_1)$ are given, it is convenient in solving the difference equations to take

$$z_0(x_0) = y(x_0), \qquad\qquad z_0(x_1) = y(x_1),$$
$$z_1(x_0) = z_2(x_0) = \cdots = 0, \qquad z_1(x_1) = z_2(x_1) = \cdots = 0.$$

As an illustrative example we select the differential equation

$$y'' + xy' + y = 2x$$

with given values $y = 1$ at $x = 0$, $y = 1$ at $x = h = 0.1$.

Here equation 44.4 becomes

$$(44.5) \quad \left(1 + \frac{n}{200}\right) y_{n+1} - (1.99)y_n + \left(1 - \frac{n}{200}\right) y_{n-1} = \frac{2n}{1000} + Ky$$

and

$$Ky = \frac{1}{12}\left[\frac{n}{100}\left(\delta^2 y_{n+1} - \delta^2 y_{n-1}\right) + \delta^4 y_n\right]$$

Method XIII

Computation 25

Differential equation $y'' + xy' + y = 2x$

Difference equation $(1 + .05x_n)y_{n+1} - (1.99)y_n + (1 - .05x_n)y_{n-1} = 0.02x_n$

x	$(1 + 0.05x_n)$	$(1 - 0.05x_n)$	$0.02x$	z_0	$\delta^2 z_0$ $10^{-7}\times$	$\delta^4 z_0$ $10^{-7}\times$	Kz_0 $10^{-8}\times$	z_1 $10^{-7}\times$	$y = z_0 + z_1$
−0.1	0.995	1.005	−0.002	0.990 0000					
0	1.000	1.000	0	1.000 0000	−100 000			0	1.000 000
0.1	1.005	.995	0.002	1.000 0000	− 79 602	2162	2160	0	1.000 000
.2	1.010	.990	.004	.992 0398	− 57 042	1242	1808	215	.992 061
.3	1.015	.985	.006	.978 3754	− 33 240	278	1429	603	.978 436
.4	1.020	.980	.008	.961 3870	− 9 160	− 672	1023	1114	.961 498
.5	1.025	.975	.010	.943 4826	14 248	−1552	593	1694	.943 652
.6	1.030	.970	.012	.927 0030	36 104	−2313	143	2287	.927 232
.7	1.035	.965	.014	.914 1338	55 647	−2920	− 323	2837	.914 418
.8	1.040	.960	.016	.906 8293	72 270	−3348	− 797	3291	.907 158
.9	1.045	.955	.018	.906 7518	85 545	−3582	−1262	3602	.907 112
1.0	1.050	.950	.020	.915 2288	95 238			3731	.915 602
1.1	1.055	.945	.022	.933 2296					

where we neglect differences of higher order than the fourth. The completed work appears in computation 25.

For additional variations of the foregoing methods see e.g. Fox and Goodwin [69], Bückner [18], Nyström [146], and Zurmühl [225]. Todd [210] points out the dangers of a blind replacement of differential by difference operators.

7

TWO-POINT BOUNDARY CONDITIONS

The methods explained in Chapter 6 were based on the assumption that all information necessary to determine a unique solution of the differential equation is known at the initial point. The present chapter deals with the more complex situation where conditions are imposed on the solution at both end points of the interval in which the solution is to be found.

45. The two-end-point problem

Two-point boundary conditions arise in all sorts of physical problems and also occur regularly in the calculus of variations.

As an illustration of exceptional simplicity we examine the deflection of an elastic beam under a transverse load. Let y denote the deflection let the beam be simply supported at the end where $x = 0$ and rigidly supported at the end where $x = L$ and let the load per unit length at the point x be $\phi(x)$. Then y satisfies the differential equation

(45.1) $$EIy^{(4)} = \phi(x)$$

and the boundary conditions

(45.2)
$$y = 0, \quad y'' = 0, \quad \text{at} \quad x = 0,$$
$$y = 0, \quad y' = 0, \quad \text{at} \quad x = L.$$

If we can obtain the general solution of the differential equation in any two-point boundary problem, we may find the desired particular solution simply by determining the arbitrary constants in the general solution so as to satisfy the given conditions. Thus in the problem just cited, if $\phi(x) = w$, a constant, we have

$$EIy = wx^4/24 + c_1x^3 + c_2x^2 + c_3x + c_4.$$

From the conditions 45.2 we discover that

$$c_1 = -wL/12, \quad c_2 = 0, \quad c_3 = wL^3/48, \quad c_4 = 0,$$

and the particular solution is

$$EIy = \frac{w}{48}(2x^4 - 3Lx^3 + L^3x).$$

As another example consider the problem of finding an arc through the points $(0, 1)$, $(1, 2)$ that minimizes

$$I = \int_0^1 \frac{(1 + y'^2)^{\frac{1}{2}} \, dx}{y}.$$

Here the Euler equation is

(45.3) $yy'' = -1 - y'^2$,

the general solution of which is

$$(x - A)^2 + y^2 = B^2,$$

and the particular solution through the given points is readily found to be

$$y = +\sqrt{5 - (x - 2)^2}.$$

These samples indicate that, when the general solution of the differential equation can be found, the two-end-point problem is not essentially more difficult than the one-end-point problem. When, however, it is necessary to resort to a numerical solution, we hit a snag not previously encountered. This is due to the fact that any step-by-step method requires the solution to be uniquely determined by the data available at the starting point. In the two-point problem there is no point at which sufficient data are known to start the solution. As a consequence no numerical method is as yet known for the two-point problem comparable in directness and simplicity to those described in Chapter 6.

In the following paragraphs we examine some of the proposed ways of attacking this problem.

There is, of course, another difficulty inherent in the nature of the two-point problem. There may exist no solution at all, or multiple solutions may occur. In many cases there appears to be no easy way to foresee what will happen in a given problem.

46. Linear equations

The procedure for linear equations is considerably simpler than for non-linear and hence will be treated first. An equation of second order will suffice to illustrate the method. The general solution of

$$y'' + Py' + Qy = R$$

with P, Q, and R functions of x has the form

$$y = Au + Bv + w$$

where A and B are constants, u and v are independent solutions of the equation obtained by setting $R = 0$, and w is a particular solution of the original equation. One condition is available at the starting point, sufficing to reduce the number of arbitrary constants by one, so that the solution is now effectively of the form

$$y = Au + w,$$

where u and w are chosen so that y satisfies the initial condition regardless of A. We determine u and w numerically by any suitable step-by-step process, and finally determine A so that y satisfies the terminal condition.

Example 1. Find the solution of

(46.1) $$y'' + xy' + y = 2x$$

which goes through the points $(0, 1)$ and $(1, 0)$.

Here we may take w as a solution of equation 46.1 such that $w(0) = 1$, and u as a solution of

(46.2) $$y'' + xy' + y = 0$$

such that $u(0) = 0$. The first quantity, w, has already been found by Method XIII in computation 25. The value of u is found by the same method in computation 26. The end condition that $y = 0$ when $x = 1$ provides the equation

$$0 = Au(1) + w(1)$$

for the determination of A. With A, u, and w now known we compute the desired values of y as shown in computation 26.

Two-end-point Problem   Computation 26

Differential equation $y'' + xy' + y = 2x$
    For values of w see computation 25.

x	z_0	$\delta^2 z_0$	$\delta^4 z_0$	Kz_0	z_1	$u = z_0 + z_1$	y
				$10^{-7}\times$	$10^{-7}\times$	$10^{-7}\times$	
−0.1	−0.100 0000						
0	0	0			0	0	1.000 000
0.1	0.100 0000	−0.001 9900	787	33	0	0.100 0000	.874 091
.2	.198 0100	− .003 9013	1530	66	33	.198 0133	.742 745
.3	.292 1187	− .005 6596	2183	100	130	.292 1317	.610 617
.4	.380 5678	− .007 1996	2707	132	321	.380 5999	.482 290
.5	.461 8173	− .008 4689	3083	164	631	.461 8804	.362 105
.6	.534 5979	− .009 4299	3290	194	1080	.534 7059	.253 991
.7	.597 9486	− .010 0619	3329	223	1681	.598 1167	.161 337
.8	.651 2374	− .010 3610	3202	248	2441	.651 4815	.086 887
.9	.694 1652	− .010 3399	2937	270	3358	.694 5010	0.032 675
1.0	.726 7531	− .010 0251			4422	.727 1953	0
1.1	.749 3159						

$$y = Au + w, \quad \text{where} \quad A = -1.259087.$$

The values of y given above are correct to within one unit in the sixth decimal.
Example 2.   Find the solution of

(46.3) $y'' + (\cosh x)y = 0$

through the points $(0, 0)$ $(2.2, 1)$.

Again we use Method XIII as shown in computation 25.   Since equation 46.3 is linear and homogeneous, any solution through the point $(0, 0)$ is of the form $y = A\phi$, when ϕ is a particular solution through $(0, 0)$.   In computation 27 we obtain a particular solution which at $x = 2.2$ takes the value $y = 0.09905$.   The value $A = 10.0959$ gives the desired solution, as shown in the last column of computation 27.

It is seen that the value of y obtained at $x = 2.2$ for the preliminary solution is rather small, resulting in a relatively large A.   Had the value of y been zero, or at least zero to several decimal places, our final solution would have been either impossible or very inaccurate.   The work was originally carried to 7 places and subsequently rounded to 5.   To save space most of the details of the actual computation have been omitted.

Two-end-point Problem   Computation 27

| Differential equation $y'' + (\cosh x)y = 0$ | | | $(0, 0)$ | $(2.2, 1)$ | $h = 0.2$ |
x	z_0	z_1	z_2	$z_0 + z_1 + z_2$	y
0	0	0	0	0	0
0.2	0.20000	0	0	0.20000	2.0192
.4	.39184	−0.00005	0	.39179	3.9555
.6	.56673	− .00020	0.00001	.56654	5.7197
.8	.71476	− .00047	.00001	.71431	7.2116
1.0	.82454	− .00082	.00003	.82375	8.3165
1.2	.88343	− .00117	.00005	.88231	8.9077
1.4	.87834	− .00133	.00008	.87709	8.8550
1.6	.79768	− .00098	.00011	.79680	8.0444
1.8	.63478	.00029	.00013	.63519	6.4128
2.0	.39297	.00289	.00011	.39598	3.9978
2.2	.09203	.00698	.00003	.09905	1.0000

47. Non-linear equations.   Trial and error

For non-linear equations the method of Article 46 fails.   One rather obvious way to tackle the two-end-point problem for non-linear equations is to start with assumed conditions at the initial point and calculate a trial solution as far as the end point.   This trial solution will fail to satisfy the given conditions at the end point, but the amount and direction of the deviation may suggest appropriate corrections to be made to the assumed initial conditions.   By repetition of this scheme of trial and correction one is frequently led to a solution of the problem.

For a differential equation of the second order where a solution is required through two given points A and B, the situation may be pictured in Fig. 11.   We start with a trial slope m_1 and compute a trial solution C_1 with initial values $y = y_0$, $y' = m_1$, at $x = x_0$. If as shown in the picture the solution C_1 hits at B_1 below B, we naturally try again with a larger slope m_2. Perhaps the new solution C_2 hits at a point B_2 above B, in which case we have the target "bracketed," and, unless the differential equation has some unexpected peculiarities, we may hope to arrive at the desired solution by successive trials. Since at best this process is long and tedious we should make as much use as possible of available information. For example, before choosing m_1 we should examine the differential equation to determine, if possible, something of the direction and magnitude of the curvature of

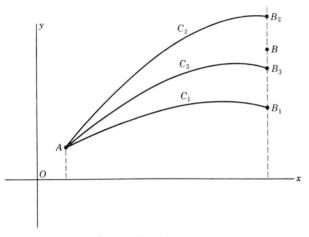

Fig. 11

the solution in the required range so as to make an intelligent guess at m_1. When two trial solutions with initial slopes m_1 and m_2 have been calculated, linear interpolation should be helpful in selecting a value for m_3. When three trials have been made, we may use quadratic interpolation, say, by means of Newton's formula with divided differences, to select an improved value of m.

All this of course is based on the assumption that the problem is well-behaved in the manner exhibited in Fig. 11. If, however, the first two trials behave as shown in Fig. 12, it is usually worthless to try interpolation until additional exploratory computations reveal more clearly the behavior of the system of integral curves through A. There may be no solution joining A and B, or there may be many such.

We see that in the simplest possible case, namely, in a single differential equation of the second order, the two-end-point problem can be troublesome. For a system of n equations in n unknowns the difficulties increase with n at an alarming rate.

Example 1. Find the solution of the equation

$$yy'' + 1 + y'^2 = 0$$

that passes through (0, 1) (1, 2).

This is the problem solved analytically in Article 45. Here we have to assume that we have no knowledge whatever of the solution and must proceed blindly.

As a first trial we start with $m = 1$. (This is obviously too small since y'' is negative for y positive.) The solution gives $y = 1.4142$ instead of $y = 2.0000$ (see computation 28). If we assume that the change in y due to a change μ in m is equal to μx, we find $\mu = 0.5858$ and accordingly compute a second trial solution

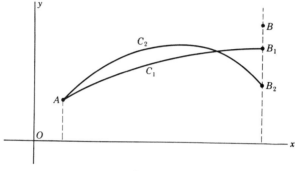

FIG. 12

with $m = 1.5858$. This gives $y = 1.7809$. Now linear interpolation yields $m = 1.9358$, and with this initial slope the third trial computation gives $y = 1.9676$. Quadratic interpolation yields $m = 2$, and the fourth trial computation gives $y = 2.0000$, the desired answer.

Two-end-point Problem   Computation 28

Differential equation $yy'' + 1 + y'^2 = 0$

x	y $m = 1.0000$	y $m = 1.5858$	y $m = 1.9358$	y $m = 2.0000$
0	1.0000	1.0000	1.0000	1.0000
0.1	1.0909	1.1433	1.1735	1.1790
.2	1.1662	1.2627	1.3169	1.3267
.3	1.2288	1.3644	1.4393	1.4526
.4	1.2806	1.4521	1.5455	1.5621
.5	1.3229	1.5283	1.6388	1.6583
.6	1.3565	1.5947	1.7213	1.7436
.7	1.3820	1.6523	1.7945	1.8193
.8	1.4000	1.7021	1.8594	1.8868
.9	1.4107	1.7448	1.9169	1.9468
1.0	1.4142	1.7809	1.9676	2.0000

Note. The values of y' and y'' required in computing each of the above solutions have been omitted to save space.

Example 2. Find a solution of the equation

$$y'' = -10y^3$$

through the points (0, 0), (1, 1).

We take $m = 1$ as a first trial and compute the first solution shown in computation 29, getting a result that falls below the desired value of 1. From it we estimate that $m = 1.55$ and compute a second trial solution which falls still farther below the desired value. From linear interpolation we next get $m = -4.61$. (Actually there is no basis whatever for using interpolation at this stage in this problem. It merely happens to give a not entirely unreasonable result.) The remaining choices of m and the corresponding solutions are shown in computation 29.

Two-end-point Problem   Computation 29

Differential equation $y'' = -10y^3$ (0, 0) (1, 1) $h = 0.1$

$m = 1$ $m = 1.55$ $m = -4.61$ $m = -5.01$ $m = -4.898$ $m = -4.910$

x	y	y	y	y	y	y
0	0.000	0.000	0.000	0.000	0.000	0.000
0.1	.100	.155	− .461	− .500	− .489	− .496
.2	.200	.309	− .907	− .982	− .961	− .963
.3	.299	.460	−1.273	−1.363	−1.338	−1.340
.4	.395	.601	−1.436	−1.497	−1.481	−1.482
.5	.485	.720	−1.316	−1.311	−1.314	−1.313
.6	.563	.801	− .974	− .903	− .925	− .921
.7	.624	.832	− .535	− .412	− .450	− .444
.8	.660	.806	− .075	.087	.039	.047
.9	.668	.728	.387	.587	.528	.537
1.0	.647	.612	.837	1.059	.996	1.006

Example 3. Consider an electron moving in a plane perpendicular to a positively charged infinite straight wire. Let x and y be the rectangular coordinates of the electron at time t referred to axes in the plane with the wire through the origin. If the unit of time is suitably chosen, the equations of motion may be written

$$x'' = -x/(x^2 + y^2),$$

$$y'' = -y/(x^2 + y^2).$$

Let us suppose that the electron leaves the point A with coordinates (1, 0) with a velocity $v = 1$. What should be its initial direction in order to reach the point B with coordinates (1, 1)?

We compute solutions of the differential equation for several choices of the initial angle θ, as shown in computation 30. The graphs of these orbits are shown in Fig. 13, and from the appearance of the orbits we infer that the electron does not reach the point B.

Two-end-point Problem   Computation 30

Differential equations $x'' = -x/(x^2 + y^2)$ $y'' = -y/(x^2 + y^2)$
At $t = 0$, $x = 1$, $y = 0$, $v = (x'^2 + y'^2)^{1/2} = 1$

	30°		40°		45°		50°		60°	
t	x	y	x	y	x	y	x	y	x	y
0	1.000	0	1.000	0	1.000	0	1.000	0	1.000	0
0.2	1.154	0.099	1.134	0.128	1.122	0.141	1.109	0.152	1.081	0.172
.4	1.274	.196	1.233	.252	1.209	.277	1.183	.300	1.125	.339
.6	1.363	.288	1.301	.370	1.265	.406	1.225	.439	1.137	.496
.8	1.424	.374	1.341	.479	1.292	.526	1.238	.569	1.119	.640
1.0	1.458	.453	1.354	.579	1.292	.635	1.225	.686	1.075	.760
1.2	1.467	.524	1.342	.669	1.267	.732	1.186	.789	1.005	.879
1.4	1.453	.587	1.306	.747	1.219	.816	1.124	.876	.914	.970
1.6	1.414	.640	1.247	.811	1.148	.884	1.040	.947	.801	1.040
1.8	1.352	.682	1.165	.861	1.055	.935	.935	.998	.670	1.085
2.0	1.266	.713	1.061	.894	.941	.967	.810	1.028	.523	1.104

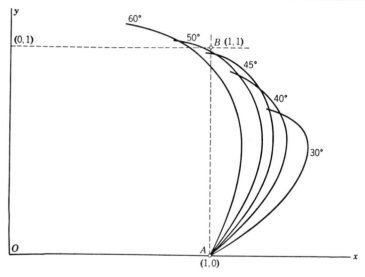

Fig. 13

48. Successive substitutions

The method explained in Article 15 for equations of the first order can be applied to the two-point boundary problem. If we seek a solution of the equation

$$y'' = f(x, y, y')$$

satisfying the conditions $y = \alpha$ at $x = a$, $y = \beta$ at $x = b$, we may start with an assumed solution $y = y_0(x)$ satisfying the boundary conditions and then construct a sequence of functions $y_1(x)$, $y_2(x)$, $\cdots$ defined by the differential equations

$$y_{n+1}''(x) = f(x, y_n(x), y_n'(x)), \qquad n = 0, 1, 2, \cdots$$

with the boundary conditions $y_{n+1}(a) = \alpha$, $y_{n+1}(b) = \beta$. Under certain rather severe limitations it can be shown that the sequence y_0, y_1, y_2, $\cdots$ converges to a solution of the two-point boundary problem. This theoretical result is of no great use in practical computation. One relies rather on the results of the computation itself to indicate whether or not convergence takes place.

In the practical application of this method we compute numerically the function

$$\bar{y}_{n+1}' = \int_a^x f(s, y_n(s), y_n'(s))\,ds$$

using some method of numerical integration. We then have

$$y_{n+1}' = \bar{y}_{n+1}' + c_1$$

and

$$y_{n+1} = \int_a^x \bar{y}_{n+1}'(s)(ds) + c_1 x + c_2.$$

The constants c_1 and c_2 are determined by the end conditions.

Any method of numerical integration which guarantees the desired accuracy may be used. Based on considerable experimentation the suggestion is offered that a convenient method is to use Gregory's formula in terms of central differences. This is

(48.1) $$y(x_{i+1}) - y(x_i) = \frac{h}{2}[\theta y'(x_{i+1}) + \theta y'(x_i)],$$

where θ is a central difference operator defined by the equation

(48.2) $$\theta = 1 - \frac{1}{12}\delta^2 + \frac{11}{720}\delta^4 - \frac{191}{60480}\delta^6 + \frac{2497}{3628\,800}\delta^8 - \cdots.$$

In spite of the fact that this formula requires values beyond the ends of the interval (which may be supplied by actual computation or by extrapolation), it furnishes one of the simplest ways of tabulating an integral on the computing machine.

Example 1. To illustrate the process of successive substitutions we apply it to Example 1 of Article 47. For the initial guess we take the straight line

$$y_0 = 1 + x.$$

Two-point boundary problem   Computation 31

Differential equation $y'' = -(1 + y'^2)/y$ Boundary points $(0, 1)$, $(1, 2)$

x	y_1''	$\theta y_1''$	$\bar{y}_1'$	$\theta \bar{y}_1'$	y_1'	y_1
0	-2.000	-1.997	2.000	1.999	1.772	1.000
0.1	-1.818	-1.815	1.809	1.808	1.580	1.168
.2	-1.667	-1.665	1.635	1.634	1.407	1.317
.3	-1.538	-1.536	1.475	1.474	1.247	1.450
.4	-1.428	-1.427	1.327	1.326	1.099	1.567
.5	-1.333	-1.332	1.189	1.188	$.961$	1.670
.6	-1.250	-1.249	1.060	1.059	$.832$	1.760
.7	-1.176	-1.175	$.939$	$.938$	$.711$	1.837
.8	-1.111	-1.110	$.825$	$.825$	$.597$	1.902
.9	-1.053	-1.053	$.717$	$.717$	$.489$	1.956
1.0	-1.000	-1.000	$.614$	$.614$	$.386$	2.000

$$c = -0.228$$

x	y_2''	$\theta y_2''$	$\bar{y}_2'$	$\theta \bar{y}_2'$	y_2'	y_2
0	-4.140	-4.074	2.000	1.985	1.991	1.000
0.1	-2.993	-2.958	1.648	1.640	1.639	1.180
.2	-2.262	-2.243	1.388	1.383	1.379	1.331
.3	-1.762	-1.750	1.189	1.185	1.180	1.458
.4	-1.408	-1.400	1.031	1.028	1.022	1.568
.5	-1.151	-1.145	$.904$	$.902$	$.895$	1.663
.6	$- .961$	$- .957$	$.799$	$.798$	$.790$	1.748
.7	$- .820$	$- .817$	$.710$	$.709$	$.701$	1.822
.8	$- .713$	$- .711$	$.633$	$.632$	$.624$	1.888
.9	$- .633$	$- .632$	$.566$	$.565$	$.557$	1.947
1.0	$- .574$	$- .573$	$.506$	$.506$	$.497$	2.000

$$c = -0.009$$

x	y_3''	$\theta y_3''$	$\bar{y}_3'$	$\theta \bar{y}_3'$	y_3'	y_3
0	-4.964	-4.810	2.000	1.978	2.008	1.000
0.1	-3.121	-3.046	1.607	1.596	1.615	1.180
.2	-2.180	-2.146	1.348	1.342	1.356	1.327
.3	-1.641	-1.624	1.159	1.155	1.167	1.453
.4	-1.304	-1.294	1.013	1.011	1.021	1.562
.5	-1.082	-1.076	$.895$	$.894$	$.903$	1.658
.6	$- .929$	$- .925$	$.795$	$.794$	$.803$	1.743
.7	$- .818$	$- .816$	$.708$	$.707$	$.716$	1.819
.8	$- .736$	$- .734$	$.630$	$.629$	$.638$	1.887
.9	$- .673$	$- .672$	$.560$	$.560$	$.568$	1.947
1.0	$- .624$	$- .623$	$.495$	$.495$	$.503$	2.000

$$c = +0.008$$

Substitution into the right-hand member of the equation

$$y_1'' = -(1 + y_0'^2)/y$$

gives $y_1'' = -2/(1 + x)$. The values of y_1'' are shown in computation 31. From these we obtain

$$\bar{y}_1' = \int_0^x y_1''(s)ds + 2,$$

the constant 2 being somewhat arbitrarily chosen simply to make $\bar{y}_1'$ positive throughout its range. Integration of $\bar{y}_1'$ gives the values of y_1 except for an additive term of the form $Ax + B$ in which A and B are chosen to make y satisfy the end conditions. The details of this step are simple and are not shown in computation 31.

A comparison of computations 28 and 31 reveals that the latter demanded more labor for less accuracy than the former. Since however both methods depend so completely on the shrewdness of the initial guess, it is difficult to make any valid comparison in the general case.

Example 2. Here we take Example 2 of Article 47, and start with $y_0 = x$, the straight line joining the end points. The results of three approximations are shown in computation 32. It is apparent that, to three approximations at least, we are getting farther and farther from a solution. Obviously our initial guess was too crude to secure convergence, at least in a reasonable number of steps.

Two-end-point Problem — Computation 32

Differential equation $y'' = -10y^3$		End points (0, 0)	(1, 1)
x	y_0	y_1	y_2	y_3
0	0	0	0	0
0.1	0.1	0.15	0.22	0.44
.2	.2	.30	.45	.88
.3	.3	.45	.67	1.31
.4	.4	.59	.90	1.70
.5	.5	.73	1.07	2.03
.6	.6	.86	1.22	2.23
.7	.7	.96	1.32	2.26
.8	.8	1.03	1.32	2.05
.9	0.9	1.05	1.22	1.62
1.0	1.0	1.00	1.00	1.00

49. Method of differential variations

In the method of Article 48 the new approximation was obtained by integrating the differential equation with the previous approximation substituted in the right-hand member. A more elegant method of correcting a trial solution is explained below.

We seek a solution of

$$(49.1) \qquad\qquad F(x, y, y', y'') = 0$$

through the points (a, α), (b, β) and start with an assumed approximate solution $\bar{y}$ through (a, α), (b, β). We may set $y = \bar{y} + \eta$ where η is the correction required to produce the correct solution y. To terms of first degree in η, η', η'', we have

(49.2) $F(x, y, y', y'') = F(x, \bar{y}, \bar{y}', \bar{y}'') + F_{y''}\eta'' + F_{y'}\eta' + F_y\eta.$

This, together with equation 49.1, gives the non-homogeneous linear differential equation

(49.3) $F_{y''}\eta'' + F_{y'}\eta' + F_y\eta + F = 0$

for the determination of the correction η. We compute a solution of equation 49.3 subject to the conditions $\eta = 0$ at $x = a$ and at $x = b$. Then an improved approximation to the solution of 49.1 is given by $y = \bar{y} + \eta$. The process is repeated and under suitable conditions will converge rapidly to the desired solution.

This is the method of differential variations. Some additional observations must be made. First, it may be impossible to solve 49.3, subject to the assigned conditions. Second, the process may not converge, or may converge too slowly for practical use. These are theoretical problems. Some computational problems are: First, the initial guess $\bar{y}$ and all subsequent corrections η must be computed to as many decimal places as are wanted in the final answer. This is due to the additive nature of the corrections, and is in contrast to the method of Article 48 where the earlier approximations may be taken to two decimal places, and later approximations taken to more and more places as the accuracy increases. Second, the convenient method of solving a linear boundary problem as explained in Articles 44 and 46 is not applicable to 49.3, since we need not only η, but also η' and η'' in order to form the new quantities y, y', y'' by the relations $y = \bar{y} + \eta$, $y' = \bar{y}' + \eta'$, $y'' = \bar{y}'' + \eta''$. Neither are the methods of Articles 42 and 43 applicable here because the coefficients are given numerically and not by analytical formulas. The computations shown here in the illustrative example were in fact done by Method VII as applied in Article 38.

Example.   To illustrate the method we take the equation

$$yy'' + 1 + y'^2 = 0$$

with the end points $(0, 1)$, $(1, 2)$ (cf. Article 45, second example; Article 47, first example; Article 48, first example).

The variational equation is

(49.4) $\bar{y}\eta'' + 2\bar{y}'\eta' + \bar{y}''\eta + (\bar{y}\bar{y}'' + 1 + \bar{y}'^2) = 0,$

Two-end-point Problem

Differential equation $yy'' + 1 + y'^2 = 0$ $(0, 1)$, $(1, 2)$ Computation 33 $h = 0.1$

x	$\bar{y}_0$	$\bar{y}_0'$	$\bar{y}_0''$	w_1	w_1'	w_1''	u_1	u_1'	u_1''	η_1	η_1'	η_1''
0	1	1	0	0	0	-2	0	1	-2	0	0	-2
0.1	1.1	1	0	$-.0091$	$-.1736$	-1.5026	.0909	.8264	-1.5026	.0818	0.6528	-1.5026
.2	1.2	1	0	$-.0333$	$-.3056$	-1.1574	.1667	.6944	-1.1574	.1333	.3888	-1.1574
.3	1.3	1	0	$-.0692$	$-.4083$	$-.9103$	.2308	.5917	$-.9103$	.1615	.1834	$-.9103$
.4	1.4	1	0	$-.1143$	$-.4898$	$-.7288$	.2857	.5102	$-.7288$	.1714	.0203	$-.7288$
.5	1.5	1	0	$-.1667$	$-.5556$	$-.5926$	.3333	.4444	$-.5926$	.1667	.1111	$-.5926$
.6	1.6	1	0	$-.2250$	$-.6094$	$-.4882$	.3750	.3906	$-.4883$	.1500	.2188	$-.4883$
.7	1.7	1	0	$-.2882$	$-.6540$	$-.4071$	.4118	.3460	$-.4071$	.1235	.3080	$-.4071$
.8	1.8	1	0	$-.3556$	$-.6914$	$-.3429$	.4444	.3086	$-.3429$	.0889	.3828	$-.3429$
.9	1.9	1	0	$-.4263$	$-.7230$	$-.2916$	.4737	.2770	$-.2916$	.0473	.4460	$-.2916$
1.0	2.0	1	0	$-.5000$	$-.7500$	$-.2500$	.5000	.2500	$-.2500$	0	.5000	$-.2500$

x	$\bar{y}_1$	$\bar{y}_1'$	$\bar{y}_1''$	w_2	w_2'	w_2''	u_2	u_2'	u_2''	η_2	η_2'	η_2''
0	1	2.0000	-4.0000	0	0	-2	-2.003	6.028	-32.125	0	0	-1
0.1	1.1818	1.6528	-3.0052	-0.0002	$+0.0022$	-1.0097	-1.524	3.829	-14.584	-0.0027	0.0014	-1.0057
.2	1.3333	1.3888	-2.3148	$-.0029$	$.0408$	$.0458$	-1.200	2.752	-7.817	.0065	.0413	$.0440$
.3	1.4615	1.1834	-1.8206	$-.0067$	$.0318$	$-.1729$	$-.958$	2.145	-4.667	.0088	.0322	$.1738$
.4	1.5714	1.0203	-1.4576	$-.0089$	$.0130$	$-.1881$	$-.764$	1.769	-3.005	.0092	.0133	$.1887$
.5	1.6667	.8889	-1.1852	$-.0094$	$.0038$	$-.1451$	$-.600$	1.520	-2.048	.0083	.0036	$.1454$
.6	1.7500	.7812	$-.9765$	$-.0083$	$-.0156$	$-.0886$	$-.457$	1.347	-1.458	.0064	$-.0154$	$.0889$
.7	1.8235	.6920	$-.8142$	$-.0064$	$-.0216$	$-.0335$	$-.329$	1.222	-1.074	.0042	$-.0214$	$.0337$
.8	1.8889	.6172	$-.6858$	$-.0042$	$-.0224$	$.0157$	$-.212$	1.128	$.814$	.0021	$-.0223$	$.0156$
.9	1.9473	.5540	$-.5832$	$-.0021$	$-.0187$	$-.0583$	-0.103	1.056	$.632$	-0.0006	$-.0186$	$.0582$
1.0	2.0000	.5000	$-.5000$	-0.0006	0.0110	$-.0944$	0	1.000	$.500$	0	.0109	$.0943$
				0	0	$-.1250$					.0001	$.1249$

x	$\bar{y}_2$	$\bar{y}_2'$	$\bar{y}_2''$
0	1	2.0014	-5.0057
0.1	1.1791	1.6116	-3.0492
.2	1.3268	1.3567	-2.1410
.3	1.4527	1.1701	-1.6319
.4	1.5622	1.0240	-1.3122
.5	1.6584	.9042	-1.0963
.6	1.7436	.8026	$-.9428$
.7	1.8194	.7143	$-.8298$
.8	1.8868	.6358	$-.7440$
.9	1.9468	.5649	$-.6775$
1.0	2.0000	.4998	$-.6249$

and just as in Article 48 let us choose the first approximation $\bar{y}$ to be

$$\bar{y} = 1 + x.$$

With this choice the equation to be solved is

$$(1 + x)\eta'' + 2\eta' + 2 = 0$$

with the conditions $\eta = 0$ at $x = 0$ and at $x = 1$. (This particular equation can be solved analytically, but usually we must resort to numerical solution.) As shown in Article 46 we first find a particular integral w of equation 49.4 with initial conditions $w = 0$ and $w' = 0$, then a solution u of the reduced equation with initial conditions $u = 0$, $u' = 1$. (In this particular example the values of $F(x, \bar{y}, \bar{y}', \bar{y}'')$ turned out to vary more smoothly near $x = 1$ than near $x = 0$ and, accordingly, in the actual numerical solution we took $x = 1$ as the initial point, because of greater simplicity in starting the computation.) Finally η is given by

$$\eta = Au + w$$

with A chosen to make η vanish at the end points.

It is evident that each step in the approximation requires two integrations of a second-order linear equation. However, as the approximation proceeds, the successive w's become smaller and require fewer significant figures, while the u's become more and more nearly equal so that eventually it is needless to recompute the u's.

The procedure is shown in considerable detail in computation 33. The work was actually carried to 5 decimal places and subsequently rounded off to 4 places.

The convergence of this and similar methods using differential variations has been investigated by Hestenes [87] and Stein [196].

50. The Ritz method

In those instances where the differential equation to be solved is the Euler equation associated with the problem of minimizing an integral

$$(50.1) \qquad J = \int_a^b F(x, y, y')dx$$

subject to the conditions

$$(50.2) \qquad y = \alpha \quad \text{at} \quad x = a, \qquad y = \beta \quad \text{at} \quad x = b,$$

the well-known Ritz method is applicable. In this method we select a basic set of linearly independent functions $u_0(x)$, $u_1(x)$, $u_2(x)$, $\cdots$, $u_n(x)$ where $u_0(x)$ satisfies equations 50.2 while $u_1(x)$, $u_2(x)$, $\cdots$, $u_n(x)$ vanish at a and at b. The desired solution y is expressed as a linear combination of the u's,

$$(50.3) \qquad y = u_0(x) + a_1u_1(x) + a_2u_2(x) + \cdots + a_nu_n.$$

It is evident that y satisfies the boundary conditions. This expression for y is then substituted into equation 50.1, the integration is performed, and we have J expressed in terms of the n quantities a_1, a_2, $\cdots$, a_n. These quantities are now determined to minimize J, and values so determined, when substituted into equation 50.3, provide an approximate solution of the problem.

Unfortunately in practice certain difficulties may arise. The required integrations may be difficult or impossible, the expression for J in terms of the a's may be too complex for us to obtain a minimum except by tedious approximations, and, of course, there is always the threat that no solution may exist. The case of principal practical importance is that where $F(x, y, y')$ is a quadratic function of y and y' and where consequently the differential equation is linear.

Example. To illustrate the Ritz method we apply it to the equation

$$y'' + xy' + y = 2x$$

with end points $(0, 1)$ and $(1, 0)$ (cf. Article 46, Example 1). Here the integral is

$$(50.4) \qquad J = \int_0^1 e^{x^2/2}(y'^2 - y^2 + 4xy)dx.$$

We may conveniently assume for the u's the polynomial functions

$$u_k = x^k(1 - x), \qquad k = 0, 1, \cdots , n$$

which are seen to satisfy the required end conditions. Then

$$(50.5) \qquad y = 1 - x + a_1x(1 - x) + a_2x^2(1 - x) + ax^3(1 - x).$$

When the values of y and y' as obtained from equation 50.5 are substituted into equation 50.4, the result is a quadratic form in a_1, a_2, a_3 with numerical coefficients expressible in terms of the integrals

$$H_k = \int_0^1 e^{x^2/2}x^k \, dx.$$

If k is odd, the value of H_k can be found explicitly by repeated integrations by parts. If k is even, integrations by parts reduce the problem to that of finding H_0, and H_0 is readily found by integrating the power series for $e^{x^2/2}$. When the integrations are performed, we get

$H_0 = 1.1950,$	$H_3 = 0.3513,$	$H_6 = 0.2116,$
$H_1 = 0.6487,$	$H_4 = 0.2874,$	$H_7 = 0.1871,$
$H_2 = 0.4538,$	$H_5 = 0.2436,$	$H_8 = 0.1677.$

In order to minimize the quadratic form in the a's, we are led to the three equations of condition:

$$
\begin{aligned}
A_{11}a_1 + A_{12}a_2 + A_{13}a_3 &= B_1, \\
(50.6) \qquad A_{21}a_1 + A_{22}a_2 + A_{23}a_3 &= B_2, \\
A_{31}a_1 + A_{32}a_2 + A_{33}a_3 &= B_3,
\end{aligned}
$$

where

$$A_{11} = H_0 - 4H_1 + 3H_2 + 2H_3 - H_4 = 0.3765,$$

$$A_{21} = A_{12} = 2H_1 - 7H_2 + 5H_3 + 2H_4 - H_5 = 0.2087,$$

$$A_{31} = A_{13} = 3H_2 - 10H_3 + 7H_4 + 2H_5 - H_6 = 0.1362,$$

$$A_{22} = 4H_2 - 12H_3 + 8H_4 + 2H_5 - H_6 = 0.1748,$$

$$A_{32} = A_{23} = 6H_3 - 17H_4 + 11H_5 + 2H_6 - H_7 = 0.1370,$$

$$A_{33} = 9H_4 - 24H_5 + 15H_6 + 2H_7 - H_8 = 0.1203,$$

$$B_1 = H_0 - H_1 - 4H_2 + 3H_3 \qquad\qquad = -0.2150,$$

$$B_2 = 2H_1 - 2H_2 - 4H_3 + 3H_4 \qquad\quad = -0.1529,$$

$$B_3 = 3H_2 - 3H_3 - 4H_4 + 3H_5 \qquad\quad = -0.1115.$$

When these values are put into equations 50.6 and the equations are solved for the a's, it turns out that

$$a_1 = -0.211, \qquad a_2 = -0.783, \qquad a_3 = 0.204.$$

We thus have as an approximate solution

$$y = (1 - x)(1 - 0.211x - 0.783x^2 + 0.204x^3).$$

The tabulated values of y are shown below.

x	y	x	y	x	y
0	1.000	0.4	0.482	0.7	0.162
0.1	0.874	0.5	0.362	0.8	0.087
0.2	0.743	0.6	0.254	0.9	0.032
0.3	0.610			1.0	0

In this example we chose the functions $u_0(x)$, $u_1(x)$, $\cdots$, $u_n(x)$ as polynomials, these being most convenient for the case in hand. In other examples it may be found expedient to use trigonometric functions, etc. Usually our choice is severely limited by the necessity of performing a lot of involved integrations.

See Ritz [162, 163].

51. Galerkin's method

A variation of the Ritz method was devised by B. G. Galerkin [75] and in most cases affords a considerable simplification in the process of finding an approximate solution. Moreover Galerkin's method makes no use of the integral (equation 50.1) and hence does not depend on a minimizing process. Without going into the analytical reasoning on which the method rests we shall simply state the basic idea. To solve a linear equation

$$L(y) - f(x) = 0$$

with boundary conditions at $x = a$ and $x = b$ we choose $u_0(x)$ satisfying the *given* conditions at a and at b, and $u_1(x)$, $u_2(x)$, $\cdots$, $u_n(x)$ satisfying corresponding *homogeneous* conditions at a and at b. To make our meaning clear, let us suppose that a particular condition is $y(a) - 2y'(b) = 3$. Then we must have

$$u_0(a) - 2u_0'(b) = 3,$$

while

$$u_k(a) - 2u_k'(b) = 0, \qquad k = 1, 2, \cdots, n.$$

Then as in the Ritz method we let

$$y = u_0(x) + a_1u_1(x) + a_2u_2(x) + \cdots + a_nu_n(x).$$

This is substituted into the differential equation and we denote the result by R, so that

$$R(x) = L[u_0(x) + \Sigma a_iu_i(x)] - f(x).$$

Galerkin's method consists in finding a set of a's (assumed for the moment to exist) such that $R(x)$ is orthogonal to each of the n functions $u_1(x)$, $u_2(x)$, $\cdots$, $u_n(x)$, i.e., such that

$$(51.1) \qquad \int_a^b R(x)u_i(x)dx = 0, \qquad i = 1, 2, \cdots, n.$$

Example. To illustrate the method we apply it to the same example used in Article 50. Just as in that case we have

$$y = 1 - x + a_1x(1 - x) + a_2x^2(1 - x) + a_3x^3(1 - x).$$

But here the integrals to be evaluated are simply the integrals

$$\int_0^1 x^k \, dx, \qquad k = 0, 1, \cdots.$$

From equation 51.1 we get the three equations

$$133a_1 + 63a_2 + 36a_3 = -70,$$

$$140a_1 + 108a_2 + 79a_3 = -98,$$

$$264a_1 + 252a_2 + 211a_3 = -210.$$

From these we find

$$a_1 = -0.2090, \qquad a_2 = -0.7894, \qquad a_3 = 0.2090,$$

and

$$y = (1 - x)(1 - 0.2090x - 0.7894x^2 + 0.2090x^3).$$

This formula gives approximately the same accuracy as that obtained by the Ritz method.

General theorems are available in the literature which, under sufficiently restrictive hypotheses, insure the possibility of carrying out the Ritz or the Galerkin method. The practical trouble with such theorems is that in order to secure reasonable generality the hypotheses have to be more restrictive than is required in many particular cases. In practical problems, if the computer can solve the conditional equations and obtain an approximate function for y, he can always substitute this y into the differential equation, and from the magnitude of the residual he can estimate the accuracy attained. If on the other hand, the conditional equations do not afford a unique solution, this fact suggests that the original problem either has no solution or is "near" a case where there is no solution.

See articles on Galerkin's method by Duncan [47, 48]. For additional material on the two-point boundary problem see Hartree [83], Nyström [147], and Fox [67].

PART II
Partial Equations

Many of the partial differential equations of mathematical physics and most of those selected in the following pages to illustrate numerical methods are special cases of the second-order linear equation

$$A U_{xx} + B U_{xy} + C U_{yy} + D U_x + E U_y + F U = G$$

in which U is the dependent variable, x and y are the independent variables, and A, B, C, D, E, F, G are constants or functions of x and y only. Partial differentiations with respect to x and y are indicated by subscripts. It is customary to call the equation *hyperbolic* if

$$B^2 - 4AC > 0,$$

parabolic if

$$B^2 - 4AC \equiv 0,$$

and *elliptic* if

$$B^2 - 4AC < 0.$$

Methods for reducing the general equation above to simpler standard forms are given, for example, in Jordan, *Cours d'analyse*, Tome 3, Articles 271-2-3. See also Goursat, *Cours d'analyse mathématique*, Tome 3, Article 479, and A. G. Webster, *Differential Equations of Mathematical Physics*, B. G. Teubner, Leipzig, 1927, p. 74.

For an up-to-date elementary treatment of these equations see F. D. Murnaghan, *Introduction to Applied Mathematics*, John Wiley & Sons, 1948 (reprinted by Dover Publications, Inc.), especially Chapters 5, 6, and 7.

8

EXPLICIT METHODS. PARABOLIC AND HYPERBOLIC EQUATIONS

Methods for the numerical solution of partial differential equations have mainly been devised for specific problems. Hence, the general theory is still but little explored. The methods so far developed center around the linear equations arising in mathematical physics, the heat equation, the wave equation, Laplace's and Poisson's equations, the biharmonic equation, and similar associated equations. Although isolated problems involving non-linear equations have been tackled with some success, the general theory for the non-linear case is still obscure.

Numerical methods tend to separate into two types: one, the *explicit* type, where the numerical solution can be calculated step by step from the given differential equation and the known initial and boundary conditions; the other, the *implicit* type, where the unknown values are bound together by a system of simultaneous equations.

In this chapter we illustrate by examples methods for solving problems of the explicit type.

52. The parabolic differential equation $U_t = c^2 U_{xx}$

This is the equation for one-dimensional flow of heat, diffusion, and similar problems. It is chosen as the first illustrative example because of its simplicity.

Associated with the differential equation are an initial condition

$$U = f(x) \quad \text{when} \quad t = 0$$

and two boundary conditions of the type

$$\alpha U + \beta U_x = g(t) \quad \text{at} \quad x = 0,$$

$$\gamma U + \delta U_x = q(t) \quad \text{at} \quad x = L.$$

To secure additional simplicity in the following analysis we treat only the case where $\beta = \delta = 0$.

In order to have a suitable representation of the differential equation by means of a difference equation we cover the region R in the first quadrant of the xt plane bounded by the lines $t = 0$, $x = 0$, $x = L$ with a net of equal rectangles each having sides $\Delta x = h$ and $\Delta t = k$. The value of Δx is chosen so that

$$\Delta x = h = L/\nu$$

where ν is an integer. We replace U_t in the differential equation by $[u(x,\ t + k) - u(x,\ t)]/k$ and U_{xx} by $[u(x + h,\ t) - 2u(x,\ t) + u(x - h,\ t)]/h^2$, multiply by k, collect terms, and arrive at the difference equation

$$(52.1) \quad u(x,\ t + k) = ru(x + h,\ t) + (1 - 2r)u(x,\ t) + ru(x - h,\ t),$$

where $r = kc^2/h^2$. This formula enables one to calculate in succession $u(x,\ k)$ from the values of $u(x,\ 0)$, $u(x,\ 2k)$ from $u(x,\ k)$, $u(x,\ 3k)$ from $u(x,\ 2k)$ and so on step by step to $u(x,\ nk)$. If the time interval to be covered is t, we must have $nk = t$, and thus the required number of steps is $n = c^2t/rh^2$.

53. Analysis of the error

Let u denote the exact solution of the difference equation 52.1 and U the exact solution of the differential equation, both solutions satisfying the given boundary and initial conditions. If the function U is substituted into the difference equation 52.1, there will be a remainder term or truncation error $T(x,\ t)$, so that we have

$$(53.1) \quad U(x,\ t + k) = rU(x + h,\ t) + (1 - 2r)U(x,\ t) + rU(x - h,\ t) + T(x,\ t).$$

We designate the error $U - u$ by e, and subtract 52.1 from 53.1 to obtain the difference equation

$$(53.2) \quad e(x,\ t + k) = re(x + h,\ t) + (1 - 2r)e(x,\ t) + re(x - h,\ t) + T(x,\ t).$$

The error $e(x,\ t)$ is that solution of 53.2 which is zero on the boundaries of R.

There are two cases to consider.

Case 1. $0 < r \leqq \frac{1}{2}$. In this case all the coefficients on the right in 53.2 are positive or zero, and their sum is unity. It is therefore evident that a dominating difference equation is

$$E(t + k) = E(t) + M$$

where $|T(x, t)| < M$. Since $E(0) = 0$ we have at once $E(nk) = nM$. Hence at time t the error is bounded by the inequality

$$(53.3) \qquad |e(x, t)| < c^2 t M/rh^2.$$

Case 2. $r > \frac{1}{2}$. It is a simple matter to verify that the complementary function in the general solution of the non-homogeneous difference equation 53.2 is made up of terms of the type

$$\sin \frac{\mu\pi x}{L} \left[1 - 2r \left(1 - \cos \frac{\mu\pi h}{L} \right) \right]^{c^2 t/rh^2}$$

in which $\mu = 1, 2, \cdots, \nu - 1$. Now, since $r > \frac{1}{2}$, the quantity

$$\left| 1 - 2r \left(1 - \cos \frac{\mu\pi h}{L} \right) \right|$$

will be greater than unity for those values of μ that make the cosine sufficiently near -1.

Thus there exist solutions $e(x, t)$ that become exponentially infinite for fixed h as t becomes infinite, or for fixed t as the mesh length h is reduced to zero. Hence, it is clear that r in equation 52.1 must not exceed $\frac{1}{2}$. Otherwise, even if there were no significant truncation error, the mere presence of round-off errors could set up exponentially increasing oscillations.

It was the noteworthy paper by Courant, Friedrichs, and Lewy [40] which called attention to the fact that the choice of the space interval places a restriction on the size of the time interval.

54. The truncation error

If U has a continuous partial derivative with respect to x of order 6 in the region under consideration, we have

$$(54.1) \quad U(x + h, t) - 2U(x, t) + U(x - h, t)$$
$$= h^2 U_{xx}(x, t) + \frac{h^4}{12} U_{x^4}(x, t) + \frac{h^6}{360} U_{x^6}(s, t)$$

where $x - h < s < x + h$. Also, if U has a continuous partial derivative with respect to t of order 3 in the region, then

$$(54.2) \quad U(x, t + k) - U(x, t) = kU_t(x, t) + \frac{k^2}{2} U_{tt}(x, t) + \frac{k^3}{6} U_{t^3}(x, \tau)$$

where $t < \tau < t + k$.

If, moreover, U is a solution of $U_t = c^2 U_{xx}$ we have

$$U_t = c^2 U_{xx}, \qquad U_{tt} = c^4 U_{x^4}, \qquad U_{t^3} = c^6 U_{x^6}.$$

We now multiply equation 54.1 by r and subtract the result from equation 54.2 obtaining (since $k = rh^2/c^2$)

(54.3) $U(x, t + k) - rU(x + h, t) - (1 - 2r)U(x, t) - rU(x - h, t)$

$$= \frac{h^4 r}{12}(6r - 1)U_{x^4}(x, t) + \frac{r^3 h^6}{6} U_{x^6}(x, \tau) - \frac{rh^6}{360} U_{x^6}(s, t).$$

It is clear from this equation that the term on the right involving h^4 drops out if $r = \frac{1}{6}$. Accordingly, we shall assume that $r = \frac{1}{6}$, or, what is the same thing, that $k = h^2/6c^2$. Then the difference equation becomes

(54.4) $U(x, t + k) = \frac{1}{6}[U(x + h, t) + 4U(x, t) + U(x - h, t)]$
$$+ T(x, t)$$

in which

$$T(x, t) = \frac{h^6 U_{x^6}(x, \tau)}{1296} - \frac{h^6 U_{x^6}(s, t)}{2160}.$$

If the constant N be chosen so that

$$|U_{x^6}(x, t)| < N \quad \text{in } R,$$

then

$$|T(x, t)| < M = h^6 N/810.$$

With this value of M the fundamental inequality (53.3) becomes

(54.5) $$|e(x, t)| < \frac{c^2 t N h^4}{135}.$$

The inequality 54.5 provides us with a rigorous bound for the error of the numerical solution in all cases where the hypotheses are satisfied. Unfortunately in probably a majority of problems there are one or more points on the boundary [often the corners $(0, 0)$ and $L, 0)$] where the hypotheses fail. The presence of isolated boundary points where the hypotheses fail gives rise to additional errors superimposed on the error shown above, but these can, by refinement of the mesh, be compressed into smaller and smaller neighborhoods of the troublesome boundary point.

55. Numerical example of parabolic equation

To illustrate the use of equation 54.4 we apply it to the approximate solution of the differential equation

$$U_t = a^2 U_{xx}$$

with boundary conditions

$$U = 0 \text{ at } x = 0, \qquad U = 0 \text{ at } x = L.$$

and initial condition

$$U = 4x(L - x)/L^2 \text{ when } t = 0.$$

Here we choose $\Delta x = h = L/10$. Then $\Delta t = k = h^2/6a^2 = L^2/600a^2$. It is convenient to arrange the computed values of $U_{m,n}$ in tabular form with m varying along the row and n varying down the column, thus

m	0	1	2	$\cdots$	10
n					
0	0	0.360	0.640	$\cdots$	0
1	0				0
2	0				0
3	0				0
.					
.					
.					

The values for $n = 0$ are given by the initial condition

$$U = 4x(L - x)/L^2$$

or, since $x = Lm/10$,

$$u = 0.04m(10 - m).$$

The boundary conditions give the value zero everywhere in the columns headed $m = 0$ and $m = 10$. The initial values are symmetrical about the column $m = 5$, and a little reflection assures one that all values of U will be symmetrical about this column, a fact that saves nearly 44 per cent of the labor. For we need calculate values of U only for $m = 1, 2, 3, 4, 5$, and can fill in the others by symmetry. The calculation carried to $n = 36$ is shown in part in computation 34. The entries for each value of n are computed from the line $n - 1$ by formula 54.4. For instance, to get $U_{3,7}$ we have

$$U_{3,7} = \tfrac{1}{6}(U_{2,6} + 4U_{3,6} + U_{4,6})$$

$$= \tfrac{1}{6}[(0.56448) + 4(0.76060) + (0.88005)]$$

$$= 0.74782.$$

For $m = 5$ we use the symmetrical property $U_{6,n} = U_{4,n}$ so that

$$U_{5,n+1} = \tfrac{1}{6}(2U_{4,n} + 4U_{5,n}).$$

The analytical solution of the foregoing differential equation with the assigned boundary conditions is found in the usual manner to be

$$(55.1) \qquad U_{m,n} = \frac{32}{\pi^3} \sum_{i=0}^{\infty} \frac{1}{(2i+1)^3} e^{\frac{-(2i+1)^2\pi^2 n}{600}} \sin \frac{(2i+1)\pi m}{10}.$$

For sake of comparison with the numerical solution of the difference equation we compute the values given by the analytical solution 55.1 for $n = 36$. These are shown in parentheses at the bottom of computation 34. A comparison of the true with the computed values reveals agreement to four decimal places.

Computation 34

Differential equation $U_t = a^2 U_{xx}$

$$U = 0 \quad \text{at} \quad x = 0 \quad \text{and at} \quad x = L \qquad U = 4x(L - x)/L^2 \quad \text{if} \quad t = 0$$

$$\Delta x = h = 0.1L \qquad \Delta t = k = h^2/6a^2$$

m	0	1	2	3	4	5
n						
0	0	0.36000	0.64000	0.84000	0.96000	1.00000
1	0	.34667	.62667	.82667	.94667	.98667
2	0	.33556	.61333	.81333	.93333	.97333
3	0	.32593	.60037	.80000	.92000	.96000
4	0	.31735	.58790	.78673	.90667	.94667
. . . . .	. . . .					
34	0	.18248	.34696	.47731	.56088	.58964
35	0	.17948	.34127	.46951	.55175	.58005
36	0	.17653	.33568	.46184	.54276	.57061
36	(0)	(.17655)	(.33572)	(.46189)	(.54280)	(.57066)

56. The hyperbolic equation $U_{tt} = a^2 U_{xx}$

This is the familiar equation associated with the propagation of waves in one dimension. As in Article 52 we cover the xt plane with rectangles having sides $\Delta x = h$ and $\Delta t = k$, designating U at the node $x = mh$, $t = nk$ by $u_{m,n}$. Upon replacing the second derivatives by corresponding second differences we obtain

$$(56.1) \quad u_{m,n+1} - 2u_{m,n} + u_{m,n-1} = \frac{k^2 a^2}{h^2} (u_{m+1,n} - 2u_{m,n} + u_{m-1,n}).$$

Now it is a familiar fact that, if $f(s)$ and $g(s)$ are any two functions of the variable s that have second-order derivatives, then

$$(56.2) \qquad U = f(x - at) + g(x + at)$$

is a solution of the equation

$$(56.3) \qquad U_{tt} = a^2 U_{xx}$$

A less publicized fact is that equation 56.2 is also a solution of the partial difference equation

$$(56.4) \qquad u_{m,n+1} = u_{m+1,n} + u_{m-1,n} - u_{m,n-1}$$

to which equation 56.1 reduces if $ka = h$. This is verified simply by substitution from 56.2 into 56.4 and elimination of h by the relation $h = ka$.

Here then we have the unusual situation where any solution of the differential equation also satisfies the associated difference equation and conversely, *regardless of mesh size*.

57. Example of the hyperbolic equation

To illustrate the numerical solution of a problem involving the hyperbolic equation we examine the transverse vibrations of a stretched string of unit length initially at rest in the undisplaced position, fastened rigidly at $x = 1$, and at $x = 0$ subject to the sinusoidal displacement

$$U = \sin \pi at.$$

The mathematical setup and several steps of the numerical solution are shown in computation 35.

At the beginning of the computation we have from the boundary conditions and initial conditions the setup

m	0	1	2	$\cdots$	10
n					
0	0	0	0	$\cdots$	0
1	0.309	0	0	$\cdots$	0
2	.588				0
3	.809				0
$\cdot$	$\cdot$				$\cdot$
$\cdot$	$\cdot$				$\cdot$
$\cdot$	$\cdot$				$\cdot$

Starting then with the line $n = 2$ and using equation (56.4), we proceed step by step to fill in the table.

This problem differs somewhat from the usual vibration problem given to illustrate vibrating strings. The numerical solution shows that this vibration is not periodic, but because of resonance the amplitudes increase linearly with time. The solution is of course exact except for round-off errors.

58. General boundary conditions

It is important to appreciate the ability of the numerical methods of Articles 55 and 57 to handle almost any type of boundary condition without modification of procedure. The analytical methods of solution depend on mathematical devices which require modification for new types of boundary conditions and which in the general case are involved and clumsy. On the other hand, we see at once that for the parabolic equation a set of boundary conditions such as

Differential equation	$U_{tt} = a^2 U_{xx}.$

Boundary conditions $U = \sin \pi at$ at $x = 0,$ $U = 0$ at $x = 1$

Initial conditions $U = 0$ when $t = 0,$ $U_t = 0$ when $t = 0$

$$\Delta x = h = 0.1, \qquad \Delta t = k = h/a$$

m \ n	0	1	2	3	4	5	6	7	8	9	10
0	0	0	0	0	0	0	0	0	0	0	0
1	0.309	0	0	0	0	0	0	0	0	0	0
2	.588	0.309	0	0	0	0	0	0	0	0	0
3	.809	.588	0.309	0	0	0	0	0	0	0	0
4	.951	.809	.588	0.309	0	0	0	0	0	0	0
5	1.000	.951	.809	.588	0.309	0	0	0	0	0	0
6	.951	1.000	.951	.809	.588	0.309	0	0	0	0	0
7	.809	.951	1.000	.951	.809	.588	0.309	0	0	0	0
8	.588	.809	.951	1.000	.951	.809	.588	0.309	0	0	0
9	.309	.588	.809	.951	1.000	.951	.809	.588	0.309	0	0
. . .	.	.	.	.	.	.	.	.	.	.	.
51	−0.309	1.176	2.545	3.666	4.427	4.755	4.618	4.029	3.045	1.764	0
52	− .588	.691	1.902	2.927	3.666	4.045	4.029	3.618	2.853	1.500	0
53	− .809	.138	1.073	1.902	2.545	2.940	3.045	2.853	2.073	1.089	0
54	− .951	− .427	.138	.691	1.176	1.545	1.764	1.500	1.089	0.573	0
55	−1.000	− .951	− .809	− .588	− .309	0	0	0	0	0	0
56	− .951	−1.382	−1.677	−1.809	−1.764	−1.854	−1.764	−1.500	−1.089	−0.573	0
57	− .809	−1.677	−2.382	−2.853	−3.354	−3.528	−3.354	−2.853	−2.073	−1.089	0
58	− .588	−1.809	−2.853	−3.927	−4.617	−4.854	−4.617	−3.927	−2.853	−1.500	0
59	−0.309	−1.764	−3.354	−4.617	−5.427	−5.706	−5.427	−4.617	−3.354	−1.764	0
60	0	−1.854	−3.528	−4.854	−5.706	−6.000	−5.706	−4.854	−3.528	−1.854	0

$U = f(x)$ for $t = 0$, $U = \phi(t)$ for $x = 0$, $U = \psi(t)$ for $x = L$ will be handled by the numerical procedure of Article 55 without any additional complication or difficulty. Likewise in the hyperbolic case the two end values may be given functions of t and the initial displacement and initial velocity may be given functions of x. The method of Article 57 still applies without modification.

EXERCISES

1. Solve the heat equation with conditions

$$U = 0 \quad \text{when} \quad t = 0,$$

$$U = 1 \quad \text{when} \quad x = 0,$$

$$U_x = 0 \quad \text{when} \quad x = L.$$

2. Solve the heat equation with the conditions

$$U = 0 \quad \text{when} \quad t = 0,$$

$$U = A \sin bt \quad \text{at} \quad x = 0 \quad \text{and at} \quad x = L.$$

3. Solve the wave equation with the conditions

$$U = 0 \qquad \text{at} \quad t = 0,$$

$$U_t = A \sin \frac{\pi x}{L} \quad \text{at} \quad t = 0,$$

$$U = \frac{L}{20} U_x \quad \text{at} \quad x = 0 \quad \text{and at} \quad x = L.$$

59. The point pattern

The first step in the derivation of a difference equation associated with a given partial differential equation is the selection of the specific pattern of points at which to take the functional values entering into the difference equation. The minimum pattern is determined by the order of the differential equation in the several variables. For instance, the equation

$$U_t = U_{xx} + U_{yyy}$$

requires a pattern with at least two points in the t direction, three in the x direction, and four in the y direction. Of course, more points may be used in order to secure a truncation error of higher degree. Thus, for the differential equation of Article 52, namely, $U_t = c^2 U_{xx}$, we may replace the time derivative by the relation (in Sheppard's notation)

$$U_t = \frac{\mu}{k} \left(\delta_t U - \frac{1^2}{3!} \delta_t{}^3 U + \frac{1^2 \cdot 2^2}{5!} \delta_t{}^5 U - \cdots \right)$$

and the second-order x derivative by

$$U_{xx} = \frac{1}{h^2} \left(\delta_x{}^2 U - \frac{2}{4!} \delta_x{}^4 U + \frac{2 \cdot 2^2}{6!} \delta_x{}^6 U - \cdots \right)$$

Indeed formulas of this type using differences of higher order (or their equivalents in terms of ordinates) provide a good way to check a computation being carried out by a simpler formula.

Actually it is usually not convenient to use one of these higher-powered formulas for the original computation because (a) their use usually leads to implicit equations, (b) they give trouble near boundaries because they overrun the boundaries, and (c) the fact that they represent difference equations of higher order than the original differential equation introduces extraneous disturbances into the solution.

For example, in place of equation 54.4 we could employ the more elaborate formula

$$(59.1) \quad U(x, t + k) - U(x, t - k) = \tfrac{1}{36}[-U(x + 2h, t) + 16U(x + h, t) - 30U(x, t) + 16U(x - h, t) - U(x - 2h, t)]$$

But clearly we cannot use equation 59.1 without modification at points adjacent to the boundary. On the other hand, we find upon applying 59.1 to computation 34 that a very satisfactory check is secured. Probably the greatest value of 59.1 is as a check formula. In many problems it is difficult to determine a rigorous estimate for the error, and we must rely upon an independent difference equation to check the results given by the difference equation actually used.

Often it is possible to use more points than the bare minimum and still not raise the order of the resulting difference equation. For instance the point pattern for 54.4 is

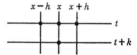

but we can equally well use the pattern

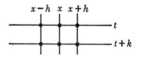

and derive for it the difference equation

$$(59.2) \quad \begin{aligned} &[U(x + h, t + k) + 10U(x, t + k) + U(x - h, t + k)] - [U(x+h, t) \\ &+ 10U(x, t) + U(x - h, t)] = 6r[U(x + h, t + k) - 2U(x, t + k) \\ &+ U(x - h, t + k)] + 6r[U(x+h, t) - 2U(x, t) + U(x - h, t),] \end{aligned}$$

where $k = rh^2$. If $r = \tfrac{1}{6}$, formula 59.2 reduces to 54.4. The truncation error of 59.2 is $O(h^6)$ except for the case $r = 1/\sqrt{20}$, where it is $O(h^8)$. A convenient check formula to use in conjunction with 54.4 is given by 59.2 when $r = \tfrac{1}{3}$. The resulting formula is

$$(59.3) \quad \begin{aligned} -U(x + h, t + k') + 14U(x, t + k') - U(x - h, t + k') \\ = 3[U(x + h, t) + 2U(x, t) + U(x - h, t)] \end{aligned}$$

where the k' in equation 59.3 is just twice the k in 54.4. Formula 59.3, being an implicit formula, is more convenient for checking a computation already under way than for performing the original calculation.

The foregoing examples give the reader an idea of various ways in which a partial differential equation can be represented by a difference equation. Other illustrations appear in ensuing articles.

60. Variable coefficients

For the parabolic equation in Article 52 it was possible to find a partial difference equation of second order such that the truncation error was $O(h^6)$. If however the coefficients are variable, we may have to be content with an error of lower order.

For example, the differential equation

$$U_t = c^2 e^x (U_{xx} + U_x)$$

and the point pattern

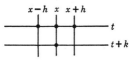

enable us to set up the difference equation

(60.1) $U(x, t + k) - U(x, t)$
$$= \frac{e^x}{6} \left[\left(1 + \frac{h}{2}\right) U(x + h, t) - 2U(x, t) + \left(1 - \frac{h}{2}\right) U(x - h, t) \right] + O(h^4),$$

where $k = h^2/6c^2$. Several steps of a calculation using 60.1 are shown in computation 36.

Computation 36

Differential equation $U_t = c^2 e^x (U_{xx} + U_x)$

Initial condition $U = 4x(1 - x)$

Boundary conditions $U = 0$ at $x = 0$ and at $x = 1$

Difference equation is 60.1, $h = 0.1$

m	0	1	2	3	4	5	6	7	8	9	10
n											
0	0	0.360	0.640	0.840	0.960	1.000	0.960	0.840	0.640	0.360	0
1	0	.351	.629	.826	.942	.978	.933	.808	.601	.314	0
2	0	.343	.617	.811	.924	.956	.906	.775	.562	.291	0
3	0	.336	.605	.796	.906	.933	.879	.742	.532	.271	0
4	0	.329	.594	.781	.887	.910	.852	.709	.505	.256	0
5	0	.323	.583	.766	.868	.887	.823	.683	.480	.243	0
. . .	. . .	. . .	. . .	. . .	. . .	. . .	. . .	. . .	. . .	. . .	. . .
16	0	.258	.463	.599	.661	.653	.582	.464	.316	.156	0
17	0	.252	.453	.585	.644	.635	.565	.450	.306	.151	0
18	0	.247	.442	.571	.628	.617	.549	.436	.296	.147	0

61. Non-linear equation

In sufficiently simple cases the foregoing methods can be applied to non-linear differential equations.

Thus for

$$\frac{\partial U}{\partial t} = \frac{\partial^2}{\partial x^2}(Ue^{-U})$$

an obvious difference equation is

$$U(x, t + k) - U(x, t) = \frac{k}{h^2}[(Ue^{-U})_{x+h} - 2(Ue^{-U})_x + (Ue^{-U})_{x-h}].$$

If U is small, this approaches the linear equation of Article 52, and, accordingly, in analogy with equation 54.4 we take $k = h^2/6$. The truncation error is $O(h^4)$ which suggests that the magnitudes of $\frac{1}{2}\Delta_t^2 U$ and of $\frac{1}{12}\Delta_x^4(Ue^{-U})$ give estimates of the accuracy attained. On this evidence the values in computation 37 appear to be accurate.

Computation 37

Differential equation $\dfrac{\partial U}{\partial t} = \dfrac{\partial^2}{\partial x^2}(Ue^{-U})$

$$U = 0 \quad \text{at} \quad x = 0, \qquad \frac{\partial U}{\partial x} = 0 \quad \text{at} \quad x = 1,$$

$$U = x \quad \text{when} \quad t = 0, \qquad h = 0.1, \quad k = h^2/6$$

t	$x = 0$	$x = h$	$x = 2h$	$x = 3h$	$x = 4h$	$x = 5h$	$x = 6h$	$x = 7h$	$x = 8h$	$x = 9h$	$x = 10h$
0	0	0.1000	0.2000	0.3000	0.4000	0.5000	0.6000	0.7000	0.8000	0.9000	1.0000
k	0	.0971	.1975	.2979	.3982	.4985	.5987	.6989	.7991	.8992	.9993
$2k$	0	.0948	.1950	.2958	.3964	.4970	.5974	.6978	.7982	.8985	.9987
$3k$	0	.0928	.1926	.2936	.3946	.4954	.5961	.6967	.7973	.8978	.9980
$4k$	0	.0911	.1902	.2914	.3927	.4938	.5948	.6956	.7964	.8970	.9973
$5k$	0	.0896	.1880	.2893	.3908	.4923	.5935	.6945	.7954	.8962	.9966
$6k$	0	.0882	.1858	.2871	.3890	.4907	.5922	.6934	.7945	.8955	.9959
$7k$	0	.0870	.1837	.2849	.3871	.4891	.5908	.6923	.7936	.8947	.9952
$8k$	0	.0859	.1818	.2828	.3852	.4874	.5894	.6912	.7926	.8939	.9944
$9k$	0	.0849	.1799	.2807	.3832	.4858	.5881	.6900	.7917	.8931	.9937

Blanch [229] has made a detailed examination of non-linear equations of parabolic type with special attention to the most economical mesh ratio. See also Crank and Nicolson [42].

62. The Laplacian operator $\nabla^2 = \dfrac{\partial^2}{\partial x^2} + \dfrac{\partial^2}{\partial y^2}$

It is convenient to introduce two symbols, say, H and X, to denote two partial difference operators intimately associated with the

Laplacian differential operator in two dimensions. They are defined by the equations

(62.1) $\quad HF(x, y) = F(x + h, y) + F(x - h, y) + F(x, y + h)$
$$+ F(x, y - h) - 4F(x, y),$$

and

(62.2) $\quad XF(x, y) = \frac{1}{2}[F(x + h, y + h) + F(x - h, y + h)$
$$+ F(x - h, y - h) + F(x + h, y - h) - 4F(x, y)].$$

They are associated with the point patterns

$$\begin{matrix} & \bullet & & & \bullet & \bullet \\ \bullet & \bullet & \bullet & \text{and} & & \bullet & , \\ & \bullet & & & \bullet & \bullet \end{matrix}$$

respectively, and may be more vividly pictured by the stencils*

(62.3) $\quad H = \begin{array}{|c|c|c|} \hline & 1 & \\ \hline 1 & -4 & 1 \\ \hline & 1 & \\ \hline \end{array},\quad X = \frac{1}{2}\begin{array}{|c|c|c|} \hline 1 & & 1 \\ \hline & -4 & \\ \hline 1 & & 1 \\ \hline \end{array},$

which show in their proper relative positions the coefficients by which the values of the operand are multiplied. Clearly X is obtained from H by a rotation through 45 degrees and a stretch with the ratio $1 : \sqrt{2}$.

Each of these operators is a first approximation to the operator ∇^2. In fact, we have

(62.4) $\quad\quad\quad\quad\quad h^2\nabla^2 U = HU + O(h^4),$

and

(62.5) $\quad\quad\quad\quad\quad h^2\nabla^2 U = XU + O(h^4).$

To establish a more accurate relationship we write down equations 62.1 and 62.2 in the symbolic forms

$$H = 2 \cosh \alpha + 2 \cosh \beta - 4,$$
(62.6)
$$X = 2 \cosh \alpha \cdot \cosh \beta - 2,$$

---

* The word "stencil" is suggested by the practice of laying a mask or stencil over the computation sheet to reveal just the numbers needed at a particular step. Some writers use the term "lozenge" to describe such a group of coefficients.

where for simplicity

$$\alpha = h\frac{\partial}{\partial x} \quad \text{and} \quad \beta = h\frac{\partial}{\partial y}.$$

We now proceed to solve the pair of equations 62.6 for α and β. The first step is to solve for cosh α and cosh β. These are then replaced in terms of $2\sinh^2 \alpha/2$ and $2\sinh^2 \beta/2$, whence we get the two equations

(62.7)
$$2\sinh \alpha/2 = \sqrt{\tfrac{1}{2}(H + \sqrt{H^2 + 8H - 8X})} = A,$$
$$2\sinh \beta/2 = \sqrt{\tfrac{1}{2}(H - \sqrt{H^2 + 8H - 8X})} = B.$$

Next from the relationship

$$\alpha = 2\sinh^{-1} A/2$$

comes the power series

$$\alpha^2 = A^2 - \frac{2A^4}{4!} + \frac{2\cdot 2^2 A^6}{6!} - \frac{2\cdot 2^2\cdot 3^2 A^8}{8!} + \frac{2\cdot 2^2\cdot 3^2\cdot 4^2 A^{10}}{10!} - \cdots.$$

From this and the similar series for β^2 we find that

(62.8)
$$h^2\nabla^2 = (A^2 + B^2) - \frac{1}{12}(A^4 + B^4) + \frac{1}{90}(A^6 + B^6) - \frac{1}{560}(A^8 + B^8) + \frac{1}{3150}(A^{10} + B^{10}) - \cdots.$$

In this equation we replace A and B by their values in terms of H and X as expressed in equations 62.7. It will be found, when the substitutions are performed and the result is arranged in decreasing powers of the operators, that

(62.9)
$$A^{2m} + B^{2m} = \sum_{k=0}^{k=[m/2]} \binom{2k}{k} \frac{\binom{m}{2k}}{\binom{m-1}{k}} H^{m-2k}(2H - 2X)^k.$$

We substitute from equation 62.9 into equation 62.8, then rearrange the terms in groups having second-order differences in the first group, fourth-order differences in the second group, etc. The result so arranged is

$$(62.10) \qquad h^2\nabla^2 = \frac{1}{3}(2H + X)$$

$$- \frac{1}{420}(13H^2 + 16HX + 6X^2)$$

$$+ \frac{1}{630}(2H^3 + H^2X + 4HX^2)$$

$$- \cdots .$$

Equation 62.10 is exact for any polynomial in x and y of degree less than $2n + 2$, provided we retain n terms on the right. There is considerable merit in two other difference operators, expressible linearly in terms of H and X as follows:

$$K = 4H + 2X,$$

$$N^2 = 2(X - H).$$

The exponents are suggested by the fact that

$$K = 6h^2\nabla^2 + O(h^4),$$

$$N^2 = h^4 \frac{\partial^4}{\partial x^2 \, \partial y^2} + O(h^6).$$

When these symbols are used to replace H and X in 62.10, we get the more elegant formula

$$(62.11) \qquad 6h^2\nabla^2 = K - \frac{K^2}{72} + \left(\frac{K^3}{3240} - \frac{KN^2}{180}\right)$$

$$- \left(\frac{K^4}{120,960} - \frac{K^2N^2}{3780} + \frac{N^4}{504}\right)$$

$$+ \cdots .$$

This formula, when truncated to n terms, is an identity when applied to any polynomial in x and y of degree less than $2n + 2$.

The stencils for K and N^2 are given by

$$(62.12) \qquad K = \begin{array}{|c|c|c|} \hline 1 & 4 & 1 \\ \hline 4 & -20 & 4 \\ \hline 1 & 4 & 1 \\ \hline \end{array}$$

and

$$(62.13) \qquad N^2 = \begin{array}{|c|c|c|} \hline 1 & -2 & 1 \\ \hline -2 & 4 & -2 \\ \hline 1 & -2 & 1 \\ \hline \end{array}$$

EXERCISES

Verify the following stencils:

1.
$$H^2 = \begin{array}{|c|c|c|c|c|} \hline & & 1 & & \\ \hline & 2 & -8 & 2 & \\ \hline 1 & -8 & 20 & -8 & 1 \\ \hline & 2 & -8 & 2 & \\ \hline & & 1 & & \\ \hline \end{array}$$

2.
$$X^2 = \frac{1}{4} \begin{array}{|c|c|c|c|} \hline 1 & 2 & & 1 \\ \hline & -8 & & -8 \\ \hline 2 & 20 & & 2 \\ \hline & -8 & & -8 \\ \hline 1 & 2 & & 1 \\ \hline \end{array}$$

3.
$$XH = HX = \frac{1}{2} \begin{array}{|c|c|c|c|c|} \hline & 1 & & 1 & \\ \hline 1 & -4 & -2 & -4 & 1 \\ \hline & -2 & 16 & -2 & \\ \hline 1 & -4 & -2 & -4 & 1 \\ \hline & 1 & & 1 & \\ \hline \end{array}$$

4. $K^2 =$

1	8	18	8	1
8	−8	−144	−8	8
18	−144	468	−144	18
8	−8	−144	−8	8
1	8	18	8	1

5. $KN^2 =$

1	2	−6	2	1
2	−32	60	−32	2
−6	60	−108	60	−6
2	−32	60	−32	2
1	2	−6	2	1

6. $N^4 =$

1	−4	6	−4	1
−4	16	−24	16	−4
6	−24	36	−24	6
−4	16	−24	16	−4
1	−4	6	−4	1

7. Using the results of 1, 2, 3, show that the stencil for the second term in 62.10 is

$-\dfrac{1}{840}$

3	16	32	16	3
16	−36	−240	−36	16
32	−240	836	−240	32
16	−36	−240	−36	16
3	16	32	16	3

8. Show that, when two terms on the right in 62.10 are retained, we have the formula

$$h^2\nabla^2 = \frac{1}{840}$$

-3	-16	-32	-16	-3
-16	176	800	176	-16
-32	800	-3636	800	-32
-16	176	800	176	-16
-3	-16	-32	-16	-3

9. Show that when two terms on the right in 62.11 are retained we have

$$h^2\nabla^2 = \frac{1}{432}$$

-1	-8	-18	-8	-1
-8	80	432	80	-8
-18	432	-1908	432	-18
-8	80	432	80	-8
-1	-8	-18	-8	-1

10. If the differential operator $\dfrac{\partial^4}{\partial x^2\,\partial y^2}$ is denoted by Q^4, show that

$$K = \frac{12\nabla^2 h^2}{2!} + \frac{12\nabla^4 h^4}{4!} + \frac{(12\nabla^6 + 24Q^4\nabla^2)h^6}{6!}$$
$$+ \frac{(12\nabla^8 + 64Q^4\nabla^4 + 80Q^8)h^8}{8!} + \cdots,$$

$$4N^2 = \frac{6Q^4 h^4}{4!} + \frac{15Q^4\nabla^2 h^6}{6!} + \frac{(28Q^4\nabla^4 + 14Q^8)h^8}{8!} + \cdots.$$

The square lattice is the one most commonly used in setting up the difference operator corresponding to Laplace's operator. See, for example, Bickley [10] and Hidaka [90]. However, equilateral triangles and regular hexagons can also be employed as is shown in detail in Southwell's book [194].

63. The equation $U_t = c^2\nabla^2 U$

By differentiation and substitution from the equation $U_t = c^2\nabla^2 U$ we get the sequence

$$U_{t^m} = c^{2m}\nabla^{2m}U, \quad \text{for} \quad m = 1, 2, \cdots .$$

Then by Taylor's series

$$U(x, y, t + k) - U(x, y, t) = kU_t + \frac{k^2 U_{tt}}{2!} + \frac{k^3 U_{ttt}}{3!} + \cdots$$

$$= \frac{1}{6} h^2\nabla^2 U + \frac{1}{72} h^4\nabla^4 U + \frac{1}{1296} h^6\nabla^6 U + \cdots$$

where we have set $k = h^2/6c^2$. Now for $h^2\nabla^2 U$ we substitute the expression in terms of differences given by equation 62.11, collect differences of like orders, and obtain finally

$$(63.1) \quad U(x, y, t + k) - U(x, y, t) = \frac{1}{36} KU(x, y, t)$$

$$+ \left(\frac{K^3}{699\,840} - \frac{KN^2}{6480}\right) U(x, y, t) + O(h^8).$$

In actual computation we drop all terms on the right in equation 63.1 except the first term and represent the resulting equation by the stencil

$$(63.2) \qquad U(x, y, t + k) = \frac{1}{36}\begin{array}{|c|c|c|} \hline 1 & 4 & 1 \\ \hline 4 & 16 & 4 \\ \hline 1 & 4 & 1 \\ \hline \end{array} U(x, y, t).$$

The terms in 63.1 that were dropped to obtain equation 63.2 can be used to estimate the magnitude of the truncation error of 63.2.

It is worthy of note that the two-dimensional operator on the right in 63.2 can be factored into the product of two similar one-dimensional. operators as indicated by the symbolic factorization

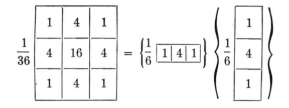

This property is of particular interest in case the computation is performed by means of computing machinery using punched cards or continuous tape where the data are available to the machine only in a linear sequence and not in a two-dimensional array. For one may scan the given region in the xy plane by vertical columns using the operator

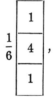

then rearrange the results into rows, and apply the operator

$$\frac{1}{6}\boxed{1\,|\,4\,|\,1}.$$

See Yowell [223].

The leading term of the truncation error in 63.2 is seen from 63.1 to be

(63.3)
$$E = \left(\frac{K^3}{699\,840} - \frac{KN^2}{6480}\right) U(x, y, t).$$

To terms of the same degree in h this can be replaced by the equation

(63.4)
$$E = \left(\frac{\Delta_t^3}{15} - \frac{\Delta_t N^2}{180}\right) U(x, y, t),$$

a form that is frequently easier to compute than equation 63.3. The values of E may be calculated as the computation proceeds, and provide a practical though non-rigorous estimate of the truncation error being committed.

Example 1. Solve the equation $U_t = c^2 \nabla^2 U$ with the condition $U = 0$ on the boundary of a square of side $6h$, and the initial value of U given by

$$U = xy(6h - x)(6h - y)/h^4.$$

Two steps of the solution are shown in computation 38.

Computation 38

Differential equation $U_t = c^2 \nabla^2 U$

Difference equation 63.2

k							
	0	0	0	0	0	0	0
	0	25	40	45	40	25	0
	0	40	64	72	64	40	0
0	0	45	72	81	72	45	0
	0	40	64	72	64	40	0
	0	25	40	45	40	25	0
	0	0	0	0	0	0	0
	0	0	0	0	0	0	0
	0	21.78	35.78	40.44	35.78	21.78	0
	0	35.78	58.78	66.44	58.78	35.78	0
1	0	40.44	66.44	75.11	66.44	40.44	0
	0	35.78	58.78	66.44	58.78	35.78	0
	0	21.78	35.78	40.44	35.78	21.78	0
	0	0	0	0	0	0	0
	0	0	0	0	0	0	0
	0	19.26	32.19	36.57	32.19	19.26	0
	0	32.19	53.78	61.11	53.78	32.19	0
2	0	36.57	61.11	69.44	61.11	36.57	0
	0	32.19	53.78	61.11	53.78	32.19	0
	0	19.26	32.19	36.57	32.19	19.26	0
	0	0	0	0	0	0	0

The test in equation 63.4 indicates that the values obtained are correct to the number of decimal places retained.

Example 2. Solve the equation $U_t = c^2 \nabla^2 U$ with the condition $U = 0$ on the boundary of a square of side $21h$, and with the initial value 1.000 inside the square.

Here we have 400 interior points at which to apply formula 63.2 in order to compute U for each value of k. When k is large, say, $k = 100$, this requires 40,000 separate calculations. In the present instance symmetry enables us to get along with only 55 calculations per step. In computation 39 the steps for $k = 15$, 16 are shown.

Computation 39

Differential equation $U_t = c^2\nabla^2 U$

k

0.116	0.211	0.276	0.313	0.330	0.337	0.339	0.339	0.340	0.340
	.386	.505	.572	.603	.615	.619	.620	.620	.620
		.661	.749	.790	.805	.810	.811	.811	.811
			.848	.895	.913	.918	.919	.920	.920
15				.944	.963	.969	.970	.971	.971
					.982	.989	.990	.991	.991
						.995	.997	.997	.997
							.999	0.999	0.999
								1	1
									1

.109	.200	.264	.301	.319	.326	.329	.330	0.330	0.330
	.367	.484	.552	.586	.599	.604	.605	.605	.605
		.638	.728	.772	.790	.796	.797	.797	.797
			.831	.881	.901	.908	.910	.910	.910
16				.934	.955	.963	.965	.965	.965
					.978	.986	.988	.988	.988
						.994	.996	.997	.997
							.998	0.999	0.999
								1	1
									1

Only interior values are given, the boundary values being always zero. Only values above the main diagonal and left of the central column are shown, since all others can be supplied by symmetry.

In computation 38 we achieved very comfortable accuracy with a relatively coarse mesh and might therefore hope for even greater accuracy in computation 39 where the mesh is much finer. An application of the error estimate (or, alternatively, a comparison with the analytical solution, which for this example is easily obtained) shows that the results in the second example are in fact much poorer than in the first. The reason is not far to seek. In Example 1 the initial value of $U(x, y, t)$, together with first and second derivatives, is continuous for all values of x and y, whereas in Example 2 the initial value itself is discontinuous all around the boundary.

A somewhat different example is the following.

Example 3. A concrete column with square cross section is subject to a periodic (sinusoidal) variation of surface temperature about the mean temperature θ_0, which we take as the origin of the temperature scale, i.e., $\theta_0 = 0$. The initial internal temperature is taken as zero also. To determine approximately the internal variations of temperature with time we solve the equation 63.2 with boundary values defined by $U_B = 100 \sin \frac{n\pi}{12}$. The value of h is taken as one fourth of the side of the square. The results are shown in computation 40.

Flow of Heat   Computation 40

$k = 0$

0	0	0	0	0
0	0	0	0	0
0	0	0	0	0
0	0	0	0	0
0	0	0	0	0

$k = 1$

26	26	26	26	26
26	0	0	0	26
26	0	0	0	26
26	0	0	0	26
26	26	26	26	26

$k = 2$

50	50	50	50	50
50	7.944	4.333	7.944	50
50	4.333	0	4.333	50
50	7.944	4.333	7.944	50
50	50	50	50	50

$k = 3$

71	71	71	71	71
71	19.771	12.265	19.771	71
71	12.265	2.808	12.265	71
71	19.771	12.265	19.771	71
71	71	71	71	71

$k = 4$

87	87	87	87	87
87	33.285	22.671	33.285	87
87	22.671	8.896	22.671	87
87	33.285	22.671	33.285	87
87	87	87	87	87

$k = 5$

96	96	96	96	96
96	46.662	34.221	46.662	96
96	34.221	17.728	34.221	96
96	46.662	34.221	46.662	96
96	96	96	96	96

For sake of brevity only five steps of computation 40 are shown. Partial results of 36 steps are displayed graphically in Fig. 14 where the curve A gives the variation of outside temperature, B the temperature at any one of the four symmetrically placed points $(0 \pm h)$, $(\pm h, 0)$, while C shows the temperature at the center $(0, 0)$.

Example 4. Consider an infinitely long heat conductor with its plane section in the xy plane shaped as shown in Fig. 15. The internal temperature is assumed to be initially zero, the boundary AA' is maintained at 100 degrees, and $BC'D'E'B'$ at 0 degrees. Because of symmetry we have $U_y = 0$ on DD' and EE'.

We apply equation 62.1 repeatedly to the ten points inside the space bounded by EE' and DD'. The boundary values on AA' and $BCC'B'$ are known while the values at points just across EE' and DD' are given by reflection in these lines. The result of several steps is shown in computation 41.

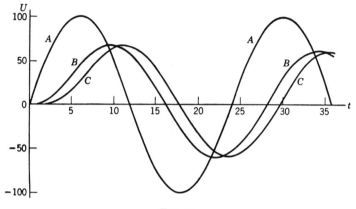

FIG. 14

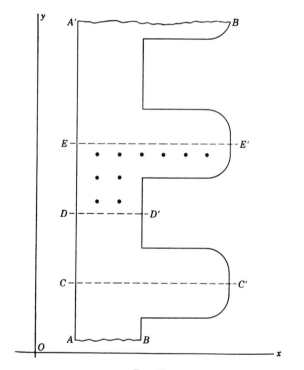

FIG. 15

Computation 41

Differential equation $U_t = c^2\nabla^2 U$

k						
0	0	0	0	0	0	0
	0	0				
	0	0				
1	0.1667	0	0	0	0	0
	.1667	0				
	.1667	0				
2	.2778	0.0278	0	0	0	0
	.2778	.0278				
	.2778	.0278				
3	.3565	.0648	0.0046	0	0	0
	.3565	.0648				
	.3565	.0648				
4	.4151	.1031	.0134	0.0006	0	0
	.4151	.1027				
	.4151	.1026				
5	.4606	.1397	.0246	.0026	0.0001	0
	.4605	.1381				
	.4605	.1376				
. . . .						
22	.6775	.3807	.1515	.0497	.0156	.0044
	.6693	.3466				
	.6632	.3314				
23	.6799	.3837	.1536	.0508	.0162	.0046
	.6714	.3490				
	.6651	.3333				

EXERCISES

1. Calculate three more stages in computation 38.
2. Calculate one more stage in computation 39.
3. Estimate the truncation error at various stages in Example 3.
4. Calculate three more stages in computation 40.

64. The equation $U_{tt} = c^2\nabla^2 U$

Let $\delta^2 U$ be defined by the equation

$$\delta^2 U = U(x, y, t + k) - 2U(x, y, t) + U(x, y, t - k).$$

Assuming the existence of a sixth derivative with respect to t, we have from equation 39.3

$$\delta^2 U = k^2(U_{tt} + \delta^2 U_{tt}/12) + O(k^6).$$

In this we replace U_{tt} by its value $c^2 \nabla^2 U$ from the differential equation and eliminate $\nabla^2 U$ by means of equation 62.11 to obtain

$$\delta^2 U = \rho^2 \left(\frac{K}{6} - \frac{K^2}{432} + \frac{\delta^2 K}{72} + \cdots \right) U$$

where $\rho = kc/h$. Operating on this equation with K, we get

$$\delta^2 K U = \frac{\rho^2}{6} K^2 U + \cdots$$

which enables us to eliminate $\delta^2 K U$ and to arrive at the equation

(64.1) $$\delta^2 U = \rho^2 \left[\frac{K}{6} - \frac{(1 - \rho^2)}{432} K^2 \right] U + O(h^6).$$

In order to insure convergence it can be shown that we must take $\rho^2 \leq (13 - \sqrt{61})/8$. Since this value is somewhat inconvenient in computation, we actually propose to use $\rho^2 = \frac{6}{10}$, a choice that entails a decrease of only about 4 per cent in the value of k and provides a better working formula. With this choice we can write 64.1 in the special form

(64.2) $$\delta^2 U = (K/10 - K^2/1800) U + O(h^6).$$

In analogy with equation 63.5 the leading term of the truncation error is

(64.3) $$E = - \delta^6 U/24 - N^2 \, \delta^2 U/180.$$

In practice, we do not use a 25-point stencil for the operator K^2, but first compute KU, and then by a second application of K we obtain $K^2 U$. This procedure avoids awkward situations at the boundaries.

Actually difference equations of the type 64.2 are seldom used to obtain approximate solutions of the wave equation. One reason perhaps is that their use requires a great deal of labor. Another is that in many practical problems the exact wave form is unimportant, and we seek instead the characteristic frequencies. Then the problem is one of characteristic numbers and is discussed in Chapter 11. But there is still another reason which we should understand clearly. Just as a microscope ceases of function for objects shorter than the wave length of visible light so difference equations cease to function for waves whose wave length is less than the mesh length of the lattice. Hence, though we may hope to represent with some accuracy the low-frequency components of a given wave motion by means of difference equations, the higher-frequency components, if present, will always be distorted. This difficulty is more pronounced for the wave equation than for the heat equation since high-frequency terms, if present initially, persist permanently in the solution of the former but are rapidly damped out in the solution of the latter.

One may speculate on the desirability of using a difference equation with some damping present as actually a better representation of some physical situations than the conventional wave equation.

Example 1. As a first illustration we consider a problem in which higher frequencies do not occur. The problem is the vibration of a square flexible membrane fastened at the edges of the square, initially displaced an amount U_0 and released with initial velocity zero. The initial displacement U_0 is deliberately chosen so as to be free of all terms of higher frequencies.

By means of equation 64.2 we compute $\delta^2 U_0$. We have

$$U_1 - 2U_0 + U_{-1} = \delta^2 U_0,$$

but, since the initial velocity is zero, we may set $U_{-1} = U_1$ so that

$$U_1 = U_0 + \tfrac{1}{2}\delta^2 U_0.$$

After U_1 has been found for all interior points, we calculate $\delta^2 U_1$ by equation 64.2 and obtain U_2 from the formula

$$U_2 = 2U_1 - U_0 + \delta^2 U_1.$$

The process for finding U_3, U_4, etc. is analogous to that for U_2. In computation 42 are shown six steps. For this carefully chosen example the accuracy is good, the error in the sixth step being about one unit in the last place retained.

Computation 42

(See Example 1)

	0	0	0	0	0	0	0
	0	1000	1732	2000	1732	1000	0
	0	1732	3000	3464	3000	1732	0
U_0	0	2000	3464	4000	3464	2000	0
	0	1732	3000	3464	3000	1732	0
	0	1000	1732	2000	1732	1000	0
	0	0	0	0	0	0	0
	0	0	0	0	0	0	0
	0	-314	-544	-629	-544	-314	0
$\dfrac{KU_0}{10}$	0	-544	-943	-1089	-943	-544	0
	0	-629	-1089	-1258	-1089	-629	0
	0	-544	-943	-1089	-943	-544	0
	0	-314	-544	-629	-544	-314	0
	0	0	0	0	0	0	0

Computation 42. (*Continued*)

	0	0	0	0	0	0	0
	0	6	10	11	10	6	0
$\dfrac{K^2U_0}{1800}$	0	10	16	19	16	10	0
	0	11	19	22	19	11	0
	0	10	16	19	16	10	0
	0	6	10	11	10	6	0
	0	0	0	0	0	0	0
	0	0	0	0	0	0	0
	0	840	1455	1680	1455	840	0
	0	1455	2520	2910	2520	1455	0
U_1	0	1680	2910	3360	2910	1680	0
	0	1455	2520	2910	2520	1455	0
	0	840	1455	1680	1455	840	0
	0	0	0	0	0	0	0
. . . . .							. . . .
	0	0	0	0	0	0	0
	0	− 962	−1667	−1925	−1667	− 962	0
	0	−1667	−2888	−3335	−2888	−1667	0
U_5	0	−1925	−3335	−3851	−3335	−1925	0
	0	−1667	−2888	−3335	−2888	−1667	0
	0	− 962	−1667	−1925	−1667	− 962	0
	0	0	0	0	0	0	0
	0	0	0	0	0	0	0
	0	302	523	605	523	302	0
$\dfrac{KU_5}{10}$	0	523	900	1048	900	523	0
	0	605	1048	1212	1048	605	0
	0	523	900	1048	900	523	0
	0	302	523	605	523	302	0
	0	0	0	0	0	0	0
	0	0	0	0	0	0	0
	0	− 5	− 9	− 11	− 9	− 5	0
$\dfrac{K^2U_5}{1800}$	0	− 9	− 15	− 19	− 15	− 9	0
	0	− 11	− 19	− 21	− 19	− 11	0
	0	− 9	− 15	− 19	− 15	− 9	0
	0	− 5	− 9	− 11	− 9	− 5	0
	0	0	0	0	0	0	0
	0	0	0	0	0	0	0
	0	− 956	−1656	−1911	−1656	− 956	0
	0	−1656	−2876	−3310	−2876	−1656	0
U_6	0	−1911	−3310	−3823	−3310	−1911	0
	0	−1656	−2876	−3310	−2876	−1656	0
	0	− 956	−1656	−1911	−1656	− 956	0
	0	0	0	0	0	0	0

Example 2. Here we take the same problem as in Example 1 except for a slightly different choice of U_0. This new choice contains terms with higher frequencies and the effect is quickly seen, for it turns out that we have errors of nearly 2 per cent. Computation 43 shows the results without the intermediate steps. For brevity only a triangle of values is shown since the remaining entries can be supplied by symmetry.

Computation 43

U_0	0	0	0	0
	0	25.00	40.00	45.00
			64.00	72.00
				81.00
U_1	0	0	0	0
	0	18.95	32.17	36.55
			54.52	61.92
				70.32
U_2	0	0	0	0
	0	6.03	13.17	16.21
			27.15	32.91
				39.73
U_3	0	0	0	0
	0	− 5.43	− 7.68	− 7.31
			−10.14	− 8.98
				− 7.25
U_4	0	0	0	0
	0	−13.00	−23.75	−27.92
			−43.00	−50.32
				−58.43
U_5	0	0	0	0
	0	−17.94	−33.26	−40.34
			−61.53	−74.38
				−90.46
U_6	0	0	0	0
	0	−19.86	−34.89	−40.54
			−61.83	−72.01
				−83.81

65. Curved boundaries

In order not to tackle too many troubles all at the same time we have so far carefully avoided the most vexing difficulty arising in two-dimensional problems, namely, how to handle boundaries that do not cut the lattice at nodal points. This boundary problem causes trouble in three ways. The first trouble is computational and stems from the fact that the simple difference formulas such as 62.4, 62.5, 62.10, 62.11, 63.1, 63.2, 64.1, and 64.2 can no longer be used at points adjacent to the boundary and must be replaced by unsymmetric and inconvenient formulas which, for a comparable number of points in the point pattern, are much less accurate. For example, equation 62.1 can be replaced by

$$(65.1) \quad H'F(x, y) =$$
$$\frac{2}{h_1 + h_3} \left[\frac{F(x + h_1, y) - F(x, y)}{h_1} + \frac{F(x - h_3, y) - F(x, y)}{h_3} \right]$$
$$+ \frac{2}{h_2 + h_4} \left[\frac{F(x, y + h_2) - F(x, y)}{h_2} + \frac{F(x, y - h_4) - F(x, y)}{h_4} \right]$$

when the associated point pattern is

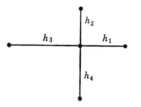

But with this formula we have

$$H'F(x, y) = \nabla^2 F(x, y) + O(h)$$

if $h_i \leq h$, $i = 1, 2, 3, 4$, and hence we do not obtain an accuracy comparable with that of equation 62.4. Likewise in analogy with equation 62.2 we have

$$(65.2) \quad X'F(x, y) = \frac{1}{h_5 + h_7} \times$$
$$\left[\frac{F(x + h_5, y + h_5) - F(x, y)}{h_5} + \frac{F(x - h_7, y - h_7) - F(x, y)}{h_7} \right]$$
$$+ \frac{1}{h_6 + h_8} \times$$
$$\left[\frac{F(x - h_6, y + h_6) - F(x, y)}{h_6} + \frac{F(x + h_8, y - h_8) - F(x, y)}{h_8} \right]$$

for the pattern

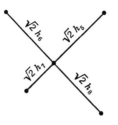

If we wish to construct a difference formula for unequally spaced points to represent $\nabla^2 F$ with an accuracy like $O(h^2)$, we find that nine conditions must be satisfied and hence at least ten points are needed. More than ten points may of course be used, in which case we may have extra parameters at our disposal.

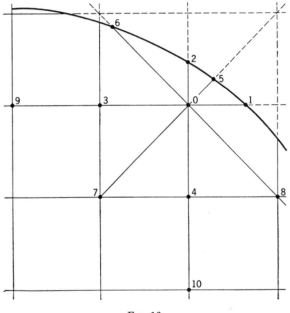

FIG. 16

To discuss in full detail the large variety of cases that might be required in practice would demand too much space. We shall be content to cite a particular formula using 11 points, four of which may be non-nodal points as shown in Fig. 16. The coordinates of the points 1, 5, 2, 6 on the boundary referred to 0 as origin are respectively $(s_1h, 0)$, (s_5h, s_5h), $(0, s_2h)$, and $(-s_6h, s_6h)$, where h is the mesh length and s_1, s_2, s_5, s_6 each lie in the interval

$$0 < s \leq 1.$$

By appropriate adjustment of the figure (rotation, reflection, etc.) and proper numbering of the points it will be found that this point pattern can be adapted to most of the cases where the boundary cuts into the standard nine-point pattern. The formula in question is

(65.3)
$$ZF = \sum_{i=1}^{10} C_i(F_i - F_0)$$

for which the coefficients C_i can be calculated one after another by means of the set of equations shown below:

Equations for the C's

$$3s_5(s_5 + 1)C_5 = s_6,$$

$$3s_6(s_6 + 1)C_6 = s_5,$$

$$C_7 = s_5{}^3C_5,$$

$$C_8 = s_6{}^3C_6,$$

$$3s_1(s_1 + 1)(s_1 + 2)C_1 = 18 - (6s_5s_6 + 2s_6 - 2s_5),$$

$$3s_2(s_2 + 1)(s_2 + 2)C_2 = 18 - (2s_5s_6 + 2s_6 + 2s_5),$$

$$3C_3 = 3s_1(s_1 + 2)C_1 - 6 + (2s_5s_6 + 2s_6 - 2s_5),$$

$$3C_4 = 3s_2(s_2 + 2)C_2 - 6 + (-2s_5s_6 + 2s_6 + 2s_5),$$

$$18C_9 = 3s_1(s_1 + 1)(s_1 - 1)C_1 + (\quad\quad - s_6 + s_5),$$

$$18C_{10} = 3s_2(s_2 + 1)(s_2 - 1)C_2 + (2s_5s_6 - s_6 - s_5).$$

To use equation 65.3 at any given nodal point P adjacent to the boundary it is first necessary to compute the coordinates of points such as 1, 5, 2, and 6 in Fig. 16 where the horizontal, vertical, and 45-degree lines radiating from P cut the boundary. Of course, if any of these intersections is farther from P than the next nodal point, we use the nodal point instead. Incidentally, if all intercepts are so placed, we may verify that 65.3 reduces to our standard formula 62.12. We do so by observing that in this situation we have $s_1 = s_2 = s_5 = s_6 = 1$, and then the equations defining the C's give us

$$C_1 = C_2 = C_3 = C_4 = \tfrac{4}{6}$$
$$C_5 = C_6 = C_7 = C_8 = \tfrac{1}{6}$$
$$C_9 = C_{10} = 0.$$

From the coordinates of the points 1, 2, 5, 6 we compute s_1, s_2, s_5 s_6, and, when these are known, the ten C's are determined by their defining equations.

We have considered the first difficulty with curved boundaries, namely, that due to the breakdown of our standard formulas. The second is due to the impossibility of representing the fine detail of a curved boundary by means of a coarse lattice. Figure 17 shows an

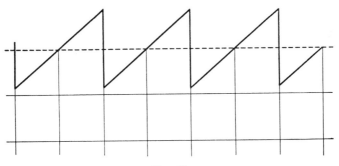

FIG. 17

example where for the given lattice the solid curve and the dotted curve are indistinguishable.

Obviously no refinement of our difference formulas is going to take care of a situation of this type, and the only answer is to choose a mesh fine enough to represent the boundary curve with as much accuracy as the problem in question requires.

The third difficulty is one which also arises in the case of straight-line boundaries and is due to discontinuities or other singularities in the assigned boundary values or initial values. For example, in the case of the heat equation it is not at all uncommon for the initial values to be discontinuous on the boundary. In this sort of situation a difference formula of high order is really no better than the simplest available difference formula. That is, we may as well use equation 65.1 at points adjacent to the boundary and equation 62.4 at interior points. In order to attain the required accuracy we have again no choice but to use a sufficiently fine mesh, since higher accuracy cannot be secured by refinement of the difference formulas. In the case of the heat equation, if the discontinuity occurs only in the initial values, it may be possible to resort to larger mesh after a few steps have been made with a fine mesh.

In some types of problems it is feasible to use a fine mesh near the boundary, with perhaps an ultrafine mesh near certain critical portions of the boundary, and a much larger mesh in the interior of the region. In the case of the wave equation and the heat equation, however, we have seen that any change in mesh size brings also a change in the time interval. If for the heat equation the mesh length is cut in half, the time interval is cut to one-fourth. Hence, variations in mesh length over a region lead to awkward computational difficulties along the boundary between subregions with different mesh lengths. Although these difficulties are not quite insuperable, they do greatly discourage the use of variable mesh lengths in this type of problem.

Example. Find the solution of $U_t = c^2 \nabla^2 U$ for the quadrant of the circle bounded by $x = 0$, $y = 0$, $x^2 + y^2 = 36$, with the boundary conditions $U = 0$ and the initial values as shown in Fig. 18.

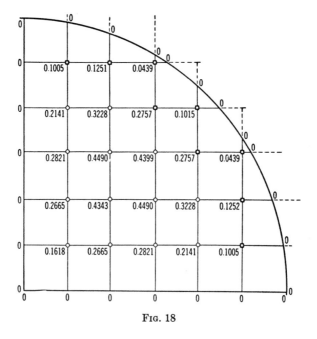

FIG. 18

At the points marked $\bigcirc$ we use equation 63.2. At those marked $\square$ we use

$$U(x, y, t + k) - U(x, y, t) = ZU(x, y, t)$$

where ZU is computed by equation 65.3. For this we need a table of the values of the s_i's and C_i's for each point. These, together with two steps of the computation, are shown in computation 44.

Computation 44

	①	②	③	④	⑤
s_1	1.0000	1.0000	0.3166	1.0000	0.4721
s_2	.9161	.6568	.1962	1.0000	.4721
s_5	.7417	.4686	.1231	.7131	.2426
s_6	1.0000	.8979	.4142	1.0000	1.0000
C_1	.7241	.8120	5.9063	.7304	2.9156
C_2	.8488	1.6630	10.8834	.7304	2.9156
C_3	.8389	1.0028	2.5603	.8580	2.0697
C_4	.9340	1.5327	3.0125	.8580	2.0697
C_5	.2580	.4349	.9986	.2729	1.1055
C_6	.1236	.0917	.0700	.1188	.0404
C_7	.1053	.0448	.0019	.0989	.0158
C_8	.1236	.0664	.0050	.1188	.0404
C_9	−.0144	−.0238	−.2966	−.0159	−.2204
C_{10}	−.0352	−.1327	−.3663	−.0159	−.2204

The encircled numbers above refer to the points □ adjacent to the boundary counted clockwise starting from the top in Fig. 18. The values for ⑥, ⑦, ⑧, and ⑨ are obtained by symmetry.

With these formulas we compute the values at $t = k$ and $t = 2k$ as shown below for the correspondingly placed points in Fig. 18.

$$t = k$$

0	0.0887	0.1105	0.0376		
0	.1895	.2858	.2435	0.0882	
0	.2497	.3974	.3894	.2435	0.0376
0	.2359	.3844	.3974	.2858	.1105
0	0.1432	0.2359	0.2497	0.1895	0.0887
0	0	0	0	0	0

$$t = 2k$$

0	0.0784	0.0976	0.0365		
0	.1677	.2528	.2150	0.0790	
0	.2210	.3517	.3445	.2150	0.0365
0	.2088	.3402	.3517	.2528	.0976
0	0.1268	0.2088	0.2210	0.1677	0.0784
0	0	0	0	0	0

Actually these results are not very good near the boundary as we see by the comparison with the correct values shown below for the analytical solution in polar coordinates:

$$U = e^{-\lambda t}J_2(ar) \sin 2\varphi$$
$$t = k$$

0	0.0889	0.1107	0.0389		
0	.1895	.2857	.2440	0.0898	
0	.2497	.3974	.3893	.2440	0.0389
0	.2359	.3844	.3974	.2857	.1107
0	0.1432	0.2359	0.2497	0.1895	0.0889
0	0	0	0	0	0

$$t = 2k$$

0	0.0787	0.0980	0.0344		
0	.1677	.2529	.2160	0.0795	
0	.2210	.3517	.3446	.2160	0.0344
0	.2088	.3402	.3517	.2529	.0980
0	0.1267	0.2088	0.2210	0.1677	0.0787
0	0	0	0	0	0

To show even more clearly the disastrous effect of curvilinear boundaries we compare below certain results, the first two sets computed with the standard formulas using values of the initial function at the lattice points including where necessary those outside the circle, the second two computed with modified formulas using boundary points. We give the ratio of $\nabla^2 U$ to U, that is

$$\rho = \nabla^2 U/U,$$

where $\nabla^2 U$ is calculated in the following different ways:

A. By using the operator H and the actual values of U at the nodal points including those outside the circle.

B. The same as A except the operator K was used.

C. By the formula 65.3 which uses points on the circle.

D. By using equation 65.1 and points on the circle.

	①	②	③	④	⑤
A	−0.7014	−0.7021	−0.7068	−0.7060	−0.7113
B	− .6892	− .6893	− .6892	− .6894	− .6896
C	− .7025	− .7005	− .8632	− .7000	− .7872
D	− .7143	− .7768	−1.6694	− .7060	− .5460

The value which all these ratios ought to have in order to give the value of $\nabla^2 U$ correctly is 0.68964. Evidently the only formula that gives reasonable results is K applied at the lattice points. The results of 65.3 are deplorable, the only consolation being that 65.1, which many authors recommend, is much worse.

EXERCISES

1. Using a three-dimensional cell block of cubes, construct possible difference operators to replace the Laplacian differential operator

$$\nabla^2 = \frac{\partial^2}{\partial x^2} + \frac{\partial^2}{\partial y^2} + \frac{\partial^2}{\partial z^2}:$$

(a) With the 7 points (x, y, z), $(x \pm h, y, z)$, $(x, y \pm h, z)$, $(x, y, z \pm h)$.

(b) With the 27 points having all the different possible combinations of coordinates $x, y, z, x \pm h, y \pm h, z \pm h$.

2. Investigate the leading term of the error for a and b in exercise 1.

3. Show that in three dimensions it is possible to set up a difference equation for

$$U_t = c^2 \nabla^2 U$$

which can be factored so that

$$U(x, y, z, t + k) = (XYZ)U,$$

where $XU = \frac{1}{6}(U(x + h) + 4U(x) + U(x - h))$ and Y and Z are similarly defied (cf equation 63.2, ff).

4. Investigate the leading term in the error for the difference equation in exercise 3.

9

LINEAR EQUATIONS AND MATRICES

Numerical methods for elliptic partial differential equations lead to systems of simultaneous algebraic equations, frequently of high order. In addition to partial differential equations many other pressing problems in industry, technology, and science lead to simultaneous systems with many variables. The importance of this problem of linear equations justifies a special chapter devoted to linear equations, matrices, and iterative methods for solving systems of high order. In most instances proofs will be omitted, since they can be found in the texts referred to below, and we shall content ourselves with numerical examples illustrating the various theorems. Applications of the results obtained in this chapter will be made in Chapters 10 and 11.

There are many excellent texts dealing with matrices. The following partial list will provide the reader with a variety of treatments and points of view: Frazer, Duncan, and Collar [72], Mac Duffee [111], Schmeidler [181], Turnbull and Aitken [213], Wedderburn [218], Zurmühl [227]. Of particular importance for the solution of linear equations is the recent (as yet unpublished) bibliography by Forsythe [253].

66. Linear equations

Associated with a set of n equations in n unknowns

$$a_{11}u_1 + a_{12}u_2 + \cdots + a_{1n}u_n = v_1,$$

$$a_{21}u_1 + a_{22}u_2 + \cdots + a_{2n}u_n = v_2,$$

(66.1)

$$\cdots \cdots \cdots \cdots \cdots \cdots \cdots \cdots$$

$$a_{n1}u_1 + a_{n2}u_2 + \cdots + a_{nn}u_n = v_n,$$

are the determinant

$$D = \begin{vmatrix} a_{11} & a_{12} & \cdots & a_{1n} \\ a_{21} & a_{22} & \cdots & a_{2n} \\ \cdots\cdots\cdots\cdots\cdots \\ a_{n1} & a_{n2} & \cdots & a_{nn} \end{vmatrix},$$

and the square matrix

$$A = \begin{bmatrix} a_{11} & a_{12} & \cdots & a_{1n} \\ a_{21} & a_{22} & \cdots & a_{2n} \\ \cdots\cdots\cdots\cdots\cdots \\ a_{n1} & a_{n2} & \cdots & a_{nn} \end{bmatrix}.$$

The former is a scalar quantity, the latter is a complex of n^2 components. Matrices will be denoted by boldfaced capital letters.

If the variables u_i are expressed linearly in terms of n other variables w_i by the equations

$$u_1 = b_{11}w_1 + b_{12}w_2 + \cdots + b_{1n}w_n,$$
$$u_2 = b_{21}w_1 + b_{22}w_2 + \cdots + b_{2n}w_n,$$

(66.2)

$$\cdots\cdots\cdots\cdots\cdots\cdots\cdots\cdots\cdots$$

$$u_n = b_{n1}w_1 + b_{n2}w_2 + \cdots + b_{nn}w_n,$$

and if these expressions for the u's are substituted into equations 66.1, the result is

$$c_{11}w_1 + c_{12}w_2 + \cdots + c_{1n}w_n = v_1,$$
$$c_{21}w_1 + c_{22}w_2 + \cdots + c_{2n}w_n = v_2,$$

(66.3)

$$\cdots\cdots\cdots\cdots\cdots\cdots\cdots\cdots\cdots$$

$$c_{n1}w_1 + c_{n2}w_2 + \cdots + c_{nn}w_n = v_n,$$

in which the coefficients c_{ij} are given by

(66.4) $$c_{ij} = a_{i1}b_{1j} + a_{i2}b_{2j} + \cdots + a_{in}b_{nj}.$$

Formula (66.4) provides the definition for the product matrix of two $n \times n$ matrices A and B. That is, if $A = [a_{ij}]$ and $B = [b_{ij}]$, then the matrix $C = [c_{ij}]$ is the product matrix

$$AB = C$$

when the elements of **C** are obtained from the elements of **A** and of **B** by means of 66.4. It is evident from this definition that **AB** is not usually equal to **BA,** and that therefore the product matrix depends on the order in which multiplication is performed. Clearly, however, the order is immaterial in the product **AA,** and it can be shown that **A(AA)** = **(AA)A,** etc. Consequently in products of this type the parentheses can be omitted and the simple notation

$$\mathbf{AA} = \mathbf{A}^2, \qquad \mathbf{AAA} = \mathbf{A}^3, \quad \text{etc.,}$$

is justified.

The matrix

$$\begin{bmatrix} 1 & 0 & \cdots & 0 \\ 0 & 1 & \cdots & 0 \\ \cdots & \cdots & \cdots & \cdots \\ 0 & 0 & \cdots & 1 \end{bmatrix}$$

with ones in the principal diagonal and zeros elsewhere is called the *unit matrix* and is denoted by **I.** The term "unit matrix" is justified by the equations

$$\mathbf{IA} = \mathbf{AI} = \mathbf{A},$$

which follow from the definition of a product.

If the $n \times n$ matrix **A** is multiplied into the $n \times n$ matrix

$$\mathbf{u} = \begin{bmatrix} u_1 & 0 & \cdots & 0 \\ u_2 & 0 & \cdots & 0 \\ \cdots & \cdots & \cdots & \cdots \\ u_n & 0 & \cdots & 0 \end{bmatrix},$$

where all elements are zero except in the first column, the rule for multiplication gives the matrix

$$\mathbf{v} = \begin{bmatrix} v_1 & 0 & \cdots & 0 \\ v_2 & 0 & \cdots & 0 \\ \cdots & \cdots & \cdots & \cdots \\ v_n & 0 & \cdots & 0 \end{bmatrix},$$

in which the v's are the quantities defined by equations 66.1. Matrices of the types **u** and **v** are called "column matrices" or simply "vectors," and are written in the shorter form

$$\mathbf{u} = \begin{bmatrix} u_1 \\ u_2 \\ \cdot \\ \cdot \\ \cdot \\ u_n \end{bmatrix}, \qquad \mathbf{v} = \begin{bmatrix} v_1 \\ v_2 \\ \cdot \\ \cdot \\ \cdot \\ v_n \end{bmatrix}, \qquad \text{etc.}$$

With this interpretation in mind we observe that equations 66.1 can be compactly written in the form

$$\mathbf{Au} = \mathbf{v}.$$

We have similarly for equations 66.2

$$\mathbf{u} = \mathbf{Bw},$$

and by elimination of **u** we get equations 66.3 in the abridged form

$$\mathbf{ABw} = \mathbf{v},$$

or, if **AB** = **C**, then

$$\mathbf{Cw} = \mathbf{v}.$$

Vectors will be denoted, as above, by lower-case boldface letters.

It is natural to define the sum of two $n \times n$ matrices as the $n \times n$ matrix whose elements are the sums of the corresponding elements of the given matrices. The extension to subtraction, to sums and differences of several matrices, and in general to linear combinations of matrices is immediate. It follows in particular that the product $c\mathbf{A}$, where c is a scalar and **A** is a matrix, is the matrix whose elements are c times the corresponding elements of **A**.

If equations 66.1 are solved by Cramer's rule, it is seen that the solution can be written in the compact form

$$\mathbf{u} = \mathbf{A}^{-1}\mathbf{v}$$

in which $\mathbf{A}^{-1}$ denotes the matrix with the ij element equal to

$$\alpha_{ji}/D.$$

Here D is the determinant of the matrix **A** and α_{ji} is the cofactor of the element a_{ji} in D. The matrix A^{-1}, which exists when and only when $D \neq 0$, is called the "inverse" of A. The appropriateness of the name and of the notation A^{-1} is clear from the inverse relations

$$Au = v,$$

$$u = A^{-1}v.$$

It is evident from the foregoing that the algebra of a set of $n \times n$ matrices is much like ordinary algebra, the commutative law in multiplication being the most conspicuous exception. For example, if $P(x)$ is a polynomial in x with scalar coefficients and if A is an $n \times n$ matrix, the rules of operation just explained enable one to calculate the $n \times n$ matrix $P(A)$.

Example. Let

$$P(x) = 2x^2 + 4x - 3$$

and

$$A = \begin{bmatrix} 1 & 3 & 1 \\ 4 & 1 & 2 \\ 2 & 4 & 3 \end{bmatrix}.$$

We then write

$$P(A) = 2A^2 + 4A - 3I,$$

and proceed to evaluate the terms of the polynomial as follows:

$$-3I = \begin{bmatrix} -3 & 0 & 0 \\ 0 & -3 & 0 \\ 0 & 0 & -3 \end{bmatrix},$$

$$4A = \begin{bmatrix} 4 & 12 & 4 \\ 16 & 4 & 8 \\ 8 & 16 & 12 \end{bmatrix},$$

$$2A^2 = \begin{bmatrix} 30 & 20 & 20 \\ 24 & 42 & 24 \\ 48 & 44 & 38 \end{bmatrix},$$

and, by addition,

$$P(A) = \begin{bmatrix} 31 & 32 & 24 \\ 40 & 43 & 32 \\ 56 & 60 & 47 \end{bmatrix}.$$

More generally it is possible to construct a polynomial in $\mathbf{A}$ with coefficients which are themselves $n \times n$ matrices. And, if the determinant associated with the denominator matrix does not vanish, we can evaluate a rational fraction in the form of the quotient of two polynomials in $\mathbf{A}$.

We conclude this article with a few theorems for reference.

1. *If two $n \times n$ matrices $\mathbf{A}_1$ and $\mathbf{A}_2$ have determinants D_1 and D_2, the determinant of the product $\mathbf{A}_1\mathbf{A}_2$ equals that of $\mathbf{A}_2\mathbf{A}_1$ and is D_1D_2.*

2. *If the determinant of $\mathbf{A}$ is $D \neq 0$, that of $\mathbf{A}^{-1}$ is $1/D = D^{-1}$.*

3. *If the determinant D of $\mathbf{A}$ is not zero, the equations*

$$\mathbf{Au} = \mathbf{v}$$

have a unique solution given by

$$\mathbf{u} = \mathbf{A}^{-1}\mathbf{v}.$$

4. *The necessary and sufficient condition that the homogeneous equations*

$$\mathbf{Au} = 0$$

have non-zero solutions is that $D = 0$.

67. Latent roots and vectors

If $\mathbf{A}$ is an $n \times n$ matrix and λ a scalar the set of n homogeneous equations

$$\mathbf{Au} = \lambda\mathbf{u}$$

is equivalent to the set

$$(\mathbf{A} - \lambda\mathbf{I})\mathbf{u} = 0,$$

and this set can have non-zero solutions if and only if the determinant of the matrix $\mathbf{A} - \lambda\mathbf{I}$ vanishes. This determinant, being of the form

$$(67.1) \qquad D(\lambda) = \begin{vmatrix} a_{11} - \lambda & a_{12} & \cdots & a_{1n} \\ a_{21} & a_{22} - \lambda & \cdots & a_{2n} \\ \cdots\cdots\cdots\cdots\cdots\cdots\cdots \\ a_{n1} & a_{n2} & \cdots & a_{nn} - \lambda \end{vmatrix},$$

is a polynomial in λ of degree exactly n and therefore vanishes for exactly n values of λ, $\lambda_1, \lambda_2, \cdots, \lambda_n$. These are called the "latent roots" of $\mathbf{A}$. Many writers use the German word *Eigenwerte*, some use the hybrid "eigenvalues" and still others the term "characteristic numbers." We shall use the words *latent roots* to refer to the roots of an algebraic equation of the type $D(\lambda) = 0$, reserving the term "characteristic numbers" for the roots of transcendental equations associated with boundary value problems, whether expressed in the form of differential or of integral equations.

The latent roots of a matrix may be real or complex, simple or multiple.

Corresponding to each latent root λ_i is a non-zero vector ϕ_i with components ϕ_{i1}, ϕ_{i2}, $\cdots$, ϕ_{in} satisfying the equations

$$\mathbf{A}\phi_i = \lambda_i \phi_i.$$

These are *latent vectors*. They are indeterminate to the extent of an arbitrary factor, i.e., if ϕ_i is a latent vector, so also is $c\phi_i$ for arbitrary c.

Associated with any matrix $\mathbf{A}$ is the *transpose* matrix $\mathbf{A}'$ obtained from $\mathbf{A}$ by changing rows to columns, that is, by replacing the element a_{ij} with the element a_{ji}. Evidently $\mathbf{A}$ and $\mathbf{A}'$ have the same determinant D. Consequently the latent roots of $\mathbf{A}'$ are the same as those of $\mathbf{A}$, but, unless $\mathbf{A}$ is symmetrical, the latent vectors will in general be different. We denote the latent vectors of $\mathbf{A}'$ by ψ_i, so that

$$\mathbf{A}'\psi_i = \lambda_i \psi_i.$$

The two sets of latent vectors ϕ_i and ψ_i have the important *orthogonal property:*

Theorem 1.   If $\lambda_i \neq \lambda_j$, then the two vectors ϕ_i and ψ_j are orthogonal, i.e., their scalar product

$$\phi_i \cdot \psi_j = \sum_{k=1}^{n} \phi_{ik}\psi_{jk}$$

vanishes.   That is,

$$\phi_i \cdot \psi_j = 0.$$

Example 1.   Let

$$\mathbf{A} = \begin{bmatrix} 23 & -9 & -2 & 0 \\ -4 & 23 & 0 & -2 \\ -8 & 0 & 23 & -9 \\ 0 & -8 & -4 & 23 \end{bmatrix}.$$

Then, by setting up the determinant $D(\lambda)$, expanding to get an algebraic equation in λ, and solving, we get

$$\lambda_1 = 13, \qquad \lambda_2 = 21, \qquad \lambda_3 = 25, \qquad \lambda_4 = 33.$$

The latent vector ϕ_i is obtained by calculating the cofactors of one row of the zero determinant $D(\lambda_i)$. Reduced to the simplest integral form, the latent vectors are

$$\begin{aligned}
\phi_1 &= (3, \quad 2, \quad 6, \quad 4), \\
\phi_2 &= (3, \quad 2, \ -6, \ -4), \\
\phi_3 &= (3, \ -2, \quad 6, \ -4), \\
\phi_4 &= (3, \ -2, \ -6, \quad 4).
\end{aligned}$$

In like manner from the cofactors of any column of $D(\lambda_i)$ are obtained the values

$$\begin{aligned}
\psi_1 &= (4, \quad 6, \quad 2, \quad 3), \\
\psi_2 &= (4, \quad 6, \ -2, \ -3), \\
\psi_3 &= (4, \ -6, \quad 2, \ -3), \\
\psi_4 &= (4, \ -6, \ -2, \quad 3).
\end{aligned}$$

From these values one may verify directly that $\phi_i \cdot \psi_j = 0$ if $i \neq j$.

The latent roots and vectors may be complex.

Example 2. Let

$$A = \begin{bmatrix} 1 & 1 \\ -1 & 1 \end{bmatrix}.$$

Then $\lambda_1 = 1 - i$, $\lambda_2 = 1 + i$, $\phi_1 = (1, -i)$, $\phi_2 = (1, i)$, $\psi_1 = (1, i)$, $\psi_2 = (1, -i)$

Note, however, that the orthogonal property holds equally well for complex cases.

The following property of latent roots has frequent applications:

Theorem 2. If $P(x)$ is a polynomial with constant (scalar) coefficients and if $\lambda_1, \lambda_2, \cdots, \lambda_n$ are the latent roots of the $n \times n$ matrix A, the latent roots of the $n \times n$ matrix $P(A)$ are given by $\mu_i = P(\lambda_i)$, $i = 1, 2, \ldots, n$, and the latent vectors of A are latent vectors of $P(A)$.

Example 3. The latent roots and vectors of

$$A = \begin{bmatrix} 6 & 1 & 1 \\ 1 & 6 & 1 \\ 3 & 3 & 6 \end{bmatrix}$$

are

λ	ϕ	ψ
4	(1, 1, -3)	(1, 1, -1)
5	(1, -1, 0)	(1, -1, 0)
9	(1, 1, 2)	(3, 3, 2).

If $P(x) = x^2 + 2x + 3$ computation gives

$$P(A) = \begin{bmatrix} 55 & 17 & 15 \\ 17 & 55 & 15 \\ 45 & 45 & 57 \end{bmatrix}$$

while

$$P(4) = 27, \qquad P(5) = 38, \qquad P(9) = 102.$$

It is a simple matter to verify that 27, 38, 102 are latent roots of $P(\mathbf{A})$ and that the ϕ and ψ given above are the latent vectors.

68. Real symmetric matrices

In a great number of important problems in pure and applied mathematics we have to deal with real symmetric matrices, i.e., the case where a_{ij} is real and equal to a_{ji}. The determinant $D(\lambda)$ is then also real (for real λ) and symmetric, and its cofactors α_{ij} are real and $\alpha_{ij} = \alpha_{ji}$.

We now state a theorem of fundamental importance in the solution of differential equations. Let

$$D_0(\lambda) = 1, \qquad D_1(\lambda) = a_{11} - \lambda, \qquad D_2(\lambda) = \begin{vmatrix} a_{11} - \lambda & a_{12} \\ a_{21} & a_{22} - \lambda \end{vmatrix},$$

and generally,

$$(68.1) \quad D_{n+1}(\lambda) = \begin{vmatrix} a_{11} - \lambda & a_{12} & \cdots & a_{1n} & a_{1,n+1} \\ a_{21} & a_{22} - \lambda & \cdots & a_{2n} & a_{2,n+1} \\ \cdots & \cdots & \cdots & \cdots & \cdots \\ a_{n1} & a_{n2} & \cdots & a_{nn} - \lambda & a_{n,n+1} \\ a_{n+1,1} & a_{n+1,2} & \cdots & a_{n+1,n} & a_{n+1,n+1} - \lambda \end{vmatrix}.$$

Then

Theorem 3. *If $D_{n+1}(\lambda)$ is real and symmetric, its latent roots are all real. Furthermore between any two distinct roots of $D_{n+1}(\lambda) = 0$ there is at least one root of $D_n(\lambda) = 0$. If λ is a multiple root of order k of $D_{n+1}(\lambda) = 0$, then λ is a root of order $k - 1$ at least of $D_n(\lambda) = 0$.*

Example 1. For the determinant

$$D_4(\lambda) = \begin{vmatrix} 4 - \lambda & -1 & 0 & 0 \\ -1 & 4 - \lambda & -1 & -1 \\ 0 & -1 & 4 - \lambda & 0 \\ 0 & -1 & 0 & 4 - \lambda \end{vmatrix}$$

the distribution of roots is

D_1			4,		
D_2		3,		5,	
D_3	2.586,		4,		5.414,
D_4	2.268,	4,		4,	5.732.

For a real symmetric matrix it is clear that the latent vectors can always be chosen so as to be real. Moreover because of symmetry we can always choose the arbitrary constants so that

$$\psi_i = \phi_i, \qquad (i = 1, 2, \cdots, n).$$

In this case, therefore, the orthogonal property is simply

$$\phi_i \cdot \phi_j = 0,$$

provided $i \neq j$.

Since non-vanishing orthogonal vectors are necessarily linearly independent, it follows at once that, if all the latent roots are distinct, all the latent vectors are linearly independent. In the case of a real symmetric matrix, however, it turns out that restriction to distinct latent roots is unnecessary, and the following important theorem holds.

Theorem 4. *A real symmetric $n \times n$ matrix has n real independent orthogonal latent vectors.*

Example 2. For the matrix

$$\mathbf{A} = \begin{bmatrix} 7 & -2 & 0 & 0 \\ -2 & 7 & -2 & -1 \\ 0 & -2 & 7 & 0 \\ 0 & -1 & 0 & 7 \end{bmatrix}$$

the latent roots and vectors are

λ	ϕ
4	(2, 3, 2, 1),
7, 7	(2, 0, −1, −2),
10	(2, −3, 2, 1).

We drop the last row and column to get the reduced matrix

$$\begin{bmatrix} 7 & -2 & 0 \\ -2 & 7 & -2 \\ 0 & -2 & 7 \end{bmatrix},$$

which has the latent root $\lambda = 7$ and corresponding latent vector $(1, 0, -1)$. Hence, $(1, 0, -1, 0)$ is a new latent vector of **A**. An orthogonal combination of the two vectors belonging to $\lambda = 7$ is $(2, 0, -1, -2)$ and $(1, 0, -2, 2)$. A complete set of orthogonal vectors for **A** is, therefore,

$$
\begin{array}{ccc}
\lambda & & \phi \\
4 & \cdots & (2, \quad 3, \quad 2, \quad 1), \\
\left.\begin{array}{c} 7 \\ 7 \end{array}\right\} & \cdots & \left\{\begin{array}{c} (2, \quad 0, -1, -2), \\ (1, \quad 0, -2, \quad 2), \end{array}\right. \\
10 & \cdots & (2, -3, \quad 2, \quad 1).
\end{array}
$$

For a real symmetric matrix $\mathbf{A}$ there are n orthogonal latent vectors ϕ_i. Since they are necessarily also linearly independent, it follows that the n-rowed determinant

$$|\phi_{ik}|$$

does not vanish. Hence, the equations

$$\sum_{i=1}^{n} \phi_{ij}\alpha_i = \beta_j$$

have a unique solution $\boldsymbol{\alpha}$ for every vector $\boldsymbol{\beta}$. Moreover, because of the orthogonality the solution is given by

$$\alpha_i = \frac{\displaystyle\sum_{j=1}^{n} \phi_{ij}\beta_j}{\displaystyle\sum_{j=1}^{n} \phi_{ij}^{2}}, \qquad (i = 1, 2, \cdots, n).$$

Hence,

Theorem 5. Every vector with n components can be expressed linearly in terms of the n latent vectors of a real symmetric $n \times n$ matrix $\mathbf{A}$.

Example 3. Suppose the $\boldsymbol{\beta}$ vector is $(1, 1, 1, 1)$. Using the latent vectors of the last example, we set up the equations

$$
\begin{aligned}
2\alpha_1 + 2\alpha_2 + \alpha_3 + 2\alpha_4 &= 1, \\
3\alpha_1 \qquad\qquad\qquad - 3\alpha_4 &= 1, \\
2\alpha_1 - \alpha_2 - 2\alpha_3 + 2\alpha_4 &= 1, \\
\alpha_1 - 2\alpha_2 + 2\alpha_3 + \alpha_4 &= 1.
\end{aligned}
$$

Multiplying these equations in turn by 2, 3, 2, 1 and adding, we have

$$\alpha_1 = \tfrac{8}{18} = \tfrac{4}{9}.$$

Similarly,

$$\alpha_2 = -\tfrac{1}{9}, \quad \alpha_3 = \tfrac{1}{9}, \quad \alpha_4 = \tfrac{2}{18} = \tfrac{1}{9}.$$

Hence, for this example

$$\boldsymbol{\beta} = \tfrac{1}{9}(4\phi_1 - \phi_2 + \phi_3 + \phi_4).$$

It is often convenient to normalize the latent vectors, i.e., to divide each one by the square root of the sum of the squares of its own components.

In this orthonormal form the following general theorem is applicable.

Theorem 6. If the n vectors $\phi_i = (x_{i1}, x_{i2}, \cdots, x_{in})$ *are orthonormal, so also are the n vectors*

$$\psi_i = (x_{1i}, x_{2i}, \cdots, x_{ni}).$$

69. Extremal properties of latent roots

Let $\mathbf{A} = [a_{ij}]$ be an $n \times n$ real symmetric matrix, and let

$$(69.1) \qquad q = \sum_{i=1}^{n} \sum_{j=1}^{n} a_{ij} u_i u_j$$

be a quadratic form in the n variables $u_1, u_2, \cdots, u_n$. Let us endeavor to determine the maxima and minima of q subject to the restriction

$$(69.2) \qquad \sum_{i=1}^{n} u_i^2 = 1.$$

The facts are stated in the following lemma

Lemma 1. If u satisfies equation 69.2 and q is defined by equation 69.1, then

$$\lambda_1 \leqq q \leqq \lambda_n,$$

where λ_1 *is the least and* λ_n *the largest latent root of* $\mathbf{A}$.

This result can be formulated somewhat differently to give the theorem:

Theorem 7. If $\mathbf{A}$ *is a real symmetric* $n \times n$ *matrix all of whose latent roots are positive, and if* $\mathbf{u}$ *is any vector, then*

$$\lambda_1 \leqq \frac{\mathbf{u} \cdot \mathbf{Au}}{\mathbf{u} \cdot \mathbf{u}} \leqq \frac{\mathbf{Au} \cdot \mathbf{Au}}{\mathbf{u} \cdot \mathbf{Au}} \leqq \lambda_n,$$

where λ_1 *is the least and* λ_n *the largest latent root of* $\mathbf{A}$.

Example 1. The matrix of the quadratic form

$$q = 7u_1^2 - 4u_1 u_2 + 7u_2^2 - 4u_2 u_3 + 7u_3^2 - 2u_2 u_4 + 7u_4^2$$

has the latent roots 4, 7, 7, 10 (see Example 2, Article 68). Hence, for all u's satisfying the condition

$$u_1^2 + u_2^2 + u_3^2 + u_4^2 = 1$$

the maximum of q is 10 and the minimum is 4.

The above result was derived for the case of a real symmetric matrix. We take now a somewhat more general case. Return to equations

66.1, and suppose that the matrix $\mathbf{A}$ is real and non-singular, but not necessarily symmetric, and that

$$(69.3) \qquad u_1{}^2 + u_2{}^2 + \cdots + u_n{}^2 = 1.$$

Subject to this condition on the u's let us find the extrema of the quantity

$$\mathbf{v} \cdot \mathbf{v} = v_1{}^2 + v_2{}^2 + \cdots + v_n{}^2$$

where the v's are those defined by equations 66.1 for the given u's. The facts are given in Lemma 2.

Lemma 2. *If $\mathbf{A}$ is real and the u's satisfy equation 69.3, then*

$$\lambda_1 \leqq \mathbf{v} \cdot \mathbf{v} \leqq \lambda_n$$

where λ_1 is the least and λ_n the largest latent root of $\mathbf{A}'\mathbf{A}$.

More generally we have the theorem:

Theorem 8. *If $\mathbf{A}$ is a real $n \times n$ matrix, the latent roots of $\mathbf{A}'\mathbf{A}$ are real and non-negative. If $\mathbf{u}$ is any n vector, then*

$$\lambda_1 \leqq \frac{\mathbf{Au} \cdot \mathbf{Au}}{\mathbf{u} \cdot \mathbf{u}} \leqq \lambda_n,$$

where λ_1 is the least and λ_n the largest latent root of $\mathbf{A}'\mathbf{A}$.

Example 2. For the equations

$$2u_1 + u_2 - u_3 = v_1$$
$$u_1 + 3u_2 + u_3 = v_2$$
$$u_1 - u_2 + 2u_3 = v_3$$

the matrix $\mathbf{A}'\mathbf{A}$ is

$$\begin{bmatrix} 6 & 4 & 1 \\ 4 & 11 & 0 \\ 1 & 0 & 6 \end{bmatrix}$$

with latent roots and latent vectors

i	λ_i	ϕ_i		
1	3.477	(0.833,	−0.443,	−0.330),
2	6.273	(−0.257,	0.218,	−0.942),
3	13.250	(0.489,	0.870,	0.067).

When the latent vectors are substituted into the original equations, the corresponding v's turn out to be

i	v_1	v_2	v_3
1	1.554	−0.826	0.616,
2	.645	− .546	−2.358,
3	1.780	3.165	− .246.

These satisfy the condition

$$v_1{}^2 + v_2{}^2 + v_3{}^2 = \lambda$$

for each latent root.

There is frequent need for an easy-to-find bound on the latent roots of a matrix

$$\mathbf{A} = [a_{ij}].$$

Let r_i and ρ_i denote the sums of the absolute values of the non-diagonal elements in the ith row and ith column, respectively. Then we have

Theorem 9. *In the complex λ plane every latent root λ_k lies in at least one of the circles* $|\lambda - a_{ii}| \leqq r_i$ *and in at least one of the circles*

$$|\lambda - a_{jj}| \leqq \rho_j.$$

Example 3. This result applied to the matrix

$$\begin{bmatrix} 2 & 1 & -1 \\ 1 & 3 & 1 \\ 1 & -1 & 2 \end{bmatrix}$$

assures us that each latent root lies in at least one of the circles

$$|\lambda - 2| \leqq 2, \qquad |\lambda - 3| \leqq 2.$$

The latent roots are 3.575, $1.71 \pm 1.35i$, and all three lie in both circles.

Limits for latent roots are treated by many writers. A few references are Gersgorin [79], Brauer [17], Karush [97], and Taussky [203, 204].

In the example of Article 66 it was shown how to construct a polynomial function of a matrix $\mathbf{A}$. In particular, one may use the polynomial

$$P(\lambda) = \lambda^n + g_1\lambda^{n-1} + \cdots + g_n = 0$$

whose roots are the latent roots of the matrix $\mathbf{A}$. We are then led to the following interesting result, known as the Cayley-Hamilton theorem.

Theorem 10. *If the latent roots of $\mathbf{A}$ satisfy the equation*

$$\lambda^n + g_1\lambda^{n-1} + \cdots + g_n = 0,$$

then

$$\mathbf{A}^n + g_1\mathbf{A}^{n-1} + \cdots + g_n\mathbf{I} = 0.$$

For example, the latent roots of

$$A = \begin{bmatrix} 3 & 1 \\ 1 & 3 \end{bmatrix}$$

satisfy the equation

$$\lambda^2 - 6\lambda + 8 = 0,$$

and we see that

$$\begin{bmatrix} 10 & 6 \\ 6 & 10 \end{bmatrix} - 6 \begin{bmatrix} 3 & 1 \\ 1 & 3 \end{bmatrix} + 8 \begin{bmatrix} 1 & 0 \\ 0 & 1 \end{bmatrix} = 0.$$

70. Methods of successive approximation

The solution of n algebraic equations in n unknowns when n is a small number is usually carried out by some method of elimination, say, that of Gauss, or of Crout or of Doolittle or of Banachiewicz. But, as n increases, the labor required by these methods of elimination increases roughly like n^3. Thus, if the solution of 4 equations can be done in 1 hour, let us say, we may expect 8 equations to require about 8 hours, 16 equations 64 hours, and 64 equations over 2 years, assuming 40-hour week, and 2 week's vacation. Consequently, when n is large, there is great demand for a shorter method than that of elimination. For certain classes of equations (and happily those associated with partial differential equations are included) there are various methods of successive approximation that can be employed, and it turns out that, although these also require much toil, the amount of labor increases roughly as n^2. For example, if 4 equations require 1 hour, by successive approximation 64 equations would need about 6 weeks.

The process of successive approximation can be described in general terms as follows:

Suppose that the system of equations to be solved is

$$(70.1) \qquad\qquad Au = b$$

in which A is a real non-singular $n \times n$ matrix. Let v_ν be an approximation to the unknown vector u at the νth stage of the process, and let the residual r_ν be defined by

$$(70.2) \qquad\qquad Av_\nu - b = r_\nu.$$

We must now select some rule of procedure for calculating a new approximation $v_{\nu+1}$ in such a manner that the corresponding sequence of residuals $r_\nu, r_{\nu+1}, \cdots$, converges to zero. As a measure of our progress in reducing the residuals $\mathbf{r}_\nu$ to zero we may use the quadratic form

$$(70.3) \qquad E(\mathbf{v}) = (\mathbf{A}\mathbf{v} - \mathbf{b}) \cdot (\mathbf{A}\mathbf{v} - \mathbf{b})$$

In particular, when $\mathbf{v} = \mathbf{v}_\nu$, we have

$$(70.4) \qquad E_\nu = E(v_\nu) = \mathbf{r}_\nu \cdot \mathbf{r}_\nu.$$

Evidently $\mathbf{r}_\nu = 0$ if and only if $E_\nu = 0$.

Let $\mathbf{w}$ be any vector, s an unknown scalar, and let us examine the quadratic form $E(\mathbf{v} - s\mathbf{w})$ as a function of s. From equation 70.3 it follows that

$$(70.5) \qquad E(\mathbf{v}) - E(\mathbf{v} - s\mathbf{w}) = 2(\mathbf{z} \cdot \mathbf{r})s - (\mathbf{z} \cdot \mathbf{z})s^2$$

in which

$$\mathbf{z} = \mathbf{A}\mathbf{w}.$$

It is convenient to define a scalar q by the equation

$$q = (\mathbf{z} \cdot \mathbf{r})/(\mathbf{z} \cdot \mathbf{z}).$$

We deduce from equation 70.5 that

$$(70.6) \qquad \begin{cases} 1) & E(\mathbf{v} - s\mathbf{w}) < E(\mathbf{v}) \quad \text{if } s \text{ is between } 0 \text{ and } 2q, \\ 2) & E(\mathbf{v} - s\mathbf{w}) = E(\mathbf{v}) \quad \text{if} \quad s = 0 \quad \text{or} \quad s = 2q \\ 3) & E(\mathbf{v} - s\mathbf{w}) > E(\mathbf{v}) \quad \text{for all other values of } s, \\ 4) & E(\mathbf{v} - s\mathbf{w}) \qquad\quad \text{has a minimum at } s = q. \end{cases}$$

The foregoing results may be formulated thus

Theorem 11.   If $\mathbf{w}$ is any vector such that $(\mathbf{A}\mathbf{w}) \cdot (\mathbf{A}\mathbf{v}_\nu - \mathbf{b}) \neq 0$, and if $v_{\nu+1}$ is defined by

$$\mathbf{v}_{\nu+1} = \mathbf{v}_\nu - s\mathbf{w},$$

then $E_{\nu+1}$ will be less than E_ν for any s between 0 and $2q$ and will be least for $s = q$.

It is apparent that the above result leaves us a good deal of latitude in the precise process of approximation to be selected since the vector $\mathbf{w}$ is almost unrestricted, and s can lie in the range between 0 and $2q$. Even this restriction is not always binding for we shall discover later that occasional choices of s outside the range may be advantageous. It should be noted that the length of $\mathbf{w}$ is immaterial and only its direction is important. Evidently the best direction for $\mathbf{w}$, the one which leads to correct value of $\mathbf{v}$ in one step, is

$$\mathbf{w} = \mathbf{A}^{-1}\mathbf{r}_\nu.$$

Since, of course, $\mathbf{A}^{-1}$ is unknown, this fact is not of immediate utility.

The various ways of solving algebraic equations by successive approximations differ principally in the manner in which the correction $z = sw$ is determined. Roughly these methods can be classified into two groups: (1) those in which the computer decides what to do next at each step of the computation, as in Articles 71, 72, 73; (2) those in which a definite routine is selected in advance and followed consistently, as in Articles 74, 75, 76, and 78. This distinction is not really clear-cut, since the definite routines may be interrupted from time to time, according to the judgment of the operator, and replaced either temporarily or permanently by some other operation, as in Articles 77 and 79. For hand computation probably the method of Article 73 is best. For high-speed automatic computing machines a fixed routine is likely to be used, probably one of the methods of Articles 74, 75, 76, or 78, possibly with occasional interruptions to accelerate the convergence as in Articles 77 and 79.

A discussion and comparison of many different procedures for solving algebraic equations, both by direct methods and also by successive approximations, is given by Bodewig [13]. See also references at the end of this chapter.

71. Methods changing one component at a time

One way to choose $\mathbf{w}$ is to take a unit vector in the direction of the kth coordinate axis. For example, we could choose the sequence of $\mathbf{w}$'s as $\mathbf{u}_1, \mathbf{u}_2, \cdots, \mathbf{u}_n, \mathbf{u}_1, \mathbf{u}_2, \cdots, \mathbf{u}_n, \mathbf{u}_1, \mathbf{u}_2, \cdots, \mathbf{u}_n$, where $\mathbf{u}_k$ denotes the unit vector in the direction of the kth coordinate axis, and for each $\mathbf{w}$ we could take the optimum s, i.e., $s = q$. In this instance

$$\mathbf{z} = \mathbf{A}\mathbf{w} = \mathbf{a}_k,$$

where $\mathbf{a}_k$ is the vector formed by the kth column of $\mathbf{A}$. Then $q = \mathbf{a}_k \cdot \mathbf{r}/\mathbf{a}_k \cdot \mathbf{a}_k$, and $\mathbf{v}_{\nu+1}$ is found from $\mathbf{v}_\nu$ by adding q to the kth component of $\mathbf{v}_\nu$. In actual practice we need not follow any prescribed order for choosing the successive components, as was implied above, but rather we try at any given stage to pick out that component whose correction seems likely to do the most good.

Example. Let the equations be

$$
\begin{aligned}
4x_1 - x_2 &&&&&= 12 \\
-x_1 + 4x_2 &- x_3 &- x_4 &&&= 5 \\
&- x_2 + 4x_3 &&- x_5 &= 7 \\
&- x_2 &+ 4x_4 &- x_5 &= 8 \\
&&- x_3 &- x_4 + 4x_5 &= 10.
\end{aligned}
$$

First of all we compute for each column of the matrix the value of $\mathbf{a}_i \cdot \mathbf{a}_i = \mathbf{z} \cdot \mathbf{z}$ for use in computing q. It is also convenient to write each of the vectors $\mathbf{a}_i$ in a column on a strip of paper to lay beside the column $\mathbf{r}_\nu$ for computing the numerator of q, and also for computing $\mathbf{r}_{\nu+1} = \mathbf{r}_\nu - q\mathbf{a}_i$. The work is conveniently arranged as shown below:

	$-\mathbf{r}_0$	$-\mathbf{r}_1$	$-\mathbf{r}_2$	$-\mathbf{r}_3$	$-\mathbf{r}_4$	$-\mathbf{r}_5$	$-\mathbf{r}_6$	$-\mathbf{r}_7$	$-\mathbf{r}_8$	$-\mathbf{r}_9$	$-\mathbf{r}_{10}$
	12	2.0	2.0	2.0	3.1	3.1	3.1	3.1	4.0	4.0	4.0
	5	7.5	7.5	8.9	4.5	6.0	6.0	6.9	3.3	4.2	4.2
	7	7.0	8.4	8.4	9.5	3.5	4.7	4.7	5.6	2.0	2.7
	8	8.0	9.4	3.8	4.9	4.9	6.1	2.5	3.4	3.4	4.1
	10	10.0	4.4	5.8	5.8	7.3	2.5	3.4	3.4	4.3	1.5
E	382					135					60
x_1	0	2.5									2.5
x_2	0			1.1					.9		2.0
x_3	0					1.5				.9	2.4
x_4	0			1.4				.9			2.3
x_5	0		1.4				1.2			.7	3.3

Here the initial vector $\mathbf{x}$ was taken as zero. The final x's after ten steps are found as the sum of the individual corrections. The work so far may be checked by substitution of the x's back into the original equations in order to obtain the last column of residuals directly. Three values of E are given to show the decrease in the quadratic form.

EXERCISE

Carry the example ten steps farther, and compute the new E.

Methods that use the corrected values as fast as they are obtained are known as Gauss-Seidel methods, to distinguish them from methods where all components of the new vector are computed entirely from the components of the old vector. Actually the distinction, though obvious in computation, is rather hard to pin down analytically until one defines precisely what is meant by a "step" in the process of successive approximations. See Seidel [187] and P. Stein [197].

72. Group changes

Quite often in a computation it becomes evident that we can make a better choice of $\mathbf{w}$ than a vector along a single component.

For instance, in the Example of Article 71 we note that the corrections are all positive so that we are increasing each component of $\mathbf{x}$ a little at a time. A rather obvious inference is that an increase in all components simultaneously ought to do some good. Let us therefore try $w = (1, 1, 1, 1, 1)$. We find $z = Aw = (3, 1, 2, 2, 2)$ whence

$$z \cdot r_{10} = 32.8, \qquad z \cdot z = 22,$$

and $q = 1.5$. This gives

$$x = x_{10} + 1.5w = (4.0, 3.5, 3.9, 3.8, 4.8)$$

and

$$r_{11} = r_{10} - qAw = (-0.5, 2.7, -0.3, 1.1, -1.5),$$

for which $E_{11} = 11.1$. This is a considerable improvement over $E_{10} = 60$.

The possibility of advantageous choices of w is limited only by the skill and insight of the computer.

73. Relaxation

When computations are to be done by hand or with the aid of desk computing machines, the method most commonly used is that which has come to be known as *relaxation*. This process is not easily explained by stating a set of rules of operation, since the main virtue of relaxation is its freedom from a rigid routine. It is best described as an art in which the computer, by examination of the residuals at any stage, makes the most effective corrections he can to reduce the residuals toward zero. No two computers, applying relaxation to the same problem, would likely follow identical steps. In general, the computer changes one component at a time, but is always on the lookout for opportunities to correct a whole group simultaneously. If one ventured to state a rule regarding relaxation, it would be that at any stage you pick out the largest residual and make a correction reducing that residual to zero (or nearly zero). How best to choose this correction is, however, not readily given by rule but really constitutes the art of relaxation.

The process is simplest when applied to a matrix for which the diagonal terms are never less than, and in some cases exceed, the sum of the absolute values of the elements in the same row and likewise in the same column. For then, if the kth component of the residual is to be reduced to zero, one can simply correct the kth component of the unknown. With practice one learns to improve on this crude method by undercorrecting or overcorrecting as may be indicated by the effect of the correction on the other components of the residual.

To illustrate the procedure we apply it to the example of Article 71, starting with the initial estimate $x = 0$. In the particular set of steps shown below the first step is a group correction where each component is changed by 4; after that only one component is changed at a time. The successive residuals r, the corrections c, and the final values of x are shown below.

$-r_0$	c	$-r_1$	c	$-r_2$	c	$-r_3$	c	$-r_4$	c	$-r_5$	c	$-r_6$
12	4	0		0		0.25		0.25	0.06	0.01		0.07
5	4	1		1.0	0.25	0		.20		.26	0.06	.02
7	4	−1		−.5		−.25		−.25		−.25		−.19
8	4	0		.5		.75	0.20	−.05		−.05		.01
10	4	2	0.5	0		0		.20		.20		.20

c	$-r_7$	c	$-r_8$	c	$-r_9$	c	$-r_{10}$	c	$-r_{11}$
	0.07		.07	0.02	−0.01		−0.01		−0.01
	−.03		−.03		−.01		0		.01
−0.05	.01		.05		.05	0.01	0.01		.01
	.01		.05		.05		0.05	0.01	.01
	.15	0.04	−.01		−.01		0		.01

$$x_1 = 4 + 0.06 + 0.02 = 4.08$$
$$x_2 = 4 + \quad .25 + \quad .06 = 4.31$$
$$x_3 = 4 - \quad .05 + \quad .01 = 3.96$$
$$x_4 = 4 + \quad .20 + \quad .01 = 4.21$$
$$x_5 = 4 + \quad .50 + \quad .04 = 4.54$$

Most of the expository papers dealing with relaxation apply it to the solution of partial differential equations. See, however, Forsythe [60], where the method is traced back to Gauss. See also Southwell [194], Grinter [82], Shortley [189], Fox [66], and Bowie [16].

EXERCISE

Continue the computation above to the point where each component of r is zero to four places.

74. A gradient method

So far we have chosen the correction vector **w** in a simple manner. A more sophisticated approach is the following. Inasmuch as our intent is to reduce the quadratic form E to zero and since at any point the function E, regarded as a point function of the point **v,** changes most rapidly in the direction of **grad** E, let us define **w** by the equation

$$\mathbf{w} = \tfrac{1}{2} \, \mathbf{grad} \, E.$$

As is known, the components of **grad** E are the partial derivatives with respect to the corresponding components of v. In this way we find that

$$w = A'r$$

where as usual r is the residual at v and A' is the transpose of A. The computational procedure is somewhat lengthy but is actually composed of the following simple steps:

1. Compute r_ν $= Av_\nu - b$ (see also step 6)
2. Compute w_ν $= A'r_\nu$
3. Compute z_ν $= Aw_\nu$
4. Compute q_ν $= z_\nu \cdot r_\nu / z_\nu \cdot z_\nu$
5. Compute $v_{\nu+1} = v_\nu - q_\nu w_\nu$
6. Compute $r_{\nu+1} = r_\nu - q_\nu z_\nu$ (see also step 1).

Let us attack the Example in Article 71 by this method, starting again from the initial guess $(0, 0, 0, 0, 0)$. A rather convenient arrangement is shown below.

v_0	$-r_0$	$-w_0$	$-z_0$	v_1	$-r_1$	$-w_1$	$-z_1$
0	12	43	179	2.52	1.50	$-$ 4.92	$-$ 51.78
0	5	$-$ 7	-101	$-$ .41	10.92	32.10	126.60
0	7	13	34	.76	5.02	3.28	$-$ 32.42
0	8	17	50	1.00	5.06	3.44	$-$ 31.78
0	10	25	70	1.47	5.88	13.44	47.04

$q_0 = 0.0587$
$E = 382$

$q_1 = .0547$
$E = 207$

v_2	$-r_2$	$-w_2$	$-z_2$	v_3	$-r_3$	$-w_3$	$-z_3$
2.25	4.34	13.37	55.46	3.03	1.10	$-$ 2.06	-24.54
1.35	3.99	$-$ 1.98	-42.17	1.23	6.46	16.30	55.62
.94	6.80	4.82	21.66	1.22	5.54	11.10	18.14
1.19	6.80	16.06	66.62	2.13	2.90	.54	-24.10
2.21	3.30	$-$ .40	-22.48	2.19	4.60	9.96	28.20

$q_2 = 0.0583$
$E = 138$

$q_3 = 0.0912$
$E = 103$

Because of the computational labor required this gradient method is not likely to find favor among hand computers. Where high-speed computing machinery is available, however, the gradient method, because of its almost universal applicability, holds out excellent promise. At present (1951) this method and various modifications of it are being tested at the National Bureau of Standards, Los Angeles, California. It is too early to state results, but the following tentative conclusions are indicated:

1. It may be better in the long run to use a value of s slightly smaller than the optimum value q. Runs with $s = 0.9q$ showed excellent results.

2. If the optimum value of s is used for a series of steps, then a much larger value is used once, and this program repeated, a great improvement in the rate of convergences can occur.

See Forsythe and Motzkin [62].

75. Iteration

In the discussion of Article 70 let us suppose that we define $\mathbf{w}$ by the equation

$$\mathbf{w} = \mathbf{Cr},$$

in which $\mathbf{C}$ is some $n \times n$ matrix. (In the gradient method we had $\mathbf{C} = \mathbf{A}'$.) Further let us suppose that s is constant from step to step. We have

$$\begin{aligned}\mathbf{v}_{\nu+1} = \mathbf{v}_\nu - s\mathbf{w}_\nu &= \mathbf{v}_\nu - s\mathbf{Cr}_\nu \\ &= \mathbf{v}_\nu - s\mathbf{CAv}_\nu + s\mathbf{Cb}.\end{aligned}$$

Also, since $\mathbf{Au} - \mathbf{b} = 0$, we have the identity

$$\mathbf{u} = \mathbf{u} - s\mathbf{CAu} + s\mathbf{Cb},$$

whence by subtraction

$$\mathbf{v}_{\nu+1} - \mathbf{u} = (\mathbf{I} - s\mathbf{CA})(\mathbf{v}_\nu - \mathbf{u}).$$

If we define the $n \times n$ matrix $\mathbf{M}$ by the equation

$$(75.1) \qquad \mathbf{M} = \mathbf{I} - s\mathbf{CA}$$

we have an iteration process using the equation

$$(75.2) \qquad \mathbf{v}_{\nu+1} = \mathbf{Mv}_\nu + \mathbf{a}$$

where $\mathbf{a} = s\mathbf{Cb}$, and

$$(75.3) \qquad \mathbf{v}_\nu - \mathbf{u} = \mathbf{M}(\mathbf{v}_{\nu-1} - \mathbf{u}) = \mathbf{M}^\nu(\mathbf{v}_0 - \mathbf{u}).$$

To investigate the convergence of this process of iteration, let us assume that the matrix $\mathbf{M}$ has n linearly independent latent vectors. As has been shown, this is always true if $\mathbf{M}$ is real and symmetric. In fact, it fails only in those cases where $\mathbf{M}$ has a k-fold latent root μ for which the rank of $\mathbf{M} - \mu\mathbf{I}$ is greater than $n - k$. Even in that case the result that we are about to obtain still holds, but the proof is more complicated and will be omitted. Assuming then that $\mathbf{M}$ has n linearly independent latent vectors $\phi_1, \phi_2, \cdots, \phi_n$, we can express any n vector, and in particular the vector $\mathbf{v}_0 - \mathbf{u}$, as a linear combination of ϕ's,

$$(75.4) \qquad \mathbf{v}_0 - \mathbf{u} = \alpha_1\phi_1 + \alpha_2\phi_2 + \cdots + \alpha_n\phi_n.$$

Since $\mathbf{M'\phi}_i = \mu_i{}'\mathbf{\phi}_i$, we have at once by repeated use of equation 75.2

(75.5) $\mathbf{v}_\nu - \mathbf{u} = \alpha_1\mathbf{\phi}_1\mu_1{}' + \alpha_2\mathbf{\phi}_2\mu_2{}' + \cdots + \alpha_n\mathbf{\phi}_n\mu_n{}'$,

This equation establishes at once

Theorem 12.   If all the latent roots of $\mathbf{M}$ *are less than unity in absolute value, the iteration* $\mathbf{v}_{\nu+1} = \mathbf{Mv}_\nu + \mathbf{a}$ *will converge no matter what initial vector* $\mathbf{v}_0$ *is chosen.*

Example.   For the set of equations

$$2u_1 + u_2 - u_3 = -0.7,$$
$$u_1 + 3u_2 + u_3 = 0.6,$$
$$u_1 - u_2 + 2u_3 = 2.6,$$

let us choose the matrix C as

$$\mathbf{C} = \begin{vmatrix} \tfrac{1}{4} & 0 & 0 \\ 0 & \tfrac{1}{4} & 0 \\ 0 & 0 & \tfrac{1}{4} \end{vmatrix}.$$

Then the equations of iteration are

$$x_1' = \tfrac{1}{4}[2x_1 - x_2 + x_3 - 0.7],$$
$$x_2' = \tfrac{1}{4}[-x_1 + x_2 - x_3 + 0.6],$$
$$x_3' = \tfrac{1}{4}[-x_1 + x_2 + 2x_3 + 2.6].$$

Here (x_1, x_2, x_3) denote the components of the vector $\mathbf{v}_\nu$, (x_1', x_2', x_3') those of the vector $\mathbf{v}_{\nu+1}$. We start with the initial choice $\mathbf{v}_0 = (0, 0, 0)$ and carry out the steps of the iteration as tabulated below.

ν	x_1	x_2	x_3
0	0	0	0
1	-0.175	0.150	0.650
2	$-.137$	.069	1.056
3	.003	$-.062$	1.230
4	.150	$-.174$	1.249
5	.256	$-.243$	1.194
6	.312	$-.273$	1.122
7	.330	$-.277$	1.065
8	.326	$-.268$	1.031
9	.313	$-.256$	1.017
10	.300	$-.246$	1.016

In this example we have

$$\mathbf{A} = \begin{bmatrix} 2 & 1 & -1 \\ 1 & 3 & 1 \\ 1 & -1 & 2 \end{bmatrix}, \qquad \mathbf{C} = \begin{bmatrix} \tfrac{1}{4} & 0 & 0 \\ 0 & \tfrac{1}{4} & 0 \\ 0 & 0 & \tfrac{1}{4} \end{bmatrix},$$

whence

$$M = I - CA = \begin{bmatrix} \frac{2}{4} & -\frac{1}{4} & \frac{1}{4} \\ -\frac{1}{4} & \frac{1}{4} & -\frac{1}{4} \\ -\frac{1}{4} & \frac{1}{4} & \frac{2}{4} \end{bmatrix}.$$

The latent roots of M are 0.11, $0.57 \pm 0.34i$ with a maximum absolute value of about 0.44. Hence, convergence is assured.

When we do not know the actual latent roots, we may appeal to Theorem 9 for an estimate of their magnitude. In the present example Theorem 9 applied to the matrix M shows that each latent root lies in at least one of the circles

$$\left|\mu - \tfrac{1}{2}\right| \leqq \tfrac{1}{2}\right| \qquad \left|\mu - \tfrac{1}{4}\right| \leqq \tfrac{1}{2}.$$

The best that we can infer from these inequalities is that no latent root exceeds unity in absolute value. Convergence is not certainly assured by this test.

76. Iteration with w = A'r

Let us consider the case of iteration where, as in the gradient method, we choose

$$w = A'r,$$

but, unlike the gradient method, we use a constant value of s throughout, so that we are iterating with the matrix

$$M = I - sA'A.$$

Theorem 13. If A is real and non-singular, it is possible to choose a constant $s = c$ so that the process of iteration with

$$M = I - cA'A$$

always converges.

To prove this we recall that the latent roots of $A'A$ are all real and positive. Let λ_1 and λ_n denote the least and largest latent root, respectively, and take for the constant c the value

$$c = 2/(\lambda_1 + \lambda_n).$$

Then the latent roots μ_i of the matrix $M = I - CA = I - cA'A$ are given by

$$\mu_i = 1 - 2\lambda_i/(\lambda_1 + \lambda_n)$$

by Theorem 2 and from this relation expressed in the form

$$\mu_i = \frac{\lambda_n - \lambda_i}{\lambda_n + \lambda_1} - \frac{\lambda_i - \lambda_1}{\lambda_n + \lambda_1}$$

we infer that μ_i lies in the interval

$$-\frac{\lambda_n - \lambda_1}{\lambda_n + \lambda_1} \leqq \mu_i \leqq \frac{\lambda_n - \lambda_1}{\lambda_n + \lambda_1}.$$

Since λ_1 and λ_n are both positive, it follows that $|\mu_i| < 1$, and, hence, by Theorem 12 the process of iteration converges.

The choice of c given above is clearly the one leading to the least possible value for the maximum $|\mu_i|$, but this choice requires that λ_1 and λ_n be known. In actual problems λ_1 and λ_n are usually not known. Hence, we usually have to sacrifice something in the speed of convergence by making a cruder choice of c for which we are at least certain that the process does converge.

Theorem 9 enables us to compute a bound σ such that $0 < \lambda_i < \sigma$. Then, if $c = 2/\sigma$, we get

$$\mu_i = 1 - 2\lambda_i/\sigma$$

and from the inequalities

$$0 < \lambda_i < \sigma$$

it follows that $|\mu_i| < 1$.

Example. Let the given equations be

$$x - 2y + z = 3,$$
$$2x + y + z = 1,$$
$$x + 2y - z = 2.$$

The multiplication by $\mathbf{A}'$ is effected practically* by multiplying each equation by the coefficient of x in that equation and adding; the same is done for y and for z. The new equations with matrix $\mathbf{A}'\mathbf{A}$ are

$$6x + 2y + 2z = 7,$$
$$2x + 9y - 3z = -1,$$
$$2x - 3y + 3z = 2.$$

Here $\sigma = 14$, and we set up the iteration equations

$$7x_{\nu+1} = x_\nu - 2y_\nu - 2z_\nu + 7,$$
$$7y_{\nu+1} = -2x_\nu - 2y_\nu + 3z_\nu - 1,$$
$$7z_{\nu+1} = -2x_\nu + 3y_\nu + 4z_\nu + 2.$$

* For n large it is better to compute $\mathbf{A}'\mathbf{A}u$ in two steps, first $\mathbf{A}u$, then $\mathbf{A}'\mathbf{A}u$.

These equations give the iterations

ν	x	y	z
0	0	0	0
1	1.0000	-0.1429	0.2857
2	1.1020	$-$ .2653	.1020
3	1.2041	$-$ .3382	$-$ .0846
4	1.2928	$-$ .4265	$-$ .2516
5	1.3784	$-$ .4982	$-$ .4045
6	1.4548	$-$ .5677	$-$ .5527

Since the correct answers are $x = 2.5$, $y = -1.5$, $z = -2.5$, it is evident that convergence in this example is slow.

The process just described is important because it converges for every real non-singular matrix $\mathbf{A}$. In actual practice however we try to avoid this method if at all possible. In the first place, for large n the calculation of $\mathbf{w} = \mathbf{A'r}$ is a considerable chore. In the second place, if $\mathbf{A}$ happens to be symmetric with positive latent roots, the simple formula

$$\mathbf{M} = \mathbf{I} - c\mathbf{A}$$

gives more rapid convergence than the elaborate one

$$\mathbf{M'} = \mathbf{I} - c'\mathbf{A'A},$$

and is consequently to be preferred whenever applicable. For under the given hypotheses, if $\lambda_1, \lambda_2, \cdots, \lambda_n$ are the latent roots of $\mathbf{A}$, we can choose c as $2/(\lambda_1 + \lambda_n)$, and get the maximum μ_i to be

$$\mu = \frac{\lambda_n - \lambda_1}{\lambda_n + \lambda_1}.$$

On the other hand, the latent roots of $\mathbf{A'A}$ are $\lambda_1^2, \lambda_2^2, \cdots, \lambda_n^2$, the best c' is

$$c' = 2/(\lambda_1^2 + \lambda_n^2)$$

and the maximum μ' is

$$\mu' = \frac{\lambda_n^2 - \lambda_1^2}{\lambda_n^2 + \lambda_1^2}.$$

For all positive values of λ_n and λ_1 it follows that

$$\mu < \mu'.$$

It should be noted, however, that we cannot always use the simpler form

$$\mathbf{M} = \mathbf{I} - c\mathbf{A}.$$

In the example of this article the latent roots are $\lambda_1 = -1.538$, $\lambda_2 = 1.269 + 1.513i$, $\lambda_3 = 1.269 - 1.513i$. With these roots it is quite impossible to set up a matrix M of the form

$$M = I - cA$$

for which all the latent roots

$$\mu_i = 1 - c\lambda_i$$

are in absolute value less than unity.

77. Solution in terms of iterates

Suppose that M denotes a real symmetric $n \times n$ matrix, with n latent roots μ_1, μ_2, $\cdots$, μ_n, and associated orthonormal latent vectors ϕ_1, ϕ_2, $\cdots$, ϕ_n. Let u be the solution of

$$u = Mu + a,$$

where $I - M$ is non-singular, let v_0 be any vector, and v_ν the νth iteration of v_0 by the formula

$$v_{\nu+1} = Mv_\nu + a.$$

Now let

$$D_n(\mu) = g_0\mu^n + g_1\mu^{n-1} + \cdots + g_n = 0, \qquad (g_0 = 1)$$

be the nth-degree equation satisfied by the n latent roots μ_1, μ_2, $\cdots$, μ_n. Using equation 75.5, we form the sum

$$(77.1) \qquad \sum_{j=0}^{n} g_{n-j}(v_{\nu+j} - u) = \sum_{i=1}^{n} \alpha_i\phi_i D_n(\mu_i)\mu_i^\nu$$

and note that the right-hand member vanishes, since $D_n(\mu_i) = 0$. Consequently, we can solve for u in the form

$$(77.2) \qquad u = \frac{\displaystyle\sum_{j=0}^{n} g_{n-j}v_{\nu+j}}{D_n(1)}$$

since no latent root is equal to unity. In particular, equation 77.2 holds for $\nu = 0$, showing that u is expressible in terms of the $n + 1$ vectors v_0, v_1, $\cdots$, v_n. We may emphasize again that 77.2 is true, no matter what non-zero initial vector v_0 is chosen.

In 77.2 we may regard ν as the independent variable and apply the difference operator Δ. Then, since u is independent of ν, it follows that

$$(77.3) \qquad \sum_{j=0}^{n} g_{n-j}\Delta v_j = 0.$$

In equation 77.3 we set $\nu = 0, 1, \cdots, n-1$ and get n linear equations for the determination of $g_1, g_2, \cdots, g_n, (g_0 = 1)$.

Hence we can express the solution in terms of $\mathbf{v}_0$ and its first n iterates as follows:

1. Calculate $\mathbf{v}_1, \mathbf{v}_2, \cdots, \mathbf{v}_n$.
2. Form the differences $\Delta\mathbf{v}_0, \Delta\mathbf{v}_1, \cdots, \Delta\mathbf{v}_{n-1}$.
3. Use equations 77.3 to obtain the g's.
4. Calculate $\mathbf{u}$ from 77.2.

Example 1. Let the given set of equations be

$$
\begin{aligned}
4u_1 - u_2 & & & & &= 4096, \\
-u_1 + 4u_2 - u_3 - u_4 & & & &= &\ 0, \\
- u_2 + 4u_3 & - u_5 &= &\ 0, \\
- u_2 & + 4u_4 - u_5 &= &\ 0, \\
- u_3 - u_4 + 4u_5 &= &\ 0,
\end{aligned}
$$

and let

$$\mathbf{M} = \mathbf{I} - \tfrac{1}{4}\mathbf{A},$$

so that the equations for iteration are

(77.4)
$$
\begin{aligned}
x_1' &= \tfrac{1}{4}(x_2 + 4096), \\
x_2' &= \tfrac{1}{4}(x_1 + x_3 + x_4), \\
x_3' &= \tfrac{1}{4}(x_2 + x_5), \\
x_4' &= \tfrac{1}{4}(x_2 + x_5), \\
x_5' &= \tfrac{1}{4}(x_3 + x_4).
\end{aligned}
$$

The initial choice of the vector $\mathbf{v}_0$ may be taken as $\mathbf{v}_0 = 0$. With this choice the iterations are computed from equation 77.4 as follows.

ν	x_1	x_2	x_3	x_4	x_5
0	0	0	0	0	0
1	1024	0	0	0	0
2	1024	256	0	0	0
3	1088	256	64	64	0
4	1088	304	64	64	32
5	1100	304	84	84	32
6	1100	317	84	84	42

The table of differences from the above is

ν	Δx_1	Δx_2	Δx_3	Δx_4	Δx_5
0	1024	0	0	0	0
1	0	256	0	0	0
2	64	0	64	64	0
3	0	48	0	0	32
4	12	0	20	20	0
5	0	13	0	0	10

Using Δx_2 and Δx_3 we get from 77.3

$$13g_0 + 48g_2 + 256g_4 = 0,$$

$$10g_0 + 32g_2 \qquad\qquad = 0,$$

the solution of these equations yields

$$g_0 = 1, \qquad g_2 = -\tfrac{5}{16}, \qquad g_4 = \tfrac{1}{128},$$

and, accordingly, the equation for the latent roots is

$$\mu^5 - \tfrac{5}{16}\mu^3 + \tfrac{1}{128}\mu = 0.$$

From this we have, as a by-product,

$$\mu_1 = 0.5339, \quad \mu_2 = 0.1655, \quad \mu_3 = 0, \quad \mu_4 = -0.1655, \quad \mu_5 = -0.5339.$$

(These are not needed to find $\mathbf{u}$.)

We conclude from 77.2 that

$$\mathbf{u} = \frac{\mathbf{v}_5 - \tfrac{5}{16}\mathbf{v}_3 + \tfrac{1}{128}\mathbf{v}_1}{1 - \tfrac{5}{16} + \tfrac{1}{128}},$$

which gives for $\mathbf{u}$

$$u_1 = 1104.54, \quad u_2 = 322.16, \quad u_3 = 92.04, \quad u_4 = 92.04, \quad u_5 = 46.02.$$

Since this process requires the solution of n simultaneous linear equations in addition to the labor of iteration and differencing, it is obviously an impractical method and arouses only academic interest. But a rather simple modification does have practical use. When ν is large in equation 75.5, it is clear that terms of the form $\alpha_i\phi_i\mu_i^\nu$ can be neglected when μ_i is small and only the terms with the larger μ_i are significant. Let us assume then that only k significant values remain, where k is small, say, $k = 1$, or $k = 2$, or $k = 3$. We then take $k + 1$ differences of the iteration and determine, not all the g's, but simply $\bar{g}_1, \bar{g}_2, \cdots, \bar{g}_k$, where the $\bar{g}$'s are the coefficients in the equation for the k largest μ's in absolute value. Then an approximation to the correct value of $\mathbf{u}$ is given by

$$(77.5) \qquad \mathbf{u} = \frac{\mathbf{v}_\nu + \bar{g}_1\mathbf{v}_{\nu-1} + \bar{g}_2\mathbf{v}_{\nu-2} + \cdots + \bar{g}_k\mathbf{v}_{\nu-k}}{1 + \bar{g}_1 + \bar{g}_2 + \cdots + \bar{g}_k}.$$

Example 2. Let the given equations be

$$
\begin{aligned}
4u_1 - u_2 \qquad\qquad - u_5 \qquad\qquad\qquad &= 400, \\
-u_1 + 4u_2 - u_3 \qquad\qquad - u_6 \qquad\quad &= 0, \\
- u_2 + 5u_3 \qquad\qquad\qquad - u_7 &= 0, \\
3u_4 - u_5 \qquad\qquad &= 900, \\
-u_1 \qquad\qquad -u_4 + 3u_5 - u_6 \qquad\quad &= 0, \\
- u_2 \qquad\qquad - u_5 + 3u_6 - u_7 &= 0, \\
- u_3 \qquad\qquad - u_6 + 4u_7 &= 0.
\end{aligned}
$$

We take the equations of iteration in the form

$$x_1' = \tfrac{1}{4}(400 + x_2 + x_5),$$

$$x_2' = \tfrac{1}{4}(x_1 + x_3 + x_6),$$

$$x_3' = \tfrac{1}{4}(x_2 - x_3 + x_7),$$

$$x_4' = \tfrac{1}{4}(900 + x_4 + x_5),$$

$$x_5' = \tfrac{1}{4}(x_1 + x_4 + x_5 + x_6),$$

$$x_6' = \tfrac{1}{4}(x_2 + x_5 + x_6 + x_7),$$

$$x_7' = \tfrac{1}{4}(x_3 + x_6),$$

and select initial estimates

$$x_1 = 300, \quad x_2 = 200, \quad x_3 = 100, \quad x_4 = 400, \quad x_5 = 300, \quad x_6 = 200, \quad x_7 = 100.$$

Iteration then gives

$\mathbf{v}$	x_1	x_2	x_3	x_4	x_5	x_6	x_7
$\mathbf{v}_0$	300	200	100	400	300	200	100
$\mathbf{v}_1$	225	150	50	400	300	200	75
$\mathbf{v}_2$	212	119	44	400	281	181	62
$\mathbf{v}_3$	200	109	34	395	268	161	56
$\mathbf{v}_4$	194	99	33	391	256	148	49
$\mathbf{v}_5$	189	94	29	387	247	138	45
$\mathbf{v}_6$	185	89	28	383	240	131	42
$\mathbf{v}_7$	182	86	26	381	235	125	40

The differences are

$-\Delta\mathbf{v}_4$	5	5	4	4	9	10	4
$-\Delta\mathbf{v}_5$	4	5	1	4	7	7	3
$-\Delta\mathbf{v}_6$	3	3	2	2	5	6	2

Altogether we have seven equations with which to determine only two unknowns, g_1, and g_2.

By least squares we obtain $g_1 = -0.05$, $g_2 = -0.53$. These values in equation 77.5 give us

$$\mathbf{v}_8 \quad 173 \quad 76 \quad 22 \quad 373 \quad 219 \quad 108 \quad 33,$$

which is close to the correct solution.

78. The correction matrix $C = F(A)$

The numerical solution of linear partial differential equations of elliptic type usually requires the solution of n linear algebraic equations

$$\mathbf{Au} = \mathbf{b}$$

with real symmetric $n \times n$ matrix $\mathbf{A}$. All the latent roots $\lambda_1, \lambda_2, \cdots,$ λ_n belonging to $\mathbf{A}$ are known to lie in the interval $0 < \lambda < L$, where L is independent of n. One way of treating this case is to choose the correction matrix **C** as a polynomial in **A** itself. This choice of **C** insures that the matrix **M**, where

$$M = I - CA$$

is also a polynomial in **A**, is real and symmetric, and has the latent roots $\mu_1, \mu_2, \cdots, \mu_n$.

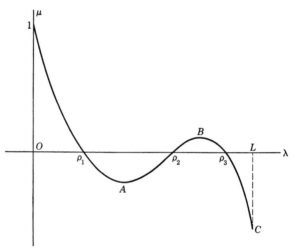

Fig. 19

It is convenient to express the polynomial in factored form

(78.1) $$\mathbf{M} = \left(\mathbf{I} - \frac{\mathbf{A}}{\rho_1}\right)\left(\mathbf{I} - \frac{\mathbf{A}}{\rho_2}\right) \cdots \left(\mathbf{I} - \frac{\mathbf{A}}{\rho_k}\right).$$

Then

(78.2) $$\mu_i = \left(1 - \frac{\lambda_i}{\rho_1}\right)\left(1 - \frac{\lambda_i}{\rho_2}\right) \cdots \left(1 - \frac{\lambda_i}{\rho_k}\right).$$

In view of Theorem 12 the one essential limitation that must be imposed on the ρ_i's is that $|\mu_i| < 1$, for every λ_i, but subject to this condition we may choose the ρ_i's as we please.

Consider the graph of μ as a function of λ defined by

(78.3) $$\mu = \left(1 - \frac{\lambda}{\rho_1}\right)\left(1 - \frac{\lambda}{\rho_2}\right) \cdots \left(1 - \frac{\lambda}{\rho_k}\right) = P(\lambda).$$

This is a polynomial curve with zeros at $\rho_1, \rho_2, \cdots, \rho_k$ and nowhere else, and always passing through the fixed point $(0, 1)$. A typical example is shown in Fig. 19. In order to secure convergence we must select ρ_1, ρ_2, and ρ_3 so that the points A, B, and C lie between the two lines $\mu = 1$ and $\mu = -1$. Obviously the best polynomial $P(\lambda)$ is the one of nth degree in which $\rho_i = \lambda_i$, $i = 1, 2, \cdots, n$, for, if we apply the **M** so defined to equation 75.5, we get at once

$$\mathbf{v}_1 - \mathbf{u} = 0$$

and hence have reached the exact solution in one application of the matrix **M,** or in n steps with the elementary matrices $(\mathbf{I} - \mathbf{A}/\rho_i)$. Since, however, the λ's are unknown, we try instead to make the μ defined by equation 78.3 as small as possible over the range $0 < \lambda < L$. This can be done by selecting $P(\lambda)$ proportional to a Čebyšev polynomial of degree k, not for the entire interval $0 < \lambda < L$, because at zero we must have $\mu = 1$, but formed for an appropriate interval $l < \lambda < L$. The constant of proportionality must be chosen to make $P(0) = 1$. For hand calculation it is inconvenient to use exactly the Čebyšev polynomials because of the additional computation involved, but they may prove valuable for machine use.

We return to equation 78.1, the polynomial expressing **M** in terms of **A**. In actual calculation we perform a single step of the iteration with **M** by performing in sequence the individual operations given by $(\mathbf{I} - \mathbf{A}/\rho_1)$, $(\mathbf{I} - \mathbf{A}/\rho_2)$, $\cdots$, $(\mathbf{I} - \mathbf{A}/\rho_k)$. In order therefore to compare justly the effect of two different polynomials $P_1(\mathbf{A})$ and $P_2(\mathbf{A})$ of different degrees we should actually compare the results of an equal number of elementary steps of the form $(\mathbf{I} - \mathbf{A}/\rho_1)$. For instance, if $P_2(\mathbf{A})$ is of second degree and $P_3(\mathbf{A})$ is of third degree, the proper polynomials to compare are obviously the two sixth-degree polynomials $P_2{}^3$ and $P_3{}^2$.

As an illustration we compare

$$P_1(\mathbf{A}) = (\mathbf{I} - 2\mathbf{A}/L)$$

with

$$P_3(\mathbf{A}) = (\mathbf{I} - 16\mathbf{A}/4L)(\mathbf{I} - 16\mathbf{A}/9L)(\mathbf{I} - 16\mathbf{A}/15L).$$

The graphs of the two polynomials

$$\mu = P_1{}^3(\lambda), \qquad \mu = P_3(\lambda)$$

are shown in Fig. 20.

The decision as to which of these two is the better evidently depends on the distribution of the latent roots $\lambda_1, \lambda_2, \cdots, \lambda_n$. If these all lie close to $L/2$, it is apparent that $P_1{}^3(\lambda)$ gives, in general, smaller values of μ and, hence, is better. In the type of problems encountered in partial differential equations, however, there are values of λ close to 0 and also close to L. For this situation the graph shows that P_3 is better. In some situations, where we do not mind having the μ for λ_n as large as the μ for λ_1, we get a better third-degree polynomial by making it symmetrical with respect to the mid-point. The polynomial $P_3(\lambda)$ in the figure was deliberately made unsymmetrical in order to damp out μ_n. To select the

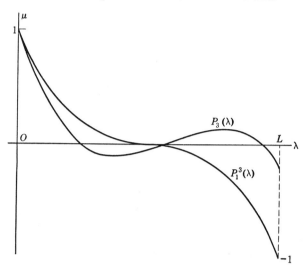

Fig. 20

very best polynomial in any individual case we have to know the greatest and least λ. Hence, in practice one takes what appears to be the best polynomial in the light of the information available.

79. Accelerating convergence

In actual practice it is suggested that the iteration be performed in a routine manner by a constant first-degree operator of the form

$$\mathbf{M} = \mathbf{I} - c\mathbf{A}$$

which gives us a sequence of iterates $\mathbf{v}_0, \mathbf{v}_1, \mathbf{v}_2, \cdots, \mathbf{v}_\nu$. The convergence generally will be slow because of latent roots μ_i near unity. If we knew the largest latent root μ_1 we could eliminate the corresponding latent vector and get an improved estimate by the formula

(79.1) $$\mathbf{v} = \frac{1}{1 - \mu_1}(\mathbf{v}_\nu - \mu_1 \mathbf{v}_{\nu-1}).$$

Then after iteration a similar formula enables us to eliminate the second latent vector, etc. Since μ_1 is only roughly known, we try to achieve the same end by the formula

(79.2) $$\mathbf{v} = (m + 1)\mathbf{v}_\nu - m\mathbf{v}_{\nu-1}.$$

This is equivalent to equation 79.1 if $m = \mu/(1 - \mu)$. We might just as well hazard a guess at m as at μ and can save labor by choosing m as an integer. If our guess for m is too small, we note that the subsequent iterates maintain the same general trend as before; if too large, the trend is reversed.

Example. Consider the example in Article 76, and at the stage $\nu = 6$ let us apply equation 79.2 with $m = 15$. This gives

$$\begin{array}{ccc} x & y & z \\ 2.6008 & -2.0214 & -2.7758, \end{array}$$

a result much closer to the true solution than the iterated values used to obtain it.

80. Determination of latent roots

One of the most difficult problems associated with matrices of high order is the determination of the latent roots. Several different methods are suggested here, but the reader must face the fact that the best methods known require much toil when n is large. This is especially true when we want to ascertain *all* the latent roots to a given degree of accuracy. Methods for the approximate evaluation of the least and largest latent roots are much simpler. Among the numerous methods that have been devised we mention the following.

(a) *Direct expansion of the determinant* $D_n(\lambda)$. This method is excellent when n is small, say not larger than 4 or 5. In general, however, it is quite impractical and will not be considered further.*

(b) *Escalator method.* One of the best ways of computing the latent roots of an $n \times n$ symmetric matrix, where n is not too large, is the "escalator method." See for example Morris [140]. With the notation of Article 68 the roots of $D_{n+1}(\lambda) = 0$ are made to depend on those of $D_n(\lambda) = 0$, the latter in turn on those of $D_{n-1}(\lambda) = 0$, etc. Therefore, to calculate the roots of $D_{n+1}(\lambda) = 0$ we start with the root of $D_1(\lambda) = 0$ (or any later known set of roots in the chain) and calculate in succession the roots of $D_2(\lambda) = 0$, $D_3(\lambda) = 0$, $\cdots$, $D_n(\lambda) = 0$. The relationship on which the escalator method is based is the following:

$$(80.1) \qquad a_{n+1,n+1} - \lambda = \sum_{k=1}^{n} \frac{N_k{}^2}{\lambda_k - \lambda}.$$

Here λ denotes a latent root of $D_{n+1}(\lambda) = 0$, and $\lambda_1, \lambda_2, \cdots , \lambda_n$ are the n roots of $D_n(\lambda) = 0$. The quantities N_k are computed by means of the formulas

$$(80.2) \qquad N_k = \sum_{j=1}^{n} a_{j,n+1} x_{kj},$$

---

* Certain special determinants satisfy recurrence relations which permit their evaluation for large n.

in which $(x_{k1}, x_{k2}, \cdots, x_{kn})$ is the normalized latent vector belonging to λ_k.

From the foregoing formulas it is evident that we must also have the latent vectors at each step in order to compute the roots for the next step. The latent vector belonging to the λ in 80.1 is computed from the formula

$$(80.3) \qquad u_q = -\sum_{k=1}^{n} x_{kq} N_k \frac{u_{n+1}}{\lambda_k - \lambda}.$$

Here $u_q(q = 1, 2, \cdots, n + 1)$ are the components (not yet normalized) of the desired latent vector.

Example. For the matrix

$$\begin{bmatrix} 4 & -1 & 0 \\ -1 & 4 & -1 \\ 0 & -1 & 4 \end{bmatrix}$$

we start with the latent roots and vectors of $D_2(\lambda)$, namely,

$$\lambda_1 = 3, \qquad \mathbf{v}_1 = (0.707, \quad 0.707),$$
$$\lambda_2 = 5, \qquad \mathbf{v}_2 = (0.707, \quad -0.707).$$

Equation 80.2 then gives $N_1{}^2 = N_2{}^2 = \tfrac{1}{2}$, and equation 80.1 becomes

$$4 - \lambda = \frac{1}{2(3 - \lambda)} + \frac{1}{2(5 - \lambda)}.$$

Its solutions are 2.586, 4, 5.414. With the aid of equation 80.3 we now get (after normalization) the three corresponding latent vectors

$$(0.500, \quad 0.707, \quad 0.500),$$
$$(0.707, \quad 0 \quad\quad, \quad -0.707),$$
$$(0.500, \quad -0.707, \quad 0.500).$$

To obtain the roots of an nth-order matrix by this procedure starting from $D_1(\lambda)$ we must compute a total of $n(n + 1)/2$ roots, and $(2n - 1)(n - 1)n/6$ components of latent vectors. For large n this becomes too ponderous for hand calculation.

(c) *Linear dependence of iterated vectors.* Let $\mathbf{v}_0$ be a non-zero vector expressed in terms of the n latent vectors of the matrix $\mathbf{A}$,

$$(80.4) \qquad\qquad \mathbf{v}_0 = \Sigma \alpha_i \phi_i.$$

Then by iteration we have

$$(80.5) \qquad \mathbf{v}_\nu = \Sigma \alpha_i \phi_i \lambda_i{}^\nu, \qquad \nu = 1, 2, \cdots, n.$$

The $n + 1$ vectors $v_0 v_1, \cdots, v_n$ are linearly dependent since they depend on n vectors $\phi_1, \cdots, \phi_n$, and we may write

(80.6) $v_n + g_1 v_{n-1} + g_2 v_{n-2} + \cdots + g_n v_0 = 0.$

From equations 80.4 and 80.5 we find that the n latent roots $\lambda_1, \lambda_2, \cdots, \lambda_n$ satisfy the nth-degree algebraic equation

(80.7) $\lambda^n + g_1 \lambda^{n-1} + \cdots + g_n = 0.$

To use this method numerically we select a vector v_0, iterate n times to get $v_1, v_2, \cdots, v_n$; then solve the n equations 80.6 (one for each component) for the coefficients $g_1, g_2, \cdots, g_n$, and finally solve the algebraic equation 80.7 for the n latent roots. The bad feature of the method is the necessity of solving the n simultaneous equations 80.6. The good feature is that by clever choice of the initial vector v_0 the task of solving these equations can sometimes be reduced to a manageable size.

Example. Find the latent roots of the matrix

$$
A = \begin{bmatrix}
4 & -1 & & -1 & & & & & & & & \\
-1 & 4 & -1 & & -1 & & & & & & & \\
& -1 & 4 & -1 & & -1 & & & & & & \\
& & -1 & 4 & & & -1 & & & & & \\
-1 & & & & 4 & -1 & & & & & & \\
& -1 & & & -1 & 4 & -1 & & -1 & & & \\
& & -1 & & & -1 & 4 & -1 & & -1 & & \\
& & & -1 & & & -1 & 4 & & & -1 & \\
& & & & -1 & & & & 4 & -1 & & \\
& & & & & -1 & & & -1 & 4 & -1 & \\
& & & & & & -1 & & & -1 & 4 & -1 \\
& & & & & & & -1 & & & -1 & 4
\end{bmatrix}
$$

where only non-zero elements have been written.

First of all we notice that, if λ is a latent root and $(x_1, x_2, x_3, x_4, x_5, x_6, x_7, x_8, x_9, x_{10}, x_{11}, x_{12})$ is the associated latent vector, then $8 - \lambda$ is also a latent root and $(x_1, -x_2, x_3, -x_4, -x_5, x_6, -x_7, x_8, -x_9, x_{10}, -x_{11}, x_{12})$ is its associated latent vector. Hence, if we use the matrix

$$M = I - \tfrac{1}{4}A$$

to perform the iterations the latent roots of M occur in pairs with opposite sign, $\mu_1 = -\mu_{12}$, $\mu_2 = -\mu_{11}$, etc. Hence, the equation

$$(80.8) \qquad D_{12}(\mu) = \mu^{12} + g_1\mu^{11} + g_2\mu^{10} + \cdots + g_{12} = 0$$

contains only even powers, and we may set $g_1 = g_3 = \cdots = g_{11} = 0$ at the outset.

Any initial vector v_0 may be used to start with. To avoid divisions and consequent loss of accuracy by round-off we keep all the work in integers. The easiest way to do this is to use $4M$ instead of M for the iterations and to take our initial vector v_0 with a single component equal to unity and all other components zero. When this is done, all the g's will obviously be integers. Accordingly we take $x_1 = 1$, all other x's equal to zero, and iterate 12 times. As already noted we need use only the even-numbered iterates. For these we find that x_2, x_4, x_5, x_7, x_9, and x_{11} are always zero so that only the remaining components need be tabulated. The result is

ν	x_1	x_3	x_6	x_8	x_{10}	x_{12}
0	1					
2	2	1	2			
4	9	9	14	4	5	
6	55	77	106	54	60	9
8	399	666	862	551	586	123
10	3188	5809	7301	5142	5373	1260
12	26787	50874	63086	46351	48074	11775

In this simple example we easily perform suitable linear combinations of the columns to get zeros above the main diagonal. One such recombination is the following

1					
2	2				
9	14	4			
55	106	48	6		
399	862	470	81	6	
3188	7301	4317	825	93	3
26787	63086	38662	7689	1014	48

and from these values it is a simple matter to determine the g's in succession, and to obtain equation 80.8

$$\mu^{12} - 16\mu^{10} + 79\mu^8 - 148\mu^6 + 96\mu^4 - 12\mu^2 = 0.$$

The solutions are

$$\mu^2 = 0,\ 0.16212,\ 1.0000,\ 2.1705,\ 3.8810,\ 8.7864.$$

Finally, the latent roots of A are

$$\lambda_1 = 1.0358, \qquad \lambda_5 = 3.5974, \qquad \lambda_9 = 5.0000,$$

$$\lambda_2 = 2.0300, \qquad \lambda_6 = 4.0000, \qquad \lambda_{10} = 5.4733,$$

$$\lambda_3 = 2.5267, \qquad \lambda_7 = 4.0000, \qquad \lambda_{11} = 5.9700,$$

$$\lambda_4 = 3.0000, \qquad \lambda_8 = 4.4026, \qquad \lambda_{12} = 6.9642.$$

(d) *Scalar products of the iterates.* A variation of method c is obtained if one forms the sequence of scalar products

$$P_0 = \mathbf{v}_0 \cdot \mathbf{v}_0, \qquad P_1 = \mathbf{v}_0 \cdot \mathbf{v}_1, \qquad P_2 = \mathbf{v}_1 \cdot \mathbf{v}_1,$$

$$P_3 = \mathbf{v}_1 \cdot \mathbf{v}_2, \qquad P_4 = \mathbf{v}_2 \cdot \mathbf{v}_2, \qquad \text{etc.},$$

and then uses the equations

$$P_n \quad + g_1 P_{n-1} + g_2 P_{n-2} + \cdots + g_n P_0 \quad = 0$$

$$P_{n+1} + g_1 P_n \quad + g_2 P_{n-1} + \cdots + g_n P_1 \quad = 0$$

$$\cdot \qquad\qquad\qquad + \quad \cdot$$

$$\cdot \qquad\qquad\qquad\qquad \cdot$$

$$\cdot \qquad\qquad\qquad\qquad \cdot$$

$$P_{2n-1} + g_1 P_{2n-2} + g_2 P_{2n-3} + \cdots + g_n P_{n-1} = 0$$

for the determination of the g's.

81. Method of iteration and elimination. Real symmetric matrix

Let us assume that the latent vectors are arranged in order of decreasing absolute value and are all distinct so that

$$|\lambda_1| > |\lambda_2| > \cdots > |\lambda_n|.$$

Starting out with an arbitrary vector $\mathbf{v}_0$ expressed in terms of the n latent vectors as in equation 80.1, we iterate ν times to get equation 80.2. It is convenient to adopt the notation

(81.1)
$$\begin{aligned}
\mathbf{v}_\nu(x) &= x\mathbf{v}_\nu - \mathbf{v}_{\nu+1}, \\
\mathbf{v}_\nu(x, y) &= y\mathbf{v}_\nu(x) - \mathbf{v}_{\nu+1}(x), \\
\mathbf{v}_\nu(x, y, z) &= z\mathbf{v}_\nu(x, y) - \mathbf{v}_{\nu+1}(x, y),
\end{aligned}$$

$$\text{etc.}$$

Note that $\mathbf{v}_\nu(x, y, z, \cdots)$ is symmetric in the arguments $x, y, z, \cdots$.

Now, since the ϕ's are orthonormal, we find on carrying out the indicated operations that

(81.2)
$$\frac{\mathbf{v}_\nu \cdot \mathbf{v}_\nu}{\mathbf{v}_{\nu-1} \cdot \mathbf{v}_\nu} = \lambda_1 \left(\frac{\alpha_1^2 + \alpha_2^2 \rho_2^{2\nu} + \cdots + \alpha_n^2 \rho_n^{2\nu}}{\alpha_1^2 + \alpha_2^2 \rho_2^{2\nu-1} + \cdots + \alpha_n^2 \rho_n^{2\nu-1}} \right)$$

where $\rho_i = \lambda_i/\lambda_1$ and is less than unity in absolute value. If $\alpha_1 \neq 0$, it is clear that the right member of equation 81.2 approaches λ_1 as ν increases. We therefore set

(81.3)
$$\frac{\mathbf{v}_\nu \cdot \mathbf{v}_\nu}{\mathbf{v}_{\nu-1} \cdot \mathbf{v}_\nu} = \lambda_1',$$

and use λ_1' as an approximation to λ_1. We can make the approximation as close as we please by taking ν large enough.

Next consider the sequence of values

$$\mathbf{v}_0(\lambda_1'), \ \mathbf{v}_1(\lambda_1'), \ \mathbf{v}_2(\lambda_1'), \ \cdots$$

which is obtained directly from the sequence $\mathbf{v}_0, \mathbf{v}_1, \mathbf{v}_2, \cdots$ by the first formula in 81.1. If all goes well, an approximate value of λ_2, say, λ_2', is supplied by the formula

$$\frac{\mathbf{v}_\nu(\lambda_1') \cdot \mathbf{v}_\nu(\lambda_1')}{\mathbf{v}_{\nu-1}(\lambda_1') \cdot \mathbf{v}_\nu(\lambda_1')} = \lambda_2'.$$

Likewise, an estimate of λ_3 is given by

$$\frac{\mathbf{v}_\nu(\lambda_1', \lambda_2') \cdot \mathbf{v}_\nu(\lambda_1', \lambda_2')}{\mathbf{v}_{\nu-1}(\lambda_1', \lambda_2') \cdot \mathbf{v}_\nu(\lambda_1', \lambda_2')} = \lambda_3'.$$

If all goes well we proceed thus step by step until all the roots have been found.

Example 1. Find the latent roots of the matrix

$$\mathbf{M} = \begin{bmatrix} 3 & 2 & 0 \\ 2 & 2 & 1 \\ 0 & 1 & 2 \end{bmatrix}.$$

Starting with the initial vector $\mathbf{v}_0 = (1, 1, 1)$ and iterating six times, we have the results

$\mathbf{v}_0$	1	1	1
$\mathbf{v}_1$	5	5	3
$\mathbf{v}_2$	25	23	11
$\mathbf{v}_3$	121	107	45
$\mathbf{v}_4$	577	501	197
$\mathbf{v}_5$	2733	2353	895
$\mathbf{v}_6$	12905	11067	4143

$$\lambda_1' = \mathbf{v}_6 \cdot \mathbf{v}_6/\mathbf{v}_5 \cdot \mathbf{v}_6 = 4.7092.$$

$\mathbf{v}_0(\lambda_1')$	-0.2908	-0.2908	1.7029
$\mathbf{v}_1(\lambda_1')$	-1.4540	$+0.5460$	3.1276
$\mathbf{v}_2(\lambda_1')$	-3.2700	1.3116	6.8012
$\mathbf{v}_3(\lambda_1')$	-7.1868	2.8844	14.9140

$$\lambda_2' = 2.1939$$

$\mathbf{v}_0(\lambda_1', \lambda_2') =$	0.8160	-1.1840	0.6222
$\mathbf{v}_1(\lambda_1', \lambda_2') =$	0.0801	-0.1137	0.0604

$$\lambda_3' = 0.0968.$$

Example 2. Find the latent roots and latent vectors of the matrix

$$M = \begin{bmatrix} 2 & 1 & 1 & 2 \\ 1 & 1 & 1 & 1 \\ 1 & 1 & 2 & 2 \\ 2 & 1 & 2 & 4 \end{bmatrix}.$$

We take the initial vector v_0 to be (1, 1, 1, 1) and iterate eight times to get the values shown below.

	1	2	3	4
v_0	1	1	1	1
v_1	6	4	6	9
v_2	40	25	40	64
v_3	273	169	273	441
v_4	1 870	1 156	1 870	3 025
v_5	12 816	7 921	12 816	20 736
v_6	87 841	54 289	87 841	142 129
v_7	602 070	372 100	602 070	974 169
v_8	4126 648	2550 409	4126 648	6677 056

$$v_7 \cdot v_8 / v_7 \cdot v_7 = 6.854101968 = \lambda_1 \text{ approx.}$$

Using this value as an approximation to λ_1, calculate v_r (λ_1) as follows:

	1	2	3	4
$v_0(\lambda_1)$	8541 01968	28541 01968	8541 01968	−21458 98032
$v_1(\lambda_1)$	11246 11808	24164 07872	11246 11808	−23130 82288
$v_2(\lambda_1)$	11640 78720	23525 49200	11640 78720	−23374 74048
$v_3(\lambda_1)$	11698 37264	23432 32592	11698 37264	−23410 32112
$v_4(\lambda_1)$	11706 80160	23418 75008	11706 80160	−23415 46800
$v_5(\lambda_1)$	11708 21888	23416 88528	11708 21888	−23415 91552

$$\frac{v_4(\lambda_1) \cdot v_5(\lambda_1)}{v_4(\lambda_1) \cdot v_4(\lambda_1)} = 1.0000\ 0000 = \lambda_2.$$

Then

	1	2	3	4
$v_0(\lambda_1, \lambda_2)$	−27050 9840	43769 4096	−27050 9840	16718 4256
$v_1(\lambda_1, \lambda_2)$	− 3946 6912	6385 8672	− 3946 6912	2439 1760
$v_2(\lambda_1, \lambda_2)$	− 575 8544	931 6608	− 575 8544	355 8064
$v_3(\lambda_1, \lambda_2)$	− 84 2896	135 7584	− 84 2896	51 4688

From these values we find

$$\lambda_3 = 0.14589\ 8033.$$

Further application of the process leads to confusing and contradictory results. Actually the figures obtained are no longer valid, all significant digits having been eliminated and only quantities arising from the errors in λ_1 and λ_2 (principally λ_1) remain. Only three roots have been found where four are known to exist. Now a root could be lost in either of two ways:

(a) If ϕ_i is orthogonal to v_0, then α_i is zero, or

(b) If λ_i is a double root, it would be found just once by this process. Actually in this example both a and b occur since $\lambda = 1$ is a double root, and one of its latent vectors is $(1, 0, -1, 0)$, which is orthogonal to $(1, 1, 1, 1)$.

It may be remarked that the values found above for the latent roots are correct to eight places of decimals. A further observation is that Examples 1 and 2 are extremely flattering to the process, since the ratios of distinct latent roots are all small, making convergence exceptionally rapid.

To find the latent vector ϕ_1 we return to v_6, v_7, v_8 and eliminate ϕ_2 and ϕ_3 by forming the vector v_6 (λ_2, λ_3). This turns out to be (not normalized)

$$\phi_1 = (3449\ 553, 2131\ 941, 3449\ 553, 5581\ 494).$$

Similarly,

$$\phi_2 = (1, 2, 1, -2)$$

$$\phi_3 = (-842\ 896, 1357\ 584, -842\ 896, 514\ 688)$$

and, as already noted, the remaining latent vector is

$$\phi_4 = (1, 0, -1, 0).$$

82. Use of orthogonality

From the inequalities

$$|\lambda_1| > |\lambda_2| > \cdots > |\lambda_n|$$

together with equation 80.2 we see that, as ν increases, the iterate v_ν tends to become proportional to the latent vector ϕ_1. Hence, for suitably large ν we assume that $v_\nu = C\phi_1$, and thus determine ϕ_1. Starting again with any initial vector v_0, we remove the component parallel to ϕ_1 by the formula

$$w_0 = v_0 - (v_0 \cdot \phi_1)\phi_1.$$

Then w_ν does not contain ϕ_1, and iteration leads to

$$w_\nu \rightarrow C\phi_2.$$

Now that ϕ_1 and ϕ_2 are known, we again form an initial vector w_0 perpendicular to ϕ_1 and ϕ_2, and by iteration arrive at ϕ_3. In this way the latent vectors and latent roots are computed one by one.

The numerical computation is not so simple as the above discussion implies, for we rarely know ϕ_1, ϕ_2, $\cdots$ exactly, and therefore the vector we iterate still contains some small components of the latent vectors that were supposedly eliminated. Iteration rapidly magnifies these components, and, hence, the elimination must be made repeatedly as the iteration progresses. How many iterations can be made between eliminations depends evidently on the ratios of the λ's.

The same difficulty arises in the method of Article 81, and only the extreme simplicity of the illustrative examples kept the resulting error within reasonable bounds. When n is large, this resurgence of components supposedly eliminated makes it necessary to perform the eliminations repeatedly. In the method of Article 81 this has the unfortunate effect of introducing factors of the type $(\lambda_i' - \lambda_j)^m$ into the components not yet eliminated, which slows down (and may even prevent) convergence. For this reason the method of Article 82 seems to be preferable to that of Article 81.

Example. Find the latent roots and vectors of the matrix

$$\mathbf{A} = \begin{bmatrix} 2 & 2 & 1 & 1 & 0 \\ 2 & 4 & 2 & 2 & 1 \\ 1 & 2 & 2 & 0 & 0 \\ 1 & 2 & 0 & 3 & 2 \\ 0 & 1 & 0 & 2 & 2 \end{bmatrix}.$$

We start with the initial vector $\mathbf{v}_0$

$$\mathbf{v}_0 = \quad 1 \quad 1 \quad 1 \quad 1 \quad 1$$

and after ten iterations (and dropping the least significant digits) arrive at

$$\mathbf{v}_9 = \quad 82\,489 \quad 151\,917 \quad 66\,721 \quad 107\,056 \quad 63\,222$$

$$\mathbf{v}_{10} = \quad 642\,591 \quad 1183\,425 \quad 519\,767 \quad 833\,936 \quad 492\,473$$

and from these get

$$\lambda_1 = 7.78989$$

$$\phi_1 = 0.36967 \quad 0.68081 \quad 0.29902 \quad 0.47975 \quad 0.28331.$$

The last digits in ϕ_1 are not reliable.

Next take a vector $\mathbf{v}_0'$ orthogonal to ϕ_1, for example

$$\mathbf{v}_0' = \quad -12 \quad -11 \quad -22 \quad 25 \quad 25.$$

This we iterate twice to get $\mathbf{v}_1'$ and $\mathbf{v}_2'$. Since the residual component of ϕ_1 in $\mathbf{v}_0'$ is magnified at every iteration, we remove it by the formula

$$\bar{\mathbf{v}}_2' = \mathbf{v}_2' - (\phi_1 \cdot \mathbf{v}_2')\phi_1.$$

Repeating the routine of two iterations and then removal of the component of ϕ_1, we finally get

$$\mathbf{v}_7' = \quad - 93\ 489 \quad - 83\ 359 \quad -165\ 622 \quad 185\ 456 \quad 183\ 044$$

$$\mathbf{v}_8' = \quad -333\ 862 \quad -297\ 702 \quad -591\ 451 \quad 662\ 249 \quad 653\ 641$$

Then

$$\lambda_2 = \quad 3.57101$$

$$\mathbf{\phi}_2 = -0.28059 \quad -0.25020 \quad -0.49708 \quad 0.55659 \quad 0.54935.$$

We start again with a vector $\mathbf{v}_0''$ orthogonal to $\mathbf{\phi}_1$ and $\mathbf{\phi}_2$

$$\mathbf{v}_0'' = \quad 7846 \quad -3219 \quad -2500 \quad -212 \quad 494,$$

iterate twice, and then remove the components of $\mathbf{\phi}_1$ and $\mathbf{\phi}_2$ by the formula

$$\bar{\mathbf{v}}_2'' = \mathbf{v}_2'' - (\mathbf{\phi}_1 \cdot \mathbf{v}_2'')\mathbf{\phi}_1 - (\mathbf{\phi}_2 \cdot \mathbf{v}_2'')\mathbf{\phi}_2.$$

Several repetitions of this ritual lead to

$$\mathbf{v}_7'' = \quad 163\ 959 \quad -31436 \quad -125\ 650 \quad 56587 \quad -101\ 599,$$

$$\mathbf{v}_8'' = \quad 195\ 983 \quad -37551 \quad -150\ 213 \quad 67650 \quad -121\ 460.$$

Then

$$\lambda_3 = 1.19539$$

$$\mathbf{\phi}_3 = 0.68561 \quad -0.13136 \quad -0.52549 \quad 0.23666 \quad -0.42490.$$

To get the two remaining roots we change to a new matrix M by the formula

$$M = 5I - A$$

and proceed as before. When all roots and latent vectors are approximately known, we continue the process of iteration and orthogonalization to secure more exact latent vectors and roots. The results as computed with a desk computer are shown below. The final digits of the latent vectors are not reliable.

$\mathbf{\phi}_1$	0.36967	0.68081	0.29902	0.47974	0.28331
$\mathbf{\phi}_2$	-0.28053	-0.25020	-0.49711	0.55662	0.54934
$\mathbf{\phi}_3$	0.68558	-0.13134	-0.52548	0.23661	-0.42497
$\mathbf{\phi}_4$	0.48921	-0.62200	0.46034	-0.01854	0.40191
$\mathbf{\phi}_5$	0.27440	0.26410	-0.41884	-0.63537	0.52520

$$\lambda_1 = 7.789\ 875$$

$$\lambda_2 = 3.571\ 014$$

$$\lambda_3 = 1.195\ 391$$

$$\lambda_4 = 0.360\ 241$$

$$\lambda_5 = 0.083\ 478$$

83. Gradient method

In the case of a real symmetric matrix one may make the step from $\mathbf{v}_\nu$ to $\mathbf{v}_{\nu+1}$ by means of the formulas

(83.1)
$$\mathbf{w}_\nu = A\mathbf{v}_\nu - \lambda(\mathbf{v}_v)\mathbf{v}_\nu,$$

where in general $\lambda(x) = x \cdot Ax / x \cdot x$, and

$$(83.2) \qquad\qquad v_{\nu+1} = v_\nu + c w_\nu,$$

in which c is a scalar at our disposal. It may be verified with the aid of equations 83.1 and 83.2 and the definition of $\lambda(x)$ that

$$(83.3) \quad \lambda(v_{\nu+1}) - \lambda(v_\nu) = c \frac{w_\nu \cdot w_\nu}{v_{\nu+1} \cdot v_{\nu+1}} \{2 + c[\lambda(w_\nu) - \lambda(v_\nu)]\}.$$

If λ_M is the largest and λ_m the least latent root of A, we know from Theorem 7 that

$$\lambda_m \leqq \lambda(x) \leqq \lambda_M,$$

no matter what vector x is chosen. Hence, $\lambda(w_\nu) - \lambda(v_\nu)$ is in absolute value not greater than $L = \lambda_M - \lambda_m$, so that, if $|c| < 2/L$, the term $2 + c[\lambda(w_\nu) - \lambda(v_\nu)]$ in equation 83.3 will be positive. Then it is evident from 83.3 that $\lambda(v_{\nu+1})$ will be greater than or less than $\lambda(v_\nu)$ according as c is positive or negative. Hence, the sequence $\lambda(v_0)$, $\lambda(v_1)$, $\lambda(v_2)$, $\cdots$ is an increasing (decreasing) sequence, provided in each step we use a positive (negative) c such that $|c| < 2/L$. It follows from Theorem 7 that the sequence converges, and converges to a latent root of A.

The process described in Article 82 may therefore be modified by the use of 83.1 and 83.2 to compute $v_{\nu+1}$ from v_ν, but with no other change.

Example. The problem worked in Article 82 was also worked by the present method at the National Bureau of Standards in Los Angeles in about six hours' computing time on the IBM card programmed calculator. The work was done to ten figures. The results, rounded to six places are shown below. Differences between the results in Articles 82 and 83 are explained in part by the fact that the iteration process in Article 82 was not carried to as many steps as in Article 83.

ϕ_1	0.369 679	0.680 813	0.299 023	0.479 744	0.283 305
ϕ_2	−0.280 571	−0.250 164	−0.497 067	0.556 613	0.549 367
ϕ_3	−0.685 567	0.131 363	0.525 522	−0.236 636	0.424 938
ϕ_4	0.489 101	−0.622 106	0.460 502	−0.018 297	0.401 705
ϕ_5	0.274 583	0.263 892	−0.418 658	−0.635 371	0.525 352

$$\lambda_1 = 7.789\,875$$

$$\lambda_2 = 3.571\,014$$

$$\lambda_3 = 1.195\,391$$

$$\lambda_4 = 0.360\,241$$

$$\lambda_5 = 0.083\,478.$$

10

IMPLICIT METHODS. ELLIPTIC
EQUATIONS

Chapter 8 dealt with those partial differential equations for which the numerical solution can be calculated step by step by means of explicit formulas. In contrast the present chapter considers those equations for which the numerical solution is contained implicitly in a system of simultaneous equations. Familiar examples of this class are Laplace's equation and Poisson's equation.

84. Laplace's equation

In Chapter 8 there was developed a difference expression representing in a formal manner the Laplacian operator. For the important special case in which the function under consideration is harmonic, i.e., satisfies $\nabla^2 u = 0$, it is advantageous to adopt a somewhat different approach in which we avail ourselves of the properties of analytic functions of a complex variable. This method of derivation uses the fact that a solution $u(x, y)$ of Laplace's equation is the real part of an analytic function

$$f(z) = u(x, y) + iv(x, y)$$

with $z = x + iy$. Suppose that $f(z)$ is analytic in a circle S with center at O. (See Fig. 21.) At any point A inside the circle $f(z)$ can be expanded in a convergent power series

$$f_A = \Sigma a_\nu (z - z_0)^\nu.$$

With the aid of polar coordinates

$$a_\nu = \rho_\nu e^{i\theta_\nu}, \qquad z - z_0 = r e^{i\phi}$$

this gives

$$f_A = \Sigma \rho_\nu r^\nu e^{i(\nu\phi + \theta_\nu)}.$$

In similar manner we get the series for $f(z)$ at the points $B(r, \phi + \pi/2)$, $C(r, \phi + \pi)$, and $D(r, \phi + 3\pi/2)$, and by addition obtain

$$f_A + f_B + f_C + f_D - 4f_0$$
$$= 4(\rho_4 r^4 e^{i(4\phi + \theta_4)} + \rho_8 r^8 e^{i(8\phi + \theta_8)} + \rho_{12} r^{12} e^{i(12\phi + \theta_{12})} + \cdots).$$

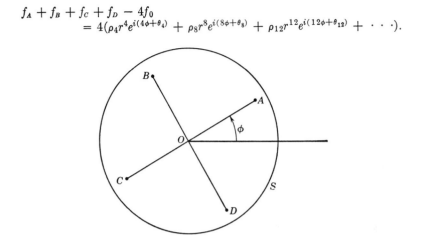

Fig. 21

In particular, since $u(x, y)$ is the real part of $f(z)$, it follows that

$$(84.1) \quad u_A + u_B + u_C + u_D - 4u_0 = 4[\rho_4 r^4 \cos(4\phi + \theta_4)$$
$$+ \rho_8 r^8 \cos(8\phi + \theta_8) + \rho_{12} r^{12} \cos(12\phi + \theta_{12}) + \cdots].$$

From this equation we derive several useful results. For example,

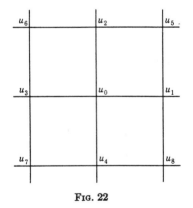

Fig. 22

referring to a portion of the net shown in Fig. 22 we note that, if $r = h$, $\phi = 0$ in equation 84.1, we have

$$(84.2) \quad u_1 + u_2 + u_3 + u_4 - 4u_0$$
$$= 4h^4 \rho_4 \cos \theta_4 + 4h^8 \rho_8 \cos \theta_8 + 4h^{12} \rho_{12} \cos \theta_{12} + \cdots$$

whereas, if $r = \sqrt{2}\,h$, $\phi = \pi/4$, the result is

$$(84.3) \quad u_5 + u_6 + u_7 + u_8 - 4u_0$$
$$= -16h^4\rho_4 \cos\theta_4 + 64h^8\rho_8 \cos\theta_8 - 256h^{12}\rho_{12} \cos\theta_{12} + \cdots.$$

If we multiply equation 84.2 by 4 and add to equation 84.3, we get the nine-point formula

$$(84.4) \quad 4u_1 + 4u_2 + 4u_3 + 4u_4 + u_5 + u_6 + u_7 + u_8 - 20u_0$$
$$= 80h^8\rho_8 \cos\theta_8 - 240h^{12}\rho_{12} \cos\theta_{12} + \cdots.$$

The left-hand members of equations 84.2, 84.3, and 84.4 are conveniently represented by means of operators H, X, and K defined by the stencils

$$(84.5) \qquad H = \begin{array}{|c|c|c|} \hline 0 & 1 & 0 \\ \hline 1 & -4 & 1 \\ \hline 0 & 1 & 0 \\ \hline \end{array},$$

$$(84.6) \qquad 2X = \begin{array}{|c|c|c|} \hline 1 & 0 & 1 \\ \hline 0 & -4 & 0 \\ \hline 1 & 0 & 1 \\ \hline \end{array},$$

and

$$(84.7) \qquad K = 4H + 2X = \begin{array}{|c|c|c|} \hline 1 & 4 & 1 \\ \hline 4 & -20 & 4 \\ \hline 1 & 4 & 1 \\ \hline \end{array}.$$

These are the same difference operators defined in Chapter 8. The stencils on the right in these equations display the coefficients by which the u's in Fig. 22 are multiplied to obtain the left-hand members of 84.2, 84.3, and 84.4, respectively.

Let us suppose that the first term in each of the series on the right in 84.2, 84.3, 84.4 does not vanish and is considerably larger in absolute value than the sum of the remaining terms. Then the three equations just cited can be written as

(84.8) $Hu = 4h^4 \rho_4 \cos \theta_4,$

(84.9) $2Xu = 16h^4 \rho_4 \cos \theta_4,$

(84.10) $Ku = 80h^8 \rho_8 \cos \theta_8.$

When h is so small that the right-hand members can be ignored in view of the accuracy desired, the foregoing equations yield three different difference equations by which Laplace's equation can be replaced:

(84.11) $Hu = 0,$

(84.12) $Xu = 0,$

(84.13) $Ku = 0.$

Equation 84.12 is similar to 84.11 rotated through 45 degrees, while 84.13 has a higher order of accuracy than the two preceding equations.

Equations 84.8 and 84.9 provide a basis for estimating the truncation error involved in using 84.11. For, if a set of v's has been obtained that satisfies $Hv = 0$, we can apply the operator X and find the values of Xv. Then the error is approximately $\frac{1}{2}Xv$. This of course is what may be called a "local error," the actual error being due to the cumulative effect over the region due to the local errors.

85. The matrix H

In many problems one needs to determine a function $u(x, y)$ that satisfies Laplace's equation everywhere in the interior of a connected region R and assumes assigned values on the boundary B of R. To attack such a problem numerically we first overlay the region R with a square lattice having mesh length h. In this initial analysis let us suppose that the boundary B cuts the lattice only at nodal points, since otherwise we encounter the vexatious difficulties mentioned in Article 65. Next, for each of the n interior points of R we apply the equation 84.11, thus obtaining a system of n simultaneous linear equations in the n unknown values $u(x, y)$. In these equations we find it convenient to change all signs, transpose to the right all known quantities, i.e., those associated with boundary points, and thus have the equations in standard form

(85.1) $\mathbf{Hu = b.}$

Here $\mathbf{H}$ denotes an $n \times n$ matrix with each element of the main diagonal equal to 4 and all others either -1 or 0.

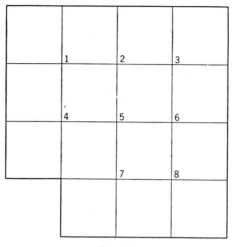

Fig. 23

For example, the matrix $\mathbf{H}$ associated with the region R in Fig. 23 is

$$\mathbf{H} = \begin{bmatrix} 4 & -1 & 0 & -1 & 0 & 0 & 0 & 0 \\ -1 & 4 & -1 & 0 & -1 & 0 & 0 & 0 \\ 0 & -1 & 4 & 0 & 0 & -1 & 0 & 0 \\ -1 & 0 & 0 & 4 & -1 & 0 & 0 & 0 \\ 0 & -1 & 0 & -1 & 4 & -1 & -1 & 0 \\ 0 & 0 & -1 & 0 & -1 & 4 & 0 & -1 \\ 0 & 0 & 0 & 0 & -1 & 0 & 4 & -1 \\ 0 & 0 & 0 & 0 & 0 & -1 & -1 & 4 \end{bmatrix}$$

We readily verify the following assertions.

1. *The matrix* $\mathbf{H}$ *depends solely on the shape of the set of lattice points within and on the boundary of region* R *and not at all on the particular boundary values.*

2. *The matrix* $\mathbf{H}$ *is symmetric for every region* R.

3. *The latent roots of* $\mathbf{H}$ *are symmetrically located with respect to* $\lambda = 4$; i.e., if λ_i is a latent root, so also is $8 - \lambda_i$. As a corollary we note that, if n is odd, $\mathbf{H}$ always has a latent root equal to 4.

To prove this statement we consider the homogeneous system

$$\mathbf{H}\phi_i = \lambda_i \phi_i,$$

where λ_i is a latent root and ϕ_i the associated latent vector. Then by subtraction from the identity $4\phi_i = 4\phi_i$ we have

$$(4\mathbf{I} - \mathbf{H})\phi_i = (4 - \lambda_i)\phi_i.$$

Now we replace $4 - \lambda_i$ by $\lambda_j - 4$ and ϕ_i by ϕ_j where the new vector ϕ_j is found from ϕ_i by the pattern of sign changes shown in Fig. 24. For any region R the pattern of sign changes is simply the one for which

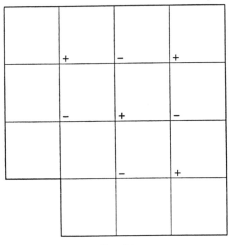

Fig. 24

signs alternate in both rows and columns. The effect in the matrix $4\mathbf{I} - \mathbf{H}$, which has zeros in the main diagonal, is to change (or leave unchanged) *all* signs in the ith row and column.

For instance, in the present example, if

$$\phi_i = (x_1, \ x_2, \ x_3, \ x_4, \ x_5, \ x_6, \ x_7, \ x_8),$$

then

$$\phi_j = (x_1, \ -x_2, \ x_3, \ -x_4, \ x_5, \ -x_6, \ -x_7, \ x_8).$$

Clearly then ϕ_j and λ_j are also a latent vector and latent root, and

$$4 - \lambda_i = \lambda_j - 4,$$

which establishes the statement.

4. *The matrix* $\mathbf{H}$ *is non-singular.* For we note from equation 84.5 that, if $Hu = 0$, the value of $u(x, y)$ lies between the greatest and least adjacent values of u. Therefore, if $Hu = 0$ at every interior point of R, it is clear that u can have neither maximum nor minimum at any interior point. Thus the value at any interior point must lie between the greatest and least boundary values. In particular, if all boundary

values are zero, so are all interior values. Hence, the homogeneous system

$$\mathbf{Hu} = 0$$

can have no non-zero solution and $\mathbf{H}$ is therefore non-singular.

For each horizontal or vertical segment of the lattice joining two adjacent points P_i and P_j we can form the square of the difference of the associated u's,

$$(u_i - u_j)^2.$$

Let S denote the positive definite quadratic form

$$S = \Sigma (u_i - u_j)^2$$

in which each difference occurs just once.

For Fig. 23 the quadratic form is

$$
\begin{aligned}
S = {}& (b_1 \ - u_1)^2 + (b_2 - u_2)^2 + (b_3 \ - u_3)^2 + (b_4 - u_3)^2 \\
& + (b_{11} - u_1)^2 + (u_1 - u_2)^2 + (u_2 \ - u_3)^2 + (u_1 - u_4)^2 \\
& + (u_5 \ - u_2)^2 + (u_6 - u_3)^2 + (b_{10} - u_4)^2 + (u_4 - u_5)^2 \\
& + (u_5 \ - u_6)^2 + (u_6 - b_5)^2 + (u_4 \ - b_9)^2 + (u_5 - u_7)^2 \\
& + (u_6 \ - u_8)^2 + (b_9 - u_7)^2 + (u_7 \ - u_8)^2 + (u_8 - b_6)^2 \\
& + (u_7 - b_8)^2 + (u_8 - b_7)^2,
\end{aligned}
$$

in which the b's denote boundary values.

Then we have the following:

5. *The solution of equations* 85.1 *minimizes* S. This is established by showing that the equations

$$S_{u_i} = 0, \qquad i = 1, 2, \cdots, n$$

reduce to 85.1.

It follows that $\mathbf{H}$ is the matrix of a positive definite quadratic form. From this together with assertion 3 we see that

6. *The latent roots of* $\mathbf{H}$ *lie in the range* $0 < \lambda < 8$.

The set of algebraic equations

$$\mathbf{Hu} = \mathbf{g},$$

where $\mathbf{g}$ is a given vector, occurs in connection with Laplace's and Poisson's equations and also in the analysis of the error. Two important properties of such equations are given below.

7. *If* $\mathbf{p}$ *is any n vector with no negative components, and if*

$$\mathbf{Hu} = \mathbf{p},$$

then $\mathbf{u}$ *has no negative components.*

The proof is conducted just as for assertion 4. If we suppose that the boundary values are all zero, we easily see that the components of **u** can have no minimum in R, and, hence, no component can be negative.

As a corollary of assertion 7 we have

8. *If* **Hu** = **a** *and* **Hv** = **b** *where* $a_i \leqq b_i$, *then* $u_i \leqq v_i$, $(i = 1, 2, \cdots, n)$.

86. Solution of Hu = b

The solution of Laplace's or Poisson's equation for a region R with given boundary values is approximated by the solution of the set of algebraic equations

(86.1)   **Hu = b,**

where **H** is the matrix formed for the region R. The solution of equations 86.1 can be carried out in practice by any of the methods described in Chapter 9, usually relaxation if the work is done by hand, or some systematic iteration process if by an automatic computing machine. However the special properties of **H** make it worth while adding some special suggestions to the general discussion in Chapter 9.

(a) *Computation sheet.* First of all it is unnecessary to write out any of the equations 86.1 or the associated matrix **H**. Instead we prepare a computation sheet ruled so that to each interior and boundary point of the actual figure there corresponds a similarly situated rectangular space on the computation sheet in which we enter the value (true or approximate as the case may be) of u at the associated point. In actual calculation corrections can be made by erasures, but for purposes of exposition we must make up a new sheet.

Example. To illustrate the methods to be described let us choose a simple example for which the analytic solutions are known and can be compared with those found numerically. The region R is chosen as a square with sides $a = 6h$, the boundary values are zero on three sides and equal to $1000 \sin \pi x/a$ on the side where $y = a$ (see Fig. 25). The computation sheet associated with Fig. 25 appears below with the known boundary values inserted.

Computation Sheet

0	500	866	1000	866	500	0
0	1	6	11	16	21	0
0	2	7	12	17	22	0
0	3	8	13	18	23	0
0	4	9	14	19	24	0
0	5	10	15	20	25	0
0	0	0	0	0	0	0

The spaces are numbered to aid subsequent explanations.

(b) *Formula for the correction.* A computer with some practical experience in relaxation associated with the operator H can develop excellent skill in estimating corrections simply by looking at the residuals. Beginners however may find it profitable to follow a more definite procedure for computing corrections c from the residuals r. One such procedure, rather simple and yet fairly effective, is to use the formula

(86.2)
$$c = Lr$$

where L is the operator defined by the stencil

(86.3)
$$L = \frac{1}{16} \begin{array}{|c|c|c|} \hline 1 & 2 & 1 \\ \hline 2 & 6 & 2 \\ \hline 1 & 2 & 1 \\ \hline \end{array} .$$

To see the effect of using L we compute the new residuals r' by

$$r' = H(u - c)$$

whence the changes in the residuals are given by

$$r' - r = -HLr$$

where the operator HL has the stencil

$$HL = \frac{1}{16} \begin{array}{|c|c|c|c|c|} \hline & 1 & 2 & 1 & \\ \hline 1 & 0 & 0 & 0 & 1 \\ \hline 2 & 0 & -16 & 0 & 2 \\ \hline 1 & 0 & 0 & 0 & 1 \\ \hline & 1 & 2 & 1 & \\ \hline \end{array} .$$

The correction at any given point therefore reduces the residual at that point to zero, leaves the adjacent residuals unchanged, and produces small changes of opposite sign in the 12 residuals two steps away. Close to the boundary of course the stencil HL is modified.

The first and sixth iteration with the operator L as applied to the example of Fig. 25 are shown in computation 45.

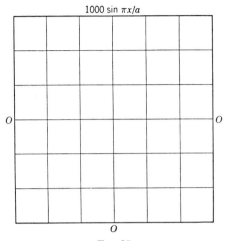

1000 sin $\pi x/a$

Fig. 25

Residuals and Corrections　　　　　　　　　　　　　　Computation 45

Initial estimate v_0 (See Article 87)							Residuals $r_0 = Hv_0$					
0	500	866	1000	866	500	0	—	—	—	—	—	— —
0	298	521	596	521	298	0	—	12	1	24	1	12 —
0	183	325	366	325	183	0	—	−1	−40	−1	−40	−1 —
0	108	190	217	190	108	0	—	2	−2	0	−2	2 —
0	61	108	122	108	61	0	—	−1	−12	−1	−12	−1 —
0	27	47	54	47	27	0	—	0	1	0	1	0 —
0	0	0	0	0	0	0	—	—	—	—	—	— —

Correction $c_0 = Lr_0$								$r_1 = r_0 + Hc_0$				
0	0	0	0	0	0	0	—	—	—	—	—	— —
0	+1.6	− 0.2	+3.7	− 0.2	+1.6	0	—	+2.1	−5.9	+1.6	−5.9	+2.1 —
0	−3.3	−13.0	−7.2	−13.0	−3.3	0	—	−1.9	−5.9	−1.7	−5.9	−1.9 —
0	−2.7	− 7.2	−7.2	− 7.2	−2.7	0	—	+ .8	− .9	+4.0	− .9	+ .8 —
0	−1.5	− 4.8	−3.2	− 4.8	−1.5	0	—	−3.1	−5.8	−6.2	−5.8	−3.1 —
0	−0.6	− 1.1	−1.2	− 1.1	−0.6	0	—	− .2	−1.2	− .6	−1.2	− .2 —
0	0	0	0	0	0	0	—	—	—	—	—	— —
.	. . .	. . .	. . .	. . .	. . .	.	.	. . .	. . .	. . .	. . .	. . . .

$c_5 = Lr_5$								$r_6 = r_5 + Hc_5$				
0	0	0	0	0	0	0	—	—	—	—	—	— —
0	−0.1	0	−0.1	0	−0.1	0	—	+0.1	−0.2	−0.2	−0.2	+0.1 —
0	− .1	−0.1	− .2	−0.1	− .1	0	—	0	0	+ .2	0	0 —
0	− .1	− .1	− .1	− .1	− .1	0	—	+0.1	−0.2	− .3	−0.2	+0.1 —
0	−0.1	− .1	− .2	− .1	−0.1	0	—	− .2	+ .1	0	+ .1	− .2 —
0	0	−0.1	−0.1	−0.1	0	0	—	− .1	− .2	+ .1	− .2	− .1 —
0	0	0	0	0	0	0	—	—	—	—	—	— —

The final values of u, obtained by applying all the successive corrections to the chosen initial values, are

0	500.0	866.0	1000.0	866.0	500.0	0
0	298.5	517.1	597.1	517.1	298.5	0
0	177.0	306.6	354.0	306.6	177.0	0
0	102.9	178.3	205.9	178.3	102.9	0
0	56.4	97.6	112.7	97.6	56.4	0
0	24.9	43.1	49.7	43.1	24.9	0
0	0	0	0	0	0	0

A comparison of these with the values calculated from the analytical solution of equations 86.1 for Fig. 25, namely,

$$u_{m,n} = 92.84 \sin \frac{m\pi}{6} \sinh (0.51202n),$$

reveals a maximum discrepancy of two units in the last digit.

(c) *Group changes.* The error at any stage of the approximation is a vector $\mathbf{e}$ which can be expressed as a linear combination of the latent vectors of the matrix $\mathbf{H}$,

(86.4) $$\mathbf{e} = \alpha_1\phi_1 + \alpha_2\phi_2 + \cdots + \alpha_n\phi_n.$$

By appeal to physical intuition we can form some idea of the character of the vector ϕ_1, and to a decreasing degree that of ϕ_2, ϕ_3, etc. For, roughly speaking, these represent the natural modes of transverse vibration of a thin elastic membrane shaped like the region R and fixed on the boundary. Now the fundamental mode has no nodal lines; hence, ϕ_1 has components all of the same sign. In the fundamental mode the membrane bulges out in a manner rather easily visualized for any simple region R. Hence, without any detailed computation we can make a rough approximation to the latent vector ϕ_1 and can remove a considerable portion of the first term in the expression for the error in equation 86.4. For we may use this estimate of ϕ_1 as a value for w in Art. 70, determine a suitable s by the methods shown there, and thus improve the approximation.

The same scheme may be applied to subregions of R wherever the residuals are predominantly of one sign. The correction is then extended over the subregion in the same manner as was done for R above.

The discussion and example of this article apply to Laplace's equation. Obviously the same matrix $\mathbf{H}$ appears also in connection with Poisson's equation

$$\nabla^2 U = f(x, y),$$

and the equation

$$\nabla^2 U + \lambda U = 0,$$

which arises in many boundary value problems.

Leibman [107] solved problems of this type using H and successive approximations of the Gauss-Seidel type. See also Shortley and Weller [191], Southwell [194], Fox [66], Shortley [189], Shortley, Weller, and Fried [192], Richardson [158], Mikeladze [117, 118], and many others.

87. The first approximation

Any set of values, no matter how unlike the true values, can be used as the initial approximation, but obviously a good initial guess saves subsequent labor. A skilled and experienced computer can look at the region and the given boundary values, sketch freehand a few equipotential curves with their orthogonal trajectories, adjust these curves to satisfy conditions of conformality, and from the resulting system of curves read off a good set of values for the initial approximation. In the hands of an expert this is probably the best way to start.

For the example of Article 86 the starting values used in computation 45 were actually found as follows. First we apply equation 84.11 with mesh size $2h$ to the four spaces numbered 7, 9, 17, 19 (see the computation sheet, Article 86), getting the equations

$$4\,u_7 - u_9 - u_{17} = 866,$$

$$4\,u_9 - u_7 - u_{19} = 0.$$

By symmetry we have $u_{17} = u_7$, $u_{19} = u_9$, and, hence, we can find all four values. When these are entered on the computation sheet, we use equation 84.12 to compute the entries for the spaces numbered 1, 3, 5, 11, 13, 15, 21, and 23. Finally the values at all the even-numbered spaces are computed by means of 84.11.

For a method of more general applicability see Frocht and Leven [74].

88. The matrix K

Instead of the operator H and the associated matrix $\mathbf{H}$ one may use the operator K and a corresponding matrix $\mathbf{K}$ in all problems where Laplace's differential operator is replaced by a finite difference operator. For the same mesh length the use of K, in general, leads to considerably more accurate results than the use of H, but, on the other hand, H is decidedly simpler in calculation than K. We propose to take advantage of both of these facts by using H in the early stages of the approximation where simplicity of calculation is our predominant concern, and by using K in the final approximations where a close fit to the true solution is the prime objective.

Using K we get, just as with H, a system of equations

(88.1) $$\mathbf{Ku} = \mathbf{b}$$

where $\mathbf{K}$ is an $n \times n$ matrix, n being the number of interior points.

For the region shown in Fig. 23 the matrix $\mathbf{K}$ is

$$\mathbf{K} = \begin{bmatrix} 20 & -4 & 0 & -4 & -1 & 0 & 0 & 0 \\ -4 & 20 & -4 & -1 & -4 & -1 & 0 & 0 \\ 0 & -4 & 20 & 0 & -1 & -4 & 0 & 0 \\ -4 & -1 & 0 & 20 & -4 & 0 & -1 & 0 \\ -1 & -4 & -1 & -4 & 20 & -4 & -4 & -1 \\ 0 & -1 & -4 & 0 & -4 & 20 & -1 & -4 \\ 0 & 0 & 0 & -1 & -4 & -1 & 20 & -4 \\ 0 & 0 & 0 & 0 & -1 & -4 & -4 & 20 \end{bmatrix}$$

The following properties are important.
1. *The matrix $\mathbf{K}$ depends only on the set of mesh points of R.*
2. *The matrix $\mathbf{K}$ is symmetric.*
3. *The latent roots of $\mathbf{K}$ lie in the range*

$$0 < \lambda < 32.$$

This is not quite obvious, being a sharper result than can be obtained from Theorem 9 of Chapter 9. It may be established by the following considerations. Imbed the region R in a rectangle R'. For R' set up the matrix $\mathbf{K'}$. Then the matrix $\mathbf{K}$ is found from $\mathbf{K'}$ by dropping from $\mathbf{K'}$ the rows and columns belonging to points in R' but not in R. Then by Theorem 3, Article 68, the latent roots for each successive matrix lie between roots of the preceding matrix. Thus, finally, the range of roots of $\mathbf{K}$ is inside the range for $\mathbf{K'}$. But the latent roots of $\mathbf{K'}$ are given explicitly by the formula

(88.2) $$\lambda = 4[9 - (2 + \cos \alpha)(2 + \cos \beta)]$$

with

$$\alpha = m_1\pi h/a, \qquad \beta = m_2\pi h/b$$

where a and b are the lengths of the sides of the rectangle R', and m_1 and m_2 are integers such that $0 < \alpha < \pi$ and $0 < \beta < \pi$. The associated latent vectors are

$$\phi = C \sin \frac{m_1\pi x}{a} \sin \frac{m_2\pi y}{b}.$$

The values of λ defined by equation 88.2 lie in the interval

$$0 < \lambda < 32,$$

and the same is consequently true for the latent roots of $\mathbf{K}$.

4. *The latent roots of $\mathbf{K}$ are not in general symmetrically placed.*

5. *The matrix $\mathbf{K}$ is non-singular.* This is proved just as for $\mathbf{H}$ (it is also implied by 3).

We can construct a quadratic form Σ similar to S containing terms

$$4(u_i - u_j)^2$$

for horizontal and vertical differences together with terms

$$(u_i - u_j)^2$$

for diagonal differences. Then

6. *The solution of equations 88.1 minimizes Σ.*

Just as in the case of $\mathbf{H}$ we can show that

7. *If $\mathbf{p}$ is any n vector with no negative components and if*

$$\mathbf{Ku} = \mathbf{p},$$

then $\mathbf{u}$ has no negative components, and the corollary

8. If $\mathbf{Ku} = \mathbf{a}$ and $\mathbf{Kv} = \mathbf{b}$, where $a_i \leqq b_i$, then $u_i \leqq v_i$ ($i = 1, 2, \cdots, n$).

The actual numerical solution of a set of equations

$$(88.1) \qquad\qquad \mathbf{Ku} = \mathbf{b}$$

is conducted in the same manner as in the case of equations

$$\mathbf{Hu} = \mathbf{a}.$$

The methods of Chapter 9 apply just as well to equation 88.1, and the suggestions of Article 86 also require little change. In place of the operator L defined in equation 86.3 we now have

$$(88.3) \qquad L' = \frac{1}{2156} \begin{vmatrix} 25 & 44 & 25 \\ 44 & 148 & 44 \\ 25 & 44 & 25 \end{vmatrix}$$

corresponding to which the changes in the residuals are given by

$$KL' = $$

0.0116	0.0668	0.1048	0.0668	0.0116
0.0668	0	0	0	0.0668
0.1048	0	−1	0	0.1048
0.0668	0	0	0	0.0668
0.0116	0.0668	0.1048	0.0668	0.0116

However, instead of equation 88.3 we suggest the somewhat less effective but much simpler operator

$$(88.4) \qquad L = \frac{1}{84} \begin{array}{|c|c|c|} \hline 1 & 2 & 1 \\ \hline 2 & 6 & 2 \\ \hline 1 & 2 & 1 \\ \hline \end{array}$$

for which

$$KL = \frac{1}{84}$$

1	6	10	6	1
6	2	−4	2	6
10	−4	−84	−4	10
6	2	−4	2	6
1	6	10	6	1

Let us return to the example of Article 86, take as the initial estimate the solution of **Hu = b** found in computation 45, and continue the approximation with the operator K. One step in the solution using L and the final solution secured after five steps are shown in computation 46.

Solution of $Ku = 0$ Computation 46

$r_0 = KV$ (From computation 45) $c_0 = Lr_0$

—	—	—	—	—	— —	0	0	0	0	0	0	0
—	21.0	38.2	44.0	38.2	21.0 —	0	−3.0	−5.2	−6.1	−5.2	−3.0	0
—	12.6	22.0	24.4	22.0	12.6 —	0	−2.7	−4.7	−5.5	−4.7	−2.7	0
—	7.0	13.9	16.4	13.9	7.0 —	0	−1.6	−2.8	−3.3	−2.8	−1.6	0
—	5.0	6.6	8.0	6.6	5.0 —	0	−1.0	−1.6	−1.8	−1.6	−1.0	0
—	2.4	4.1	3.2	4.1	2.4 —	0	−0.5	−0.7	−0.8	−0.7	−0.5	0
—	—	—	—	—	— —	0	0	0	0	0	0	0

$r_1 = r_0 + Kc_0$ $c_1 = Lr_1$

—	—	—	—	—	— —	0	0	0	0	0	0	0
—	−2.7	− 2.4	− 5.0	− 2.4	−2.7 —	0	+0.1	+0.1	+0.2	+0.1	+0.1	0
—	+3.8	+ 6.8	+ 5.6	+ 6.8	+3.8 —	0	− .7	−1.1	−1.2	−1.1	− .7	0
—	+7.3	+13.7	+14.6	+13.7	+7.3 —	0	−1.2	−2.1	−2.3	−2.1	−1.2	0
—	+3.3	+ 6.0	+ 8.2	+ 6.0	+3.3 —	0	− .8	−1.4	−1.7	−1.4	− .8	0
—	+ .8	+ 4.5	+ 3.2	+ 4.5	+ .8 —	0	−0.3	−0.7	−0.8	−0.7	−0.3	0
—	—	—	—	—	— —	0	0	0	0	0	0	0

0	500	866	1000	866	500	0
0	295.2	511.2	590.3	511.2	295.2	0
0	173.1	299.9	346.3	299.9	173.1	0
0	99.7	172.7	199.4	172.7	99.7	0
0	54.1	93.7	108.3	93.7	54.1	0
0	23.7	41.1	47.5	41.1	23.7	0
0	0	0	0	0	0	0

A comparison of these values with those given by the analytical solution of Laplace's equation,

$$u = 1000 \sin \frac{\pi x}{a} \sinh \frac{\pi y}{a} \bigg/ \sinh \pi,$$

shows agreement to within one unit in the last digit retained.

89. Further refinement

If, having carried out the computations of Articles 86 and 88 and having obtained the solution of equations 88.1, we are still not satisfied with the accuracy obtained we may replace the lattice by a new one having a finer mesh, say, with the new h half the old h, and solve the new set of equations **Ku** = **b** for the new lattice. The solution already obtained should provide excellent initial values at certain points, and the remaining values can be found by simple interpolation formulas.

On the other hand, one may seek further improvement, not by refining the mesh, but by using more terms in formula 62.11. To illustrate the procedure, suppose that the problem is to solve Poisson's equation

$$\nabla^2 U = f(x, y)$$

with $U = 0$ on the boundary of a region R. From 62.11 we have

(89.1) $$KU - \frac{K^2 U}{72} + \cdots = 6h^2 f(x, y)$$

First we solve the equation

(89.2) $$KU_0 = 6h^2 f(x, y)$$

as explained in Article 88. Then we write equation 89.1 in the form

(89.3) $$KU_{n+1} = 6h^2 f(x, y) + K^2 U_n/72,$$

let $n = 0, 1, 2, \cdots$ in succession, and, if the process converges, arrive at a value of U that approximately satisfies 89.1. To be at all practical this process must not merely converge but must converge rapidly, since otherwise the numerical work is prohibitive. If further accuracy is desired more terms of 62.11 may be used on the right in equation 89.3. However, the point is soon reached where it is better to use a smaller mesh than to add terms to 89.3. Just which is better in any specific case, more terms or a finer mesh, is a matter of good judgment based on experience, on the analysis of error as explained in the next article, and on the computational facilities available.

For other applications of this idea the reader is referred to Fox [65].

90. Analysis of errors

The errors to be examined are of two kinds: (a) the error committed by using an approximate solution obtained by relaxation or iteration in place of the exact solution of $\mathbf{Hu} = \mathbf{a}$ or $\mathbf{Ku} = \mathbf{b}$ and (b) the error committed by using the solutions of the difference equations $\mathbf{Hu} = \mathbf{a}$ or $\mathbf{Ku} = \mathbf{b}$ in place of the true solution of Laplace's or Poisson's equations.

To estimate the magnitude of errors of type a we recall that any method of approximation is designed to reduce the residuals to zero, but that in actual practice we never succeed in making the residuals rigorously vanish. We wish, therefore, to find a bound for the error committed by using an approximation for which the residuals are all less in absolute value than some constant m. To be precise let $\mathbf{u}$ be the true solution, v the approximate solution, and e the error, so that

$$Hu = a,$$
$$Hv = a + r,$$

and

$$H(v - u) = He = r,$$

where $|r_i| < m$, $(i = 1, 2, \cdots, n)$. Enclose the region R in a circle C with center (α, β) and radius ρ, and construct the vector ε with components equal to the values of

$$(90.1) \qquad E = \frac{m}{4h^2} [\rho^2 - (x - \alpha)^2 - (y - \beta)^2]$$

at the interior nodal points in R. Applying the operator H to E, we find that

$$H\varepsilon = m.$$

It follows from proposition 8 of Article 85 that

$$e_i < \epsilon_i.$$

Reversing the reasoning we have also

$$e_i > - \epsilon_i,$$

so that

$$|e_i| < \epsilon_i.$$

From equation 90.1 it is seen that $\epsilon_i \leqq m\rho^2/4h^2$. Therefore, finally we get

Theorem 1. If the maximum residual does not exceed a positive quantity m, the maximum error in the approximate solution of

$$Hu = a$$

does not exceed $m\rho^2/4h^2$.

By an almost identical proof it is shown that

Theorem 2. If the maximum residual does not exceed a positive number μ, the maximum error in the approximate solution of

$$Kv = b$$

does not exceed $\mu\rho^2/24h^2$, where ρ is the radius defined above.

Turning now to errors of type b, we first of all define an auxiliary function $F(t)$ by the formula

$$(90.2) \quad F(t) = U(x + th, y) + U(x, y + th) + U(x - th, y) + U(x, y - th) - h^2 \nabla^2 U.$$

Assuming the existence of fourth partial derivatives of U with respect to x and y for all values in the ranges $x \pm h$ and $y \pm h$, we may apply Taylor's series with remainder and get

$$(90.3) \qquad F(1) = F(0) + F'(0) + \frac{F''(0)}{2!} + \frac{F'''(0)}{3!} + \frac{F^{(4)}(s)}{4!}$$

where $0 < s < 1$. When equation 90.3 is evaluated with the aid of 90.2, it gives

$$HU - h^2 \nabla^2 U = \frac{1}{4!} F^{(4)}(s)$$

or

$$(90.4) \qquad \left| HU - h^2 \nabla^2 U \right| \leqq M_4 h^4 / 6,$$

where M_4 is the maximum of the absolute values of the fourth derivatives of U in the circle of radius h with center at (x, y).

Suppose that the problem is to obtain the solution of

$$(90.5) \qquad \nabla^2 U = f(x, y)$$

in a region R with assigned values of U on the boundary, and that as an approximation we are using v, where

$$(90.6) \qquad Hv = h^2 f(x, y)$$

in R, and $v = U$ on the boundary. From equations 90.4, 90.5, and 90.6 we get

$$\left| HU - Hv \right| \leqq M_4 h^4 / 6.$$

Thus it appears that we can use $M_4 h^4 / 6$ in place of m in Theorem 1, and arrive at

Theorem 3. *If the maximum of the absolute values of the fourth order partial derivatives of U in R (including all areas covered by circles of radius h with centers at interior nodal points) is M_4, then the maximum error resulting from using the difference operator H in place of the differential operator $h^2 \nabla^2$ does not exceed*

$$M_4 h^2 \rho^2 / 24.$$

This result applies to Laplace's and Poisson's equations when the boundary values are assigned.

In a similar manner we derive

Theorem 4. *For the case of Laplace's equation with assigned boundary values the maximum error committed by using K in place of $6h^2 \nabla^2$ does not exceed*

$$\rho^2 h^6 M_8 / 12{,}096.$$

Here M_8 denotes the maximum of the absolute values of the eighth-order partial derivatives of U in R (including areas covered by squares of side $2h$ with centers at interior nodal points).

This result does not necessarily apply to Poisson's equation

$$\nabla^2 U = f(x, y),$$

since in this case the error depends on the character of $f(x, y)$. However, in the important special case where $f(x, y)$ is a constant, the reader may show without difficulty that the result of Theorem 4 holds also for Poisson's equation.

In the numerical calculation there is no difficulty in applying Theorems 1 and 2, since h is known, m can be taken as the largest residual (in absolute value), and ρ is the radius of the smallest circle inclosing R. But Theorems 3 and 4 cannot be applied directly since we have no direct way of computing M_4 or M_8, quantities depending on derivatives of the unknown function U. In the case where U is harmonic, a rough estimate, not rigorous but practically helpful, can be made as follows. Since U is harmonic

$$h^4 \frac{\partial^4 U}{\partial x^4} = h^4 \frac{\partial^4 U}{\partial y^4} = -h^4 \frac{\partial^4 U}{\partial x^2\,\partial y^2},$$

and the last expression, as we saw in Article 62, corresponds to the finite difference expression $N^2 U$.

Thus we obtain a general idea of the magnitude of $h^4 M_4$ by calculating at each point of R the value of

$$h^4 M_4 = (2H - 2X)U = -N^2 U = \begin{vmatrix} -1 & 2 & -1 \\ 2 & -4 & 2 \\ -1 & 2 & -1 \end{vmatrix} U.$$

Hence, for practical computation we replace the rigorous bound in Theorem 3 by the non-rigorous but easily found estimate

$$\frac{\rho^2 \max\left|N^2 U\right|}{24h^2}.$$

In similar fashion the rigorous bound in Theorem 4 may be replaced by the estimate

$$\frac{\rho^2 \max\left|N^4 U\right|}{21096h^2}.$$

Note that these numerical estimates apply only at interior points. Boundary values can be estimated from these only by the dangerous expedient of extrapolation.

In many, if not most, practical problems the hypotheses of Theorem 3 or Theorem 4 are not fulfilled, owing to discontinuities or other singularities in the boundary values. In such cases it generally happens that the *maximum* error of the numerical solution obtained by the matrix **H** or the matrix **K** does not approach zero as the mesh length h approaches zero. Nevertheless the approximate solution obtained numerically will, in most cases encountered in practice, converge to the true solution at points inside the region. Without attempting a rigorous discussion we shall explain in general terms what may happen.

Take the case of Laplace's equation for example, where singularities occur only on the boundary B. Suppose for simplicity that there is a discontinuity in the given boundary values at a point P on B, but otherwise the hypotheses of Theorem 3 are fulfilled on $R + B$. No matter how fine the mesh is made, the error at points of R adjacent to P will remain finite. Suppose that ϵ is a given positive constant and that $A(\epsilon, h)$ is the area in R for which the error is greater in absolute value than ϵ for a given h. When h is sufficiently small, $A(\epsilon, h)$ will consist of a portion of R close to the point P, and, as h is decreased, the area $A(\epsilon, h)$ will shrink up toward the point P. Any fixed interior point of R will eventually lie outside of $A(\epsilon, h)$, and at such a point the error will then be less than ϵ.

A table providing an estimate for the magnitude of the area $A(\epsilon, h)$ in certain simple cases is given in Art. 92.

Sometimes singularities on the boundary can be removed beforehand by analytical means. A simple case where this is possible is treated in Art. 91.

EXERCISES

1. In the 8×12 rectangle shown below the values of W are given by the equations

$W = 100x(12 - x)$ on AB and CD

$W = 100y(8 - y)$ on AC and BD.

Determine W to satisfy Laplace's equation within the rectangle. Take $h = 1$.

2. An approximate answer to the problem in exercise 1 is given in computation 47. Apply internal tests to determine how good an approximation has been secured.

Solution of Dirichlet Problem   Computation 47

x	0	1	2	3	4	5	6	7	8	9	10	11	12
y													
0	0	1100	2000	2700	3200	3500	3600	3500	3200	2700	2000	1100	0
1	700	1301	1909	2428	2819	3060	3142	3060	2819	2428	1909	1301	700
2	1200	1519	1912	2287	2588	2780	2846	2780	2588	2287	1912	1519	1200
3	1500	1669	1937	2223	2466	2626	2681	2626	2466	2223	1937	1669	1500
4	1600	1721	1949	2205	2428	2577	2629	2577	2428	2205	1949	1721	1600
5	1500	1669	1937	2223	2466	2626	2681	2626	2466	2223	1937	1669	1500
6	1200	1519	1912	2287	2588	2780	2846	2780	2588	2287	1912	1519	1200
7	700	1301	1909	2428	2819	3060	3142	3060	2819	2428	1909	1301	700
8	0	1100	2000	2700	3200	3500	3600	3500	3200	2700	2000	1100	0

91. Removal of singularities on the boundary.

For simplicity we consider only the case where a portion of the boundary of the region R is a straight line and where a discontinuity in boundary values or their derivatives occurs at a point P on this line segment. By a change of axes we can move P to the origin and make the straight-line segment of the boundary coincide with the x axis, the region immediately above the axis being interior to R.

We suppose that the boundary values $U(B)$ on the axis are given by

$$U(B) = \begin{cases} f_1(x) & \text{if} \quad x > 0, \\ f_2(x) & \text{if} \quad x < 0, \end{cases}$$

and that in their respective regions $f_1(x)$ and $f_2(x)$ have fourth derivatives less in absolute value than some positive constant M.

Let

$$f_1(0) - f_2(0) = a_0,$$
$$f_1'(0) - f_2'(0) = a_1,$$
$$f_1''(0) - f_2''(0) = a_2,$$
$$f_1'''(0) - f_2'''(0) = a_3,$$
$$f_1^{(4)}(0) - f_2^{(4)}(0) = a_4.$$

Also for brevity in writing let us introduce the symbols

$$p_0 = 1,$$

$$p_1 = x, \qquad\qquad\qquad q_1 = y,$$

$$p_2 = (x^2 - y^2)/2!, \qquad\qquad q_2 = xy,$$

$$p_3 = (x^3 - 3xy^2)/3!, \qquad\quad q_3 = (3x^2y - y^3)/3!,$$

$$p_4 = (x^4 - 6x^2y^2 + y^4)/4!, \qquad q_4 = (4x^3y - 4xy^3)/4!,$$

$$T = \frac{1}{\pi}\tan^{-1}\frac{y}{x}, \qquad\qquad L = \tfrac{1}{2}\log(x^2 + y^2).$$

Note that the pair of functions on the same line is a conjugate pair of harmonic functions. With these we construct the function

$$W(x, y) = (a_0p_0 + a_1p_1 + a_2p_2 + a_3p_3 + a_4p_4)T$$
$$- (a_1q_1 + a_2q_2 + a_3q_3 + a_4q_4)L.$$

It is a simple matter to verify that $W(x, y)$ is harmonic except at the origin. If now we let y approach zero in the upper half plane we find that, if $x > 0$,

$$W(x, 0) \equiv 0$$

whereas, if $x < 0$,

$$W(x, 0) = a_0 + a_1x + a_2x^2/2! + a_3x^3/3! + a_4x^4/4!.$$

Finally, consider the function

$$V = U + W.$$

Clearly V is harmonic in R, takes on the values $U(B) + W(B)$ on B and at the origin is continuous together with the first four derivatives.

92. Area of infection

For various reasons one may decide not to bother with the analytical procedure of Article 91, but simply to use a finer mesh in the vicinity of the discontinuity and thus confine the error to a small neighborhood. The serious objection to meshes of different sizes for different parts of the same region that we encountered in Article 65 is not so important here where we have no time interval to worry about.

If we decide to refine the mesh in the vicinity of a discontinuity we need some criterion by which to judge when our results are satisfactory.

The following table has been computed to give a means of estimating how a discontinuity in the boundary values contaminates the adjacent values.

$$10^5 \left(V - \frac{2}{\pi} \tan^{-1} \frac{y}{x} \right)$$

0	0	0	0	0	0	0	0	0	0	0	0	0	0	0	0	0	0	0	0	0
0	0	718	456	253	145	88	58	38	26	19	13	10	8	6	5	3	2	2	1	1
0	−718	0	238	229	173	124	88	63	45	33	25	18	14	11	9	6	5	3	2	1
0	−456	−238	0	100	119	107	88	69	53	40	31	24	19	14	11	8	7	4	2	1
0	−253	−229	−100	0	51	68	68	60	50	40	33	26	21	17	14	10	8	5	3	1
0	−145	−173	−119	−51	0	29	42	44	41	35	30	25	21	17	14	11	8	6	4	2
0	−88	−124	−107	−68	−29	0	17	26	29	28	26	22	19	17	14	11	9	7	4	2
0	−58	−88	−88	−68	−42	−17	0	11	17	19	20	19	17	15	12	10	8	6	4	2
0	−38	−63	−69	−60	−44	−26	−11	0	8	12	14	14	14	12	11	9	8	6	3	1
0	−26	−45	−53	−50	−41	−29	−17	−8	0	6	8	9	10	10	10	8	7	5	3	1
0	−19	−33	−40	−40	−35	−28	−19	−12	−6	0	4	7	7	8	8	6	5	5	3	1
0	−13	−25	−31	−33	−30	−26	−20	−14	−8	−4	0	3	5	6	6	5	5	4	2	1
0	−10	−18	−24	−26	−25	−22	−19	−14	−9	−7	−3	0	2	3	5	4	4	4	2	1
0	−8	−14	−19	−21	−21	−19	−17	−14	−10	−7	−5	−2	0	2	3	3	3	3	2	1
0	−6	−11	−14	−17	−17	−17	−15	−12	−10	−8	−6	−3	−2	0	1	2	3	3	2	1
0	−5	−9	−11	−14	−14	−14	−12	−11	−10	−8	−6	−5	−3	−1	0	1	2	2	1	0
0	−3	−6	−8	−10	−11	−11	−10	−9	−8	−6	−5	−4	−3	−2	−1	0	1	1	0	0
0	−2	−5	−7	−8	−8	−9	−8	−8	−7	−5	−5	−4	−3	−3	−2	−1	0	0	0	0
0	−2	−3	−4	−5	−6	−7	−6	−6	−5	−5	−4	−4	−3	−3	−2	−1	0	0	0	0
0	−1	−2	−2	−3	−4	−4	−4	−3	−3	−3	−2	−2	−2	−2	−1	0	0	0	0	0
0	−1	−1	−1	−1	−2	−2	−2	−1	−1	−1	−1	−1	−1	−1	0	0	0	0	0	0

The function V satisfies the difference equation 85.1 in the region bounded by the lines OX and OY, on OX assumes the boundary value 0, on OY the boundary value 1. On the other hand, $\frac{2}{\pi} \tan^{-1} \frac{y}{x}$ satisfies Laplace's equation and the same boundary values. The difference $V - \frac{2}{\pi} \tan^{-1} \frac{y}{x}$, therefore, represents the error. To use the table let us suppose that the magnitude of the discontinuity is D, and the largest permissible error is e. First calculate $e/D = E$. Then draw a curve through the table, separating tabular values greater than E from those less than E. This curve shows how far the error penetrates into the region, and from it we can estimate how much finer to take the mesh in this neighborhood.

Note that the same table can be extended to the left by reflection in the y axis and thus serves also in the case where there is a jump of two units as x passes through the origin.

Example 1. In a certain set of boundary values there is a discontinuity of magnitude 12 at a point B on a straight boundary. We have an over-all mesh length h and wish to introduce a finer mesh with length h' so that the discontinuity at B will not produce an error exceeding 0.05 at distances from B greater than h. How should we choose h'?

Here $D = 6$ and $E = 0.00833$. Examination of the table reveals no entry as large as E. Hence a finer mesh is unnecessary.

Example 2. Suppose that the magnitude of the discontinuity is 100 and the admissable error at distances exceeding h is 0.005. How large should we take h'?

Here $E = 0.005/50 = 0.0001$. The row and column for which the error is less than 0.0001 is numbered 17. Hence we take

$$h' \leq h/17.$$

For computational convenience we would probably take $h' = h/20$.

93. Poisson's Equation

Many practical problems require the solution of the equation

(93.1)
$$\nabla^2 U = f(x, y)$$

in the interior of a region R while U must take on assigned values on the boundary of R. As usual we overlay R with a square lattice having mesh length h and replace equation 93.1 by an appropriate difference equation. The simplest such equation is

(93.2)
$$HU = h^2 f(x, y)$$

which follows readily from equation 62.4. The local error for this difference equation is seen to be $O(h^4)$.

It is possible to secure a somewhat more accurate difference equation by formula 62.11, which gives us

(**93.3**)
$$KU - K^2U/72 = 6h^2 f(x, y) + O(h^6).$$

For computational purposes this equation can be considerably improved if the function $f(x, y)$ is such that $\nabla^2 f(x, y)$ exists everywhere in R. For in that case we have approximately

$$K^2U = 36h^4\nabla^4 U = 36h^4\nabla^2 f(x, y)$$

and we can replace equation 93.3 by

(**93.4**)
$$KU = 6h^2 f(x, y) + \frac{h^4}{2}\nabla^2 f(x, y)$$

with an error having as its leading term the expression

$$[(K^3/3240) - (KN^2/180)]U.$$

In problems where $f(x, y)$ is harmonic, equation 93.4 simplifies to

(93.5)
$$KU = 6h^2 f(x, y).$$

Example. Let us solve the equation

$$\nabla^2 U = W(x, y)$$

where $W(x, y)$ is the function given numerically in computation 47, and where the boundary values of U are given by

$$U = \frac{100}{12}(24x^3 - x^4 - 1728x)$$

on AB and CD while

$$U = \frac{100}{12} (16y^3 - y^4 - 512y)$$

on AC and BD (see also the example of Article 94).

The work is carried out as usual by relaxation with the operator H in the earlier steps and the operator K later on where greater accuracy is demanded. Lack of space forbids showing the details, but the result, assumed sufficiently accurate, appears below in computation 48.

The values of U are shown only for the upper left hand quarter of the region R, since all remaining entries can be supplied by symmetry. In computation 48 appear also the values of $KU/6$; for a correct solution these should be exactly equal to the values of W in computation 47.

Computation 48

Poisson's Equation $\nabla^2 U = W$
(W is given in computation 47)

0	-14208.00	-27333.33	-38475.00	-46933.33	-52208.00	-54000.00
-4142.00	-16898.00	-28767.60	-38923.05	-46681.89	-51541.56	-53196.00
-7600.00	-19183.05	-30017.03	-39348.20	-46518.75	-51028.03	-52566.30
-9875.00	-20707.91	-30870.70	-39661.24	-46442.48	-50719.28	-52180.36
-10666.67	-21241.95	-31173.78	-39776.67	-46422.14	-50617.72	-52051.88

Values of $KU/6$

	1301.1	1908.7	2428.0	2818.6	3059.5	3141.7
	1519.0	1912.0	2286.8	2587.9	2779.4	2845.8
	1669.7	1937.4	2223.5	2465.6	2625.4	2681.5
	1720.4	1949.5	2205.6	2427.4	2576.4	2629.3

94. The biharmonic equation

If

$$(94.1) \qquad \nabla^2\nabla^2 U = 0$$

we readily find from the expansion of K^2U in powers of h that

$$(94.2) \qquad K^2U = \frac{h^{10}}{42} \left[\frac{\partial^{10}U}{\partial x^8\,\partial y^2} + \frac{\partial^{10}U}{\partial x^2\,\partial y^8} \right] + \cdots.$$

Accordingly, it is natural to replace the differential equation 94.1 by the fourth-order difference equation

$$(94.3) \qquad K^2U = 0.$$

If equation 94.3 is to be satisfied at each interior point of a region R, and if in addition the values of U are known on the boundary and on the layer of points adjacent to the boundary, we have exactly enough information to determine a unique solution of 94.3.

When the boundary information is of the type just described the problem can be solved by the stencil

$$K^2 = \begin{array}{|c|c|c|c|c|}
\hline
1 & 8 & 18 & 8 & 1 \\
\hline
8 & -\,8 & -144 & -\,8 & 8 \\
\hline
18 & -144 & 468 & -144 & 18 \\
\hline
8 & -\,8 & -144 & -\,8 & 8 \\
\hline
1 & 8 & 18 & 8 & 1 \\
\hline
\end{array}$$

together with the method of relaxation.

In certain problems of elasticity where the region R is bounded by horizontal and vertical straight lines we may encounter the case where not two sets of values of U, but the values of U_{xx} and U_{yy} are known on the boundary. In this case one may arrive at the solution in two steps, as follows:

First let

$$\nabla^2 U = W.$$

Then, since U_{xx} and U_{yy} are known on the boundary the value of W is also known there and since

$$\nabla^2 \nabla^2 U = \nabla^2 W = 0$$

at interior points, we have a straightforward Dirichlet problem for finding W, and can obtain W by relaxation in the usual manner.

Second, having found W, we must solve the Poisson equation

$$\nabla^2 U = W$$

with boundary values of U obtained by integrating U_{xx} twice along horizontal boundaries and U_{yy} twice along vertical boundaries. The constants of integration can be chosen so as to make U continuous at corners.

Example. In an 8×12 rectangle the values of U_{xx} and U_{yy} are given as follows:

On AB and CD, $U_{xx} = 100x(12 - x)$, $U_{yy} = 0$.
On AC and BD, $U_{xx} = 0$, $U_{yy} = 100y(8 - y)$.
To find U satisfying

$$\nabla^2 \nabla^2 U = 0$$

inside the rectangle and fulfilling the given boundary conditions, we set up the Dirichlet problem

$$\nabla^2 W = 0 \quad \text{in } R,$$

$$W = 100x(12 - x) \quad \text{on } AB \text{ and } CD, \qquad W = 100y(8 - y) \quad \text{on } AC \text{ and } BD.$$

This problem was solved in computation 47, Article 90.

To proceed we must first obtain the boundary values for U by double integration. Thus we get

$$U = \frac{100}{12} (24x^3 - x^4 - 1728x)$$

on AB and CD while

$$U = \frac{100}{12} (16y^3 - y^4 - 512y)$$

on AC and BD. Here the constants of integration were chosen to make U vanish at the four corners of the rectangle. Using the boundary values supplied by these equations and the values of W found in computation 47, we solve Poisson's equation

$$\nabla^2 U = W.$$

This is precisely the problem solved in computation 48, Article 93. Hence the results obtained in computation 48 furnish the answer to the example of this section.

For general discussions of the material of this chapter see, e.g., Grinter [82], Southwell [194], O'Brien, Hyman, and Kaplan [149], Young [221]; errors are treated by Collatz [26] and Gersgorin [78]; Gilles [81] uses interlacing nets for problems of elasticity; Frocht [73] deals with composite rectangular areas; and Fox [64, 68] treats more general boundary conditions.

11

CHARACTERISTIC NUMBERS

One of the most important and at the same time one of the most difficult problems associated with numerical solutions of differential equations is the determination of characteristic numbers. Most of the numerical methods available reduce essentially to the problem of finding the latent roots of a matrix, a problem treated in some detail in Chapter 9. In the present chapter are given additional examples for the cases of one and two dimensions. We consider first the case of one independent variable. The basic idea appears in a paper by Lanczos [104]. See also Milne [136].

95. Formulation of the problem

Let

$$L(u) = P_0(x)u'' + P_1(x)u' + P_2(x)u,$$

where primes mean differentiation with respect to x. The problem before us is to find those characteristic values of λ for which the differential system

$$L(u) + \lambda^2 u = 0$$

(95.1)
$$u + gu' = 0 \quad \text{at} \quad x = a$$

$$u + Gu' = 0 \quad \text{at} \quad x = b$$

possesses non-zero solutions in the interval $a \leq x \leq b$.

Problems of this type arise in many different ways in mathematical physics. For example, they may occur in connection with the heat equation

$$V_t = L(V),$$

if we assume particular solutions of the form

$$V = u(x)e^{-\lambda^2 t},$$

228

or in connection with the wave equation

$$V_{tt} = L(V),$$

if we assume particular solutions of the form

$$V = u(x) \cos \lambda t,$$

or they may occur after separation of variables in more general partial differential equations.

The important fact is that the characteristic numbers are completely determined by the system 95.1 regardless of the particular physical problems from which equations 95.1 may have been derived. Consequently, in order to obtain numerical solution we may assume a fictitious partial differential system which leads to 95.1. The most satisfactory fictitious system for our purpose appears to be

$$(95.2) \qquad V_{tt} = L(V),$$

with boundary conditions

$$(95.3) \qquad \begin{aligned} V + gV_x &= 0 \quad \text{at} \quad x = a, \\ V + GV_x &= 0 \quad \text{at} \quad x = b, \end{aligned}$$

and initial conditions

$$(95.4) \qquad \begin{aligned} V &= F(x) \quad \text{when} \quad t = 0, \\ V_t &= 0 \quad \text{when} \quad t = 0. \end{aligned}$$

Following standard procedure, we assume that the solution of equations 95.2, 95.3, and 95.4 is

$$(95.5) \qquad V = \sum_k c_k u_k(x) \cos \lambda_k t,$$

where the λ_k's are the desired characteristic numbers, the $u_k(x)$'s are the characteristic functions (normalized in some definite manner) and the c_k's are constants depending on $F(x)$.

It will be assumed with respect to the coefficient functions in $L(u)$ that for all values of x in the interval $a \leqq x \leqq b$

(a) $P_0(x) > 0$,

(b) $\dfrac{d}{dx} P_0(x)$ exists and is continuous,

(c) $P_1(x)$ and $P_2(x)$ are continuous.

We find it desirable to change the independent variable x to a new variable s by the relation

$$dx/\sqrt{P_0(x)} = ds.$$

This change of variable carries $L(u)$ into what we shall call the "normal form"

$$L(u) = u'' + p(s)u' + q(s)u,$$

where primes now mean differentiation with respect to s and where $p(s)$ and $q(s)$ are seen to be continuous in the interval $a' \leqq s \leqq b'$ corresponding to $a \leqq x \leqq b$. It will be assumed from now on that $L(u)$ has already been put in the normal form.

Let the interval $a \leqq x \leqq b$ be divided into $n + 1$ equal subintervals of length

$$h = (b - a)/(n + 1),$$

and let the value of $V(x, t)$ at the point $x = a + ih$ and at time $t = jh$ be designated by V_{ij}. Then in the differential equation

$$V_{tt} = V_{xx} + p(x)V_x + q(x)V$$

we make the substitutions

$$V_{tt} = (V_{i,j+1} - 2V_{ij} + V_{i,j-1})/h^2,$$
$$V_{xx} = (V_{i+1,j} - 2V_{ij} + V_{i-1,j})/h^2,$$
$$V_x = (V_{i+1,j} - V_{i-1,j})/2h,$$

and after simplification obtain the *approximate* difference equation

(95.6) $V_{i,j+1} + V_{i,j-1} = V_{i+1,j} + V_{i-1,j}$
$$+ \frac{hp_i}{2}(V_{i+1,j} - V_{i-1,j}) + h^2 q_i V_{ij},$$

in which

$$p_i = p(a + ih), \qquad q_i = q(a + ih).$$

If we assume the existence and continuity of the second derivatives of $p(x)$ and $q(x)$, and suppose that the error of equation 95.6 is expanded in powers of h, we find that the leading term of the error is

(95.7) $\dfrac{h^4}{12}[(p^2 + 2p' + 2q)V_{xx} + (p'' + pp' + 2q' + 2pq)V_x$
$$+ (q'' + pq' + q^2)V],$$

showing that the error of equation 95.6 is of fourth order in h.

96. Derivation of the characteristic equation

For simplicity in the following explanation let us suppose that in the boundary conditions $g = G = 0$, so that $V_{0j} = 0$, $V_{n+1,j} = 0$. For $t = 0$ (or $j = 0$) the values of V_{i0} are arbitrary [being equal to the arbitrary function $F(x)$]. Hence we choose these in the most convenient manner, say, $V_{10} \neq 0$, $V_{i0} = 0$, $i > 1$. Then we have for V_{ij} the array of values

	i	0	1	2	3	$\cdots$	n	$n+1$
	j							
	0	0	V_{10}	0	0	$\cdots$	0	0
	1	0	V_{11}	V_{21}	0		0	0
(96.1)	2	0	V_{12}	V_{22}	V_{32}	$\cdots$	0	0
	3	0	V_{13}	V_{23}	V_{33}		0	0
	$\cdots$	$\cdots$	$\cdots$	$\cdots$	$\cdots$	$\cdots$	$\cdots$	$\cdots$
	$n-1$	0	V_{1n-1}	V_{2n-1}	V_{3n-1}		V_{nn-1}	0
	n	0	V_{1n}	V_{2n}	V_{3n}		V_{nn}	0

As just explained, the entries of the first row are arbitrary and are chosen as shown for convenience. To obtain the second row we use equation 95.6, together with the relation

$$V_{i,1} = V_{i,-1},$$

derived from the condition $\partial V/\partial t = 0$ at $t = 0$. All remaining rows are computed by 95.6.

So far no restriction has been placed on the magnitude of the interval h. We now impose the condition that h is chosen small enough that

$$|\tfrac{1}{2}hp_i| < 1,$$

for all values of i.

It will then be seen from array 96.1, together with 95.6, that the entry in the principal diagonal in the second column is given by

$$V_{21} = \tfrac{1}{2}(1 - \tfrac{1}{2}hp_2)V_{10},$$

whereas for subsequent columns the entries in the principal diagonal are

$$V_{32} = (1 - \tfrac{1}{2}hp_3)V_{21},$$
$$V_{43} = (1 - \tfrac{1}{2}hp_4)V_{32},$$

etc.

Since by our choice $V_{10} \neq 0$, no entry in the principal diagonal is zero, and the n-square matrix obtained from 96.1 by dropping the first and last columns and the last row is not singular. Then there will exist n constants $A_0, A_1, \cdots, A_{n-1}$ such that

$$\sum_{j=0}^{n} A_j V_{ij} = 0 \quad \text{for} \quad i = 1, 2, \cdots, n,$$

(where $A_n = 1$). See Theorem 10, Article 69.

Now from equation 95.5 we have

$$V_{ij} = \sum_{k=1}^{n} c_k u_{ki} \cos \lambda_k hj,$$

where $u_{ki} = u_k(a + ih)$. Therefore,

$$\sum_{k=1}^{n} c_k u_{ki} \sum_{j=0}^{n} A_j \cos \lambda_k hj = 0, \qquad i = 1, 2, \cdots, n.$$

It can be shown that these equations imply

$$\sum_{j=0}^{n} A_j \cos \lambda_k hj = 0, \qquad k = 1, 2, \cdots, n.$$

It is apparent, therefore, that the desired characteristic numbers λ_k are given by

$$\lambda_k = \mu_k/h, \qquad k = 1, 2, \cdots, n,$$

where $\mu_1, \mu_2, \cdots, \mu_n$ are roots of the characteristic equation

$$(96.2) \quad \cos n\mu + A_{n-1} \cos (n - 1)\mu + \cdots + A_1 \cos \mu + A_0 = 0.$$

The λ_k's thus found are correct for the *difference* equation with the given boundary conditions but in general are only approximations to the λ_k's belonging to the differential system 95.1.

In order to obtain numerical values for the characteristic function u_{ki} associated with λ_k, we substitute

$$V_{ij} = u_{ki} \cos \mu_k j,$$

in 95.6.   Since

$$V_{i,j+1} + V_{i,j-1} = 2V_{ij} \cos \mu_k,$$

the result, after removal of the factor $\cos \mu_k j$, may be put in the form

(96.3) $(1 + \frac{1}{2}hp_i)u_{k,i+1} = (2 \cos \mu_k - h^2q_i)u_{ki} - (1 - \frac{1}{2}hp_i)u_{k,i-1}.$

This is an ordinary difference equation of second order.   We may choose any non-zero value for $u_{k,1}$, say,

$$u_{k1} = 1.$$

Since $u_{k0} = 0$, we obtain $u_{k,2}$ from equation 96.3, then $u_{k,3}$, etc.   The fact that $u_{k,n+1}$ should turn out to be zero serves as a check on the calculations.

Example 1.   As a very simple example, let us take the system

$$u'' + \lambda^2 u = 0,$$
$$u = 0 \quad \text{at} \quad x = 0,$$
$$u = 0 \quad \text{at} \quad x = 1.$$

Let $h = \frac{1}{6}$.   For this system equation 95.6 becomes

$$V_{i,j+1} + V_{i,j-1} = V_{i+1,j} + V_{i-1,j},$$

and, if we take $V_{1,0} = 2$, the array 96.1 turns out to be

0	2	0	0	0	0	0
0	0	1	0	0	0	0
0	−1	0	1	0	0	0
0	0	−1	0	1	0	0
0	0	0	−1	0	1	0
0	0	0	0	−1	0	0

From this we have immediately

$$A_0 = A_2 = A_4 = 0,$$
$$A_1 = A_3 = 1,$$

and the characteristic equation is

$$\cos 5\mu + \cos 3\mu + \cos \mu = 0,$$

or in factored form

$$\cos 3\mu(2 \cos \mu + 1) = 0.$$

From the second form we read off the roots at once.

$$\mu_1 = \frac{\pi}{6}, \quad \mu_2 = \frac{2\pi}{6}, \quad \mu_3 = \frac{3\pi}{6}, \quad \mu_4 = \frac{4\pi}{6}, \quad \mu_5 = \frac{5\pi}{6},$$

whence

$$\lambda_k = k\pi, \qquad k = 1, 2, 3, 4, 5.$$

To obtain values of the characteristic functions, we resort to equation 96.3 which here becomes

$$u_{k,i+1} = 2 \ (\cos \mu_k)u_{k,i} - u_{k,i-1},$$

where $\cos \mu_1 = \frac{1}{2} \sqrt{3}$, $\cos \mu_2 = \frac{1}{2}$, $\cos \mu_3 = 0$, $\cos \mu_4 = -\frac{1}{2}$, $\cos \mu_5 = -\frac{1}{2} \sqrt{3}$. Upon calculating the u_{ki} with this difference equation we have

i =	0	1	2	3	4	5	6
u_{1i} =	0	1	$\sqrt{3}$	2	$\sqrt{3}$	1	0
u_{2i} =	0	1	1	0	-1	-1	0
u_{3i} =	0	1	0	-1	0	1	0
u_{4i} =	0	1	-1	0	1	-1	0
u_{5i} =	0	1	$-\sqrt{3}$	2	$-\sqrt{3}$	1	0

This particular example is remarkable in that the values obtained for the λ_k's and for the u_{ki}'s not only are exact for the difference equation but also are exact for the differential system from which we started. We know, of course, that the correct λ's are given by

$$\lambda_k = k\pi.$$

We readily see that the values obtained numerically for u_{ki} are also the values of the true characteristic functions

$$u_1(x) = 2 \sin \pi x,$$

$$u_2(x) = \frac{2}{\sqrt{3}} \sin 2\pi x,$$

$$u_3(x) = \sin 3\pi x,$$

$$u_4(x) = \frac{2}{\sqrt{3}} \sin 4\pi x,$$

$$u_5(x) = 2 \sin 5\pi x.$$

When $p(x)$ or $q(x)$ are different from zero, the results will no be longer exact but only approximate.

Computation 49

Example 2.

Differential equation $\quad u'' + 2u' + \lambda^2 u = 0$

Boundary conditions $\quad u = 0$ at $x = 0$ and at $x = 1$

Interval $\quad h = \frac{1}{16}$

Difference equation $\quad V_{i,j+1} = \frac{1}{16}(15V_{i-1,j} + 17V_{i+1,j}) - V_{i,j-1}$

Boundary conditions $\quad V_{o,j} = V_{16,j} = 0$

0	1	2	3	4	5	6	7	8	9	10	11	12	13	14	15	16
0	115292	0	0	0	0	0	0	0	0	0	0	0	0	0	0	0
0	0	54043	0	0	0	0	0	0	0	0	0	0	0	0	0	0
0	-57871	0	50665	47499	0	0	0	0	0	0	0	0	0	0	0	0
0	0	-54465	0	0	44530	41747	0	0	0	0	0	0	0	0	0	0
0	1.7592	0	-51259	-48241	0	0	39138	36692	0	0	0	0	0	0	0	0
0	0	4.1232	0	0	-45400	-42726	0	0	34399	32249	0	0	0	0	0	0
0	2.6216	0	6.9578	10.147	0	0	-40208	-37838	0	0	30233	28344	0	0	0	0
0	0	5.7273	0	0	13.590	17.199	0	0	-35608	-33508	0	0	26572	24911	0	0
0	3.4636	0	9.1926	12.910	0	0	20.902	24.634	0	0	-31532	-29672	0	0	23354	0
0	0	7.2869	0	0	16.788	20.748	0	0	28.344	31.985	0	0	-27921	-26274	0	0
0	4.2787	0	11.356	15.573	0	0	24.723	28.658	0	0	35.522	38.924	0	0	0	0
0	0	8.7900	0	0	19.856	24.136	0	0	32.508	36.233	0	0	0	0	0	0
0	5.0606	0	13.431	18.116	0	0	28.354	32.463	0	0	0	0	0	0	0	0
0	0	10.225	0	0	22.772	27.339	0	0	0	0	0	0	0	0	0	0
0	5.8035	0	15.403	20.519	0	0	0	0	0	0	0	0	0	0	0	0
0	0	11.581	0	0	0	0	0	0	0	0	0	0	0	0	0	0

Characteristic equation

$F(\mu) = \cos 15\mu + (1.05469) \cos 13\mu + (1.10275) \cos 11\mu + (1.14366) \cos 9\mu$
$+ (1.17695) \cos 7\mu + (1.20225) \cos 5\mu + (1.21928) \cos 3\mu$
$+ (1.22784) \cos \mu = 0$

k	μ_k	λ_k (approx.)	λ_k (true)	Relative error
	Degrees			
1	11.800	3.2952	3.2969	−0.00052
2	22.769	6.3583	6.3623	− .00063
3	33.917	9.4715	9.4777	− .00065
4	45.112	12.5976	12.6061	− .00067
5	56.325	15.7289	15.7398	− .00069
6	67.546	18.8625	18.8761	− .00072
7	78.772	21.9974	22.0139	− .00075
8	90	25.1327	25.1526	− .00079
9	101.228	28.2681	28.2920	− .00084
10	112.454	31.4030	31.4318	− .00092
11	123.675	34.5366	34.5720	− .00102
12	134.888	37.6679	37.7124	− .00118
13	146.083	40.7940	40.8529	− .00144
14	157.231	43.9072	43.9937	− .00197
15	168.200	46.9703	47.1345	− .00348

Example 3.　　　　　　　　　　　　　　　　　　　　　　Computation 50

Differential equation	$4xu'' + \lambda^2 u = 0$
Boundary conditions	$u = 0$ at $x = 1$ and at $x = 4$
Change variable	$x = s^2$
Normal equation	$u'' - \dfrac{1}{s} u' + \lambda^2 u = 0,\ u = 0$ at $s = 1$ and at $s = 2$
Interval	$h = \frac{1}{8}$
Difference equation	$V_{i,j+1} = \dfrac{1}{2i}[(2i + 1)V_{i-1,j} + (2i - 1)V_{i+1,j}] - V_{i,j-1}$
Boundary conditions	$V_{8,j} = V_{16,j} = 0$

8	9	10	11	12	13	14	15	16
0	13178880	0	0	0	0	0	0	0
0	0	6918912	0	0	0	0	0	0
0	−6644352	0	7233408	0	0	0	0	0
0	0	−7023744	0	7534800	0	0	0	0
0	10816	0	−7384104	0	7824600	0	0	0
0	20202	0	−7728000	0	8104050	0	0	0
0	8263.667	0	28497	0	−8057475	0	8374185	0
0	0	15547	0	35937.5	0	−8374185	0	0

Characteristic equation

$$F(\mu) = \cos 7\mu + (1.03333) \cos 5\mu + (1.05506) \cos 3\mu + (1.06578) \cos \mu = 0$$

k	μ_k	λ_k (approx.)	λ_t (true)	Relative error
	Degrees			
1	22.873	3.1936	3.1966	-0.00094
2	45.164	6.3060	6.3124	$-$ .00101
3	67.568	9.4343	9.4445	$-$ .00108
4	90	12.5664	12.5812	$-$ .00118
5	112.432	15.6984	15.7199	$-$ .00137
6	134.836	18.8267	18.8595	$-$ .00174
7	157.127	21.9391	21.9997	$-$ .00275

97. Case of two dimensions

A similar treatment applies in the case of two dimensions, but unfortunately the great advantage accruing from the triangular matrix 96.1 is largely lost even for the simplest two-dimensional cases. Also because of the greater number of points required to secure reasonable accuracy the order of the matrix is generally much greater than in the case of one dimension.

Let us consider, for example, the problem of finding approximately the characteristic numbers ρ_i associated with the differential equation

$$(97.1) \qquad \nabla^2 U + \rho^2 U = 0$$

where U is subject to the condition $U = 0$ on the boundary B of a plane region R. This problem arises in particular in connection with vibrations of a membrane stretched over the region R and fixed in the boundary points B of R. Here we have to solve the equation

$$(97.2) \qquad W_{tt} = a^2 \nabla^2 W$$

with the condition $W = 0$ on B. If we replace equation 97.2 in the simplest possible manner by a difference equation set up for a square lattice in the xy plane and for equal time intervals of length k, we get

$$(97.3) \qquad W(t + k) - 2W(t) + W(t - k) = (k^2 a^2/h^2)HW(t),$$

in which H is the operator defined in equation 62.1. It may be verified that a permissible choice of k is that for which $k^2 a^2/h^2 = \frac{1}{2}$. If we now assume that

$$W = U(x, y) \cos a\rho t,$$

the differential equation 97.2 for W reduces to the differential equation 97.1 for U, while the difference equation 97.3, upon the removal of the factor $\frac{1}{2} \cos a\rho t$, and the substitution $k = h/a\sqrt{2}$, reduces to

$$(97.4) \qquad 4\left[\cos\frac{h\rho}{\sqrt{2}} - 1\right] U = HU.$$

We are, therefore, led to approximate the characteristic numbers ρ_i by means of the relation

$$(97.5) \qquad \cos\frac{h\rho}{\sqrt{2}} = 1 - \lambda/4,$$

in which λ is a latent root of the matrix $\mathbf{H}$ defined in Article 85. In actual practice we find it advantageous to introduce a new matrix $\mathbf{M}$ defined by the equation

$$(97.6) \qquad \mathbf{M} = 4\mathbf{I} - \mathbf{H}.$$

The latent roots μ_i of $\mathbf{M}$ are symmetrical with respect to zero, and lie in the interval $-4 < \mu < 4$, as is apparent from the relation

$$\mu = 4 - \lambda,$$

and the discussion in Article 85. Hence, the procedure proposed for the determination of characteristic numbers is first to cover the region R with a square lattice of side h, then in some manner to calculate the latent roots $\mu_1, \mu_2, \mu_3, \cdots$ of the corresponding matrix $\mathbf{A}$. The characteristic numbers are now given approximately by

$$(97.7) \qquad h\rho_i = \sqrt{2}\left[\cos^{-1}(\mu_i/4)\right].$$

If the region R is unrestricted as to shape, it is difficult to determine the error committed by the use of formula 97.7. However, in the case where R is a rectangle whose sides α and β are integral multiples of the grid length h, we have an explicit formula for the true values, namely,

$$(97.8) \quad \rho^2 = \pi^2\left(\frac{m^2}{\alpha^2} + \frac{n^2}{\beta^2}\right), \quad (m = 1, 2, \cdots, \quad n = 1, 2, \cdots),$$

and also an analytical formula for the corresponding approximation given by 97.7, namely,

$$(97.9) \qquad \cos h\rho/\sqrt{2} = \tfrac{1}{2}\left(\cos\frac{m\pi h}{\alpha} + \cos\frac{n\pi h}{\beta}\right).$$

When we solve equation 97.8 for ρ^2 as a series of powers of h, we get

$$(97.10) \quad \rho^2 = \pi^2 \left(\frac{m^2}{\alpha^2} + \frac{n^2}{\beta^2} \right) - \frac{\pi^4 \left(\dfrac{m^2}{\alpha^2} - \dfrac{n^2}{\beta^2} \right)^2 h^2}{24}$$
$$- \frac{\pi^6 \left(\dfrac{m^2}{\alpha^2} + \dfrac{n^2}{\beta^2} \right) \left(\dfrac{m^2}{\alpha^2} - \dfrac{n^2}{\beta^2} \right)^2 h^4}{720} - \cdots,$$

a result which shows that for the case of the rectangle the approximate value converges to the true value as the mesh length is reduced to zero.

It is instructive also to compare the actual numerical values given by formula 97.8 with those obtained from formula 97.9. To this end Table IV presents the first 20 characteristic values for the case of a 2×3 rectangle together with the approximate values found by 97.9 for three different mesh sizes.

Table IV.　Characteristic Numbers for a Rectangle 2×3　Comparison of True Values with Approximations Using a Square Lattice

			Values of ρ^2	
	True	8×12 grid 77 points	16×24 grid 345 points	24×26 grid 805 points
1	3.564	3.558	3.562	3.562
2	6.854	6.845	6.853	6.853
3	10.966	10.762	10.917	10.939
4	12.337	12.189	12.277	12.323
5	14.256	14.176	14.237	14.245
6	19.739	19.739	19.739	19.739
7	20.013	19.394	19.840	19.950
8	23.303	22.082	23.011	23.163
9	26.593	25.718	26.382	26.493
10	27.416	27.250	27.374	27.392
11	29.883	28.160	29.448	29.694
12	32.076	31.648	31.949	32.032
13	37.285	36.418	37.077	37.200
14	39.752	39.690	39.741	39.747
15	40.575	36.438	39.598	40.155
16	41.946	38.064	41.006	41.544
17	43.864	40.355	43.043	43.422
18	49.348	46.805	48.760	49.101
19	49.348	46.805	48.760	49.101
20	49.622	49.542	49.600	49.600

True values computed by
$$\rho^2 = \frac{\pi^2}{36} (4m^2 + 9n^2)$$

Approximate values computed by
$$\sin^2 \frac{h\rho}{\sqrt{8}} = \frac{1}{4} \left(2 - \cos \frac{m\pi h}{3} - \cos \frac{n\pi h}{2} \right)$$

98. Calculation of the latent roots of M

In the preceding article it was indicated that approximate values for characteristic numbers associated with equation 97.1 and the boundary condition $U = 0$ can be found by equation 97.7 in terms of the latent roots μ_1 of the matrix **M**. These latent roots may be computed by some one of the various methods explained in Articles 80, 81, 82, and 83. However, there are certain simplifications available for the matrix **M** that are not possible for matrices in general,

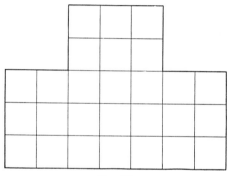

Fig. 26

one of the most important being due to the symmetry of the latent roots of **M**.

This fact is particularly helpful if we use method c of Article 80, for, instead of solving n linear equations to find the g's, we need solve only $n/2$ or $(n - 1)/2$ equations. Likewise we have to solve an algebraic equation of degree $n/2$ or $(n - 1)/2$ instead of nth degree.

These simplifications, as well as a suggested model for carrying out the computation, are illustrated by the following example.

Example 1. Find the latent roots of **H** for the region R shown in Fig. 26. Here, as in Article 86a, we make up a computation sheet with a space corresponding to each interior point of R, thus:

Computation Sheet

		8)	9)		
		7)	10)		
2)	3)	6)	11)	14)	15)
1)	4)	5)	12)	13)	16)

For convenience of reference the spaces of the computation sheet are numbered the same as corresponding components of u.

As the initial vector of the iteration we take $u_1 = 1$, all other u's = 0. This choice simplifies the solution of the linear equations for the g's, for we find upon carrying out the required 16 iterations that all the odd-numbered components are zero for the odd-numbered iterations and all the even-numbered components are zero for the even-numbered iterations. It suffices then to set up the equations for the g's from the even (or odd) iterations only. These give the following table of values for the odd-numbered components.

ν	u_1	u_3	u_5	u_7	u_9	u_{11}	u_{13}	u_{15}
0	1							
2	2	2	1					
4	9	12	9	3		4	1	
6	51	79	72	38	10	54	20	6
8	332	575	591	393	150	571	252	106
10	2 405	4 535	5 042	3 787	1 657	5 587	2 701	1 287
12	18 922	37 873	44 263	35 426	16 475	52 887	26 892	13 563
14	157 853	328 302	395 550	327 138	156 689	492 623	257 842	133 799
16	1367 860	2911 473	3571 333	3003 890	1460 277	4550 342	2421 920	1275 905

By rather simple recombinations of the columns of the above array we get rid of all numbers above the diagonal.

One such recombination is the following:

ν	u_1	u_3	u_5	u_7	u_9	u_{11}	u_{13}	u_{15}
0	1							
2	2	1						
4	9	9	1					
6	51	72	20	1				
8	332	591	252	29	1			
10	2 405	5 042	2 701	368	85	159		
12	18 922	44 263	26 893	3 724	1 652	3 405	10 445	
14	157 853	395 550	257 842	34 789	22 101	47 073	221 290	1
16	1367 860	3571 333	2421 920	316 310	249 783	540 219	3040 505	22

The g's can now be calculated one by one, and we get the algebraic equation

$$\mu^{16} - 22\mu^{14} + 175\mu^{12} - 632\mu^{10} + 1059\mu^8 - 787\mu^6 + 233\mu^4 - 16\mu^2 = 0,$$

and from it we obtain the solutions

$$\mu^2 = 0, \ 0.0960, \ 0.5932, \ 0.7120, \ 1.9035, \ 4.4088, \ 5.1408, \ 9.1457.$$

From these values we compute by equation 97.7 the following approximations:

i	$h^2\rho^2$
1	1.02
2	1.87
3	2.07
4	2.97
5	3.69
6	3.79
	etc.

The method used in Example 1 has the advantage that it gives us all the latent roots, but, when the number of interior points is large, that method is too laborious for practical calculation. We are then usually content to calculate a few of the latent roots approximately by the methods of Articles 81 and 82. For this purpose it speeds the convergence to start the iteration $\mathbf{v}_\nu$ with an initial $\mathbf{v}_0$ as near as possible to the fundamental latent vector ϕ_1. Since ϕ_n, the latent vector belonging to the largest λ, differs from ϕ_1 only in having a checkerboard of negative signs, we can also usually take $\mathbf{v}_0$ to be nearly orthogonal to ϕ_n. The process of iteration then quickly gives us a good estimate for μ_1 by the formula

$$(98.1) \qquad \mu_1 = \mathbf{v}_\nu \cdot \mathbf{v}_\nu / \mathbf{v}_\nu \cdot \mathbf{v}_{\nu-1}.$$

If we cannot readily make $\mathbf{v}_0$ orthogonal to ϕ_n, it is better to determine μ_1 by the approximate equation

$$(98.2) \qquad \mu_1{}^2 = \mathbf{v}_\nu \cdot \mathbf{v}_\nu / \mathbf{v}_\nu \cdot \mathbf{v}_{\nu-2}.$$

From μ_1 we obtain the least and largest λ. If desired we can use the orthogonal properties, as in Article 82, to carry out further iterations with vectors free from ϕ_1 and ϕ_n and thus calculate approximations for μ_2, μ_3, etc.

Example 2. For the problem treated in Example 1 of this article we start the

		1		1	
		1		1	
1	1	1	1	1	1
1	1	1	1	1	1

iteration with v_0 chosen as shown below. This choice is suggested by the fact that ϕ_1 has components all of the same sign. From symmetry it is clear that the chosen initial vector is orthogonal to ϕ_{16}. Accordingly, we iterate a suitable number of times (actually 5 in this case) and compute μ_1 by equation 98.1, getting

$$\mu_1 = 3.02415,$$

and hence

$$\lambda_1 = 0.97585, \qquad \lambda_{16} = 7.02415.$$

99. Use of the operator K

Since the operator $K/6h^2$ is usually a much more accurate substitute for the Laplacian operator ∇^2 than H/h^2, it is natural to hope that the latent roots of the matrix $\mathbf{K}$ of Article 88 lead to better estimates for the characteristic numbers than do the latent roots of $\mathbf{H}$. On the other hand, the great computational aid provided by symmetry in the roots of $\mathbf{H}$ is lacking in the case of $\mathbf{K}$.

In Article 63 we found that the heat equation

$$(99.1) \qquad\qquad U_t = c^2 \nabla^2 U$$

can be approximated by the difference equation 63.2 with an error which is seen from equation 63.1 to be $O(h^6)$. No simple difference equation of corresponding accuracy was obtained for the case of the wave equation (see Article 64, especially equation 64.1). Hence, we shall endeavor to relate the characteristic numbers ρ_i to the latent roots λ_i of **K** through equation 99.1 rather than through equation 97.2. To do this we assume a solution of 63.2 of the form

$$U = e^{-c^2 \rho^2 t} V(x, y),$$

recall that the time interval k is equal to $h^2/6c^2$, and arrive at the relation

$$(99.2) \qquad\qquad h^2 \rho^2 = -6 \log_e (1 - \lambda/36)$$

in which λ is a latent root of the matrix **K**. It is of interest to compare the values of ρ computed with equation 99.2 for the case of the same rectangles used for the comparison in Article 97. The values of λ required in 99.2 are obtained for any rectangle by equation 88.2. Table V gives the comparison.

Table V. Characteristic Numbers for a Rectangle 2×3 Comparison of True Values with Approximations Using a Square Lattice

		Value of ρ^2		
	True	8×12 grid	16×24 grid	24×36 grid
1	3.564	3.564	3.564	3.564
2	6.854	6.853	6.854	6.854
3	10.966	10.959	10.966	10.966
4	12.337	12.330	12.336	12.337
5	14.256	14.249	14.255	14.256
6	19.739	19.727	19.738	19.739
7	20.013	19.970	20.011	20.013
8	23.303	23.215	23.298	23.302
9	26.593	26.504	26.588	26.592
10	27.416	27.366	27.413	27.415
11	29.883	29.709	29.873	29.881
12	32.076	31.982	32.071	32.075
13	37.285	37.105	37.275	37.283
14	39.752	39.621	39.745	39.751
15	40.575	40.021	40.546	40.569
16	41.946	41.392	41.916	41.940
17	43.864	43.310	43.836	43.859
18	49.348	48.789	49.318	49.342
19	49.348	48.789	49.318	49.342
20	49.622	49.360	49.607	49.619

True values computed by $\quad \rho^2 = \dfrac{\pi^2}{36} (4m^2 + 9n^2)$

Approximate values computed by

$$h^2 \rho^2 = -6 \log_e \left[\left(\frac{2 + \cos \dfrac{m \pi h}{3}}{3} \right) \left(\frac{2 + \cos \dfrac{n \pi h}{2}}{3} \right) \right]$$

It may be noted that for the first ten characteristic numbers the approximation using K with 77 interior points is as good on the average as that furnished by H with 805 points. The implication is obvious that the use of K requires less labor in the long run than the simpler operator H.

Corresponding to formula 97.10 we have in the present case

$$(99.3) \qquad \rho^2 = \pi^2 \left(\frac{m^2}{\alpha^2} + \frac{n^2}{\beta^2} \right) - \frac{\pi^6 \left(\dfrac{m^6}{\alpha^6} + \dfrac{n^6}{\beta^6} \right) h^4}{540} - \frac{\pi^8 \left(\dfrac{m^8}{\alpha^8} + \dfrac{n^8}{\beta^8} \right) h^6}{6048}$$

Hence, for a rectangle the error of equation 97.7 is $O(h^2)$ while that of 99.2 is $O(h^4)$.

100. Characteristic numbers for curved boundaries

Cases where the bounding curve cuts the lattice at points other than nodal points present the same kind of difficulties encountered in Chapters 8 and 10. Three procedures are suggested here for coping with these difficulties. The first is based on Theorem 3 of Article 68 on the separation of the roots of $D_{n+1}(\lambda)$ by those of $D_n(\lambda)$. Suppose that $\lambda_1' \leqq \lambda_2' \leqq \lambda_3' \leqq \cdots \leqq \lambda_n'$ are the latent roots of the matrix $\mathbf{H}$ (or $\mathbf{K}$) for a given set R_n' of n interior nodal points. Let $\lambda_1'' \leqq \lambda_2'' \leqq \lambda_3'' \leqq \cdots \leqq \lambda_{n+p}''$ be the latent roots of the matrix $\mathbf{H}$ (or $\mathbf{K}$) for a larger region R_{n+p}'' which includes all the points of R_n'. Then

$$(100.1) \qquad \lambda_i'' \leqq \lambda_i', \qquad i = 1, 2, \cdots, n.$$

The proof is conducted in much the same manner as for property 3, Article 88. To apply this result practically we compute λ_1', λ_2', λ_3', $\cdots$ for the largest block of squares that can be fitted inside the curved region R, and again compute λ_1'', λ_2'', λ_3'', $\cdots$ for the smallest block of squares that contains R. We assume without proof that the latent roots λ_i which we should use in equation 97.5 or in equation 99.2 satisfy the equalities

$$\lambda_i'' \leqq \lambda_i \leqq \lambda_i'.$$

Hence, we may for example assume that

$$\lambda_i = (\lambda_i'' + \lambda_i')/2$$

with an error that does not exceed $(\lambda_i' - \lambda_i'')/2$.

A second procedure is based on the following observation. If ρ_i is a characteristic number belonging to a region R with area A, while ρ_i' belongs to R' with area A', where R and R' are geometrically similar, then

$$(100.2) \qquad A\rho_i^2 = A'\rho_i'^2.$$

This suggests that, having decided on the size of the mesh length h, we construct a region R' of squares as nearly as possible similar in shape to the curved region R. Having computed the value ρ_i' by the methods of Article 97, 98, or 99, we then use equation 100.2 to compute an approximation to ρ_i.

A third method is to overlay the region R with a square net in the usual manner and then to compute values for the successive iterants v_r at external nodal points adjacent to the boundary by some type of extrapolation. The extrapolation uses interior values of v_r computed by equation 97.6 or by equation 63.2 together with the zero value at points where the boundary of R cuts the lattice.

Example. We illustrate all three of the above procedures by applying each in turn to the problem of finding the smallest characteristic number belonging to a circle of unit radius. To avoid laborious calculations we take a comparatively coarse net with $h = 0.2$, and make use of symmetry to reduce the amount of computation required to perform the successive iterations. Regions R' and R'' are shown in Figs. 27 and 28, respectively. The initial vector v_0 was chosen as

$$v_0 = 25[1 - x^2 - y^2].$$

We use formula 63.2, and after eight iterations (omitted here to save space) we compute λ' for Fig. 27 and λ'' for Fig. 28 by using the approximate formula

$$1 - \lambda/36 = v_8 \cdot v_8/v_7 \cdot v_8.$$

It turns out that

$$\lambda' = 1.5889, \qquad \lambda'' = 1.2981.$$

These give

$$\rho' = 2.602, \qquad \rho'' = 2.347,$$

and the average of ρ' and ρ'' is

$$\rho = 2.474.$$

To illustrate the second method we apply it to the values of ρ' and ρ'' just obtained for the regions R' and R'' in Figs. 27 and 28, respectively. The areas of F', F'', and the circle are 3.04, 3.52, and π, respectively, so that the conversion factors are $\sqrt{3.04/\pi}$ and $\sqrt{3.53/\pi}$. From these and ρ' and ρ'' we get the estimates 2.484 and 2.560.

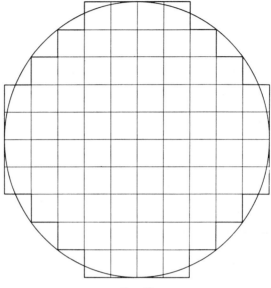

FIG. 27

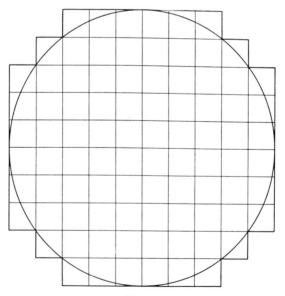

FIG. 28

To apply the third procedure we have to perform an iteration using the extrapolation formula at each step to get the exterior values. The extrapolation was done by means of a second-degree polynomial approximation along horizontal lines for the points with coordinates (0.8, 0.8) (1, 0.6) (1, 0.4) and (1, 0.2). The other extrapolated quantities were supplied by symmetry. The value of ρ obtained by this method with formula 99.2 proved to be

$$\rho = 2.4126.$$

Summarizing the foregoing results we have

> First method,   2.602, 2.347, average 2.474.
> Second method, 2.484, 2.560, average 2.522.
> Third method,   2.413
> Correct (least zero of $J_0(x)$)   2.4048

The following references deal with many widely differing methods for finding characteristic numbers: Aronszajn [2, 3], Collatz [31, 32, 37], Donsker and Kac [46], Flanders and Shortley [57], Karush [97], Lanczos [104], Lyusternik [110], Morris [140], Rayleigh [157], von Sanden [176], Schulte [183], Southwell [193].

With these examples we close the discussion of partial differential equations. As stated in the beginning we have here no comprehensive theory but only a collection of special cases which, it is hoped, may serve to light the way toward the solution of other problems.

APPENDIX A

ROUND-OFF ERRORS

In the discussions of the error involved in various methods of integration the effect of errors due to round-off has been ignored. A satisfactory method of dealing with this type of error is complicated by the fact that to apply a *rigorous* bound at every step generally leads to an overall bound which is much larger than the actual error. Hence, one is tempted not to use a rigorous bound but to arrive at a *statistical* estimate based on the assumption that round-off errors are distributed at random in an interval of unit length in the last digit retained. Working on this basis, Rademacher [156] investigated the accumulation of errors in a process of integration, nearly the same as Method II. But Huskey [92] showed that in certain instances the assumption of randomness can lead to a serious underestimate for the actual error. However, Forsythe [58] devised a method of round-off which actually is random even in cases where the customary method of round-off is not, and Rademacher's results apply.

Forsythe's method may be used for machine computation as follows: Instead of adding a 5 in the most significant position of the digits to be dropped (ordinary round-off), one adds a *random* decimal digit to each of the digital positions to be dropped. Lists of random numbers which may be used for this purpose are given, e.g., by Kendall and Smith [240] or by Dixon and Massey [45].

In practical applications it seems probable that hand computations will usually not be so extensive as to require elaborate precautions beyond the retention of a suitable number of digits in excess of those to be finally accepted. High-speed machine calculation, especially with ultrashort steps, is quite another matter, and some method such as Forsythe's may be indispensable.

APPENDIX B

LARGE-SCALE COMPUTING MACHINES

Just as the advent of desk calculators has influenced traditional computing methods (e.g., in the elimination of the common logarithms as a standard computational tool), it is reasonable to expect that the current development of large-scale digital computing machines will have an even greater impact on numerical procedures.

With respect to the solution of differential equations it seems to be generally believed that the new machines will use the simplest possible methods and attain desired accuracy by means of many very short steps, for ordinary equations, or by an ultrafine mesh, for partial equations. This forecast is probably correct as long as the machines are employed on problems of about the same overall magnitude as have up to the present been worked by hand (for example, the calculation of a trajectory in ballistics). But it is my belief that the big machines will tend to be employed on bigger and bigger problems, where the danger of accumulated error due to many steps, and the limitations on fineness of mesh imposed by limited high-speed memory capacity, will both oppose extravagantly short steps and ultrafine mesh. It may also be pointed out that for complicated systems of differential equations the substitution routine which must be carried through for each step will greatly outweigh the integration routine per step, even where the latter is based on fairly powerful methods of quadrature. Hence, the use of very simple integration methods may not really save a great amount of coding time or of machine operating time. For partial differential equations the number of values that must be kept in the memory and used in every iteration increases so rapidly as the mesh size is reduced that many difficulties will be encountered in going to a very fine mesh with any of the machines now in operation.

For all these reasons it is probable that the existence of these new machines will intensify the search for methods simple enough to be coded for practical use but powerful enough to produce adequate. accuracy without unreasonable reduction in step length or mesh size.

It may not be amiss to mention briefly a few machines, or types of machines that are being or can be used for the solution of differential equations.

The first to be effectively developed were the *analog differential analyzers,* associated with the name of Dr. Vannevar Bush. Machines of this type perform the required integrations by the continuous operation of a mechanical or electrical "integrator." Input and output are by means of curves or other physical representation of the data. Since analog machines do not operate on numbers as such, the numerical methods of this text have no application to their operation. Quite recently there have appeared *digital differential analyzers,* which are high-speed electronic machines. The integrators in these machines are accumulators that in effect perform the integrations by summations, using very short steps. They possess one advantage over analog machines in that they can be built to operate with such a large number of digits that their accuracy and capacity (range of values of the variables) can meet any reasonable requirements, while analog machines are subject to decided physical limitations.

Perhaps the most celebrated machine is the *Electronic Numerical Integrator and Calculator* (ENIAC) at Aberdeen Proving Ground. This machine is not so highly specialized as the differential analyzers just mentioned but nevertheless was designed with the solution of differential equations in mind, especially those associated with exterior ballistics. In performing its integrations the ENIAC uses a method much like Method II.

We turn now to *general-purpose calculators,* not designed especially for the solution of differential equations, but versatile enough to handle, by proper coding, any ordinary equation, system of ordinary equations, or partial equations whose solution can be approximated by known numerical methods. A number of such machines are now in operation, but only those will be mentioned which I have seen. Of these the *Automatic Sequence Controlled Calculator* at the Computation Laboratory of Harvard University is the oldest. It operates with mechanical calculators, but uses electric relays for storage (memory) and electric circuits for controlling operations and transferring data. Ordinary differential equations are integrated by Method VII.

At the National Bureau of Standards in Washington, D. C., is the *National Bureau of Standards Eastern Automatic Computer* (SEAC), which is an automatic-sequenced electronic calculator operating in the binary system and using mercury delay line storage. (Recently an auxiliary cathode-ray tube memory has been added.)

The *National Bureau of Standards Western Automatic Computer* (SWAC) at Los Angeles is also an automatic-sequenced electronic calculator operating in the binary system and using a cathode-ray tube storage. The digits of the binary numbers are handled simultaneously in parallel circuits, in contrast to the SEAC where digits follow one another in series. The SWAC is the fastest machine yet built.

In a trial run with Method II the SWAC calculated automatically four hundred steps in the numerical integration of the simultaneous system

$$dx/dt = yz, \quad dy/dt = -xz, \quad dz/dt = -\tfrac{1}{2}xy$$

in a few minutes' computing time. The computed values of x, y, and z were correct to approximately eight decimal digits.

On another test run the SWAC solved a Dirichlet problem by iteration using the operator

$$\left(1 + \frac{1}{16}K\right) = \frac{1}{16}
\begin{array}{|c|c|c|}
\hline
1 & 4 & 1 \\
\hline
4 & -4 & 4 \\
\hline
1 & 4 & 1 \\
\hline
\end{array}$$

and the Liebmann process. The region R was a rectangle with 108 interior points. The set of points was scanned by the machine 32 times, and the values obtained for the function at the interior points were correct to about four decimal digits.

Any operator of the form

$$\left(1 + \frac{1}{\theta}K\right)$$

might have been used. To assure convergence of a straight iteration (not Liebmann process) we must take $\theta \geq 16$. The choice $\theta = 16$ gives the most rapid convergence and at the same time luckily provides an operator all of whose multipliers are integral powers of 2, a convenience for a binary machine.

A large volume of computation is being currently done with *punched-card computing machinery*. Such machines have the great advantage of an almost unlimited storage or memory, coupled with the disadvantage that this memory can only be used with speed and efficiency in problems for which the required values can be arranged in some serial order. This presents certain difficulties in scanning two- or three-dimensional arrays, as in the case of Laplace's equation. For this problem the factored form of the operator $(1 + K/36)$, as shown in Article 63, can be used to advantage as was shown in detail by Yowell [223].

APPENDIX C

THE MONTE CARLO METHOD*

It has lately become fashionable to apply a rather picturesque name, the "Monte Carlo method," to any procedure that involves the use of sampling devices based on probabilities to approximate the solution of mathematical or physical problems. The Monte Carlo method has, so far, largely been used in connection with functional equations and quadratures. No comprehensive technical exposition of the method has as yet appeared.

The use of functional equations in connection with probability problems arising in games of chance dates back to the earliest beginnings of probability theory, and was familiar to de Moivre, Lagrange, and Laplace. The connection between probabilities and the differential equations of mathematical physics is also an old story to theoretical physicists. During the early years of the present century, it was studied particularly extensively by Einstein, Smoluchowski, Lord Rayleigh, Langevin, and many others. In 1899 Lord Rayleigh proved that a random walk† in one dimension yields an approximate solution of a parabolic differential equation. The relationship between random flights (i.e., random walks in more than one dimension) and the first boundary value problem, or Dirichlet problem, for elliptic difference and differential equations was established by Courant, Friedrichs, and Lewy [40] in 1928.

The novelty that the method possesses lies chiefly in its point of view. With few exceptions, the authors cited above proceed from a problem in probabilities to a problem in functional equations, whose solution is then obtained, or at least proved to exist, by classical methods and furnishes the answer to the probability problem. In the Monte Carlo method, the situation is reversed. The probability problem (whose solution can always be approximated by repeated trials) is regarded as the tool for the numerical analysis of a functional equation. Or alternatively, in a physical problem which, classically,

---

* The material of this appendix is taken almost word for word from Curtiss [43] and Yowell [222].

† For a detailed discussion of "random walks" see Feller [241].

would call for an analytic model, the equivalent probability problem is regarded as an adequate model in itself, and derivation of an analytic equivalent is considered to be superfluous. The use of probability theory in this way to solve physical problems was apparently first suggested by von Neumann and Ulam.

But it is worth noting that the Monte Carlo method is not at all novel to statisticians, as was pointed out by Professor John Wishart at a symposium on the Monte Carlo method held in 1949 at Los Angeles, California. For more than fifty years, when statisticians have been confronted with a difficult problem in distribution theory, they have resorted to what they have sometimes called "model sampling." This process consists of setting up some sort of urn model or system, or drawings from a table of random numbers, whereby the statistic whose distribution is sought can be observed over and over again and the distribution in question is usually a multiple integral over a peculiar region in many dimensions; so, in such cases, "model sampling" is clearly a Monte Carlo method of numerical quadrature. In fact, the distribution of "Student's t" was first determined in this way. Many other examples can be found by leafing through the pages of *Biometrika* and the other statistical journals.

In the old days of hand computation the use of random methods for the solution of differential equations was a visionary idea—novel and clevel, but impractical. Suddenly, thanks to large-scale computing machinery, such methods not only are possible but may, in fact, prove to be the best methods for tackling certain problems as well.

As yet, however, Monte Carlo methods, in spite of considerable theoretical study, have not been thoroughly tested by actual numerical computation. One computation at least was carried out by Yowell [222], who set up a two-dimensional Dirichlet problem to be solved by the Monte Carlo method using IBM computing machinery. The following extracts taken from Yowell's report describe the procedure in general terms, technical details of coding being omitted. In the case of Laplace's equation in two dimensions, the random process is merely a two-dimensional random walk. In such a walk, uniform steps are made along one or the other of the coordinate directions, the choice being purely random, and in either a positive or a negative direction, the choice once again being random. If such a walk is started at a point inside the boundary of a region, it will eventually end up at the boundary. The number of steps will vary for two different walks starting at the same point, but the average number of steps can be computed. Suppose, now, we make a large number of walks from a single interior point. Each time we come to a boundary, we stop, record the value of the function on the boundary, and return to our starting point to repeat the process. If, after a large number of walks, we average all the boundary values we have noted, that average will approximate the value of the function at the starting point and will converge to that value as the number of walks increase to infinity.

This process was tested on the 604 Electronic Calculating Punch. A region bounded by the four lines $x = y = 0$ and $x = y = 10$ was selected, and a unit step was made in the random walk. The boundary values were $f(10, y) = f(x, 10) = 0$, $f(0, y) = 100 - 10y$, $f(x, 0) = (10 - x)^2$.

The random variables were introduced into the problem according to the evenness or oddness of the digits in a table of random digits prepared and furnished us by Rand Corporation.

The operating procedure was as follows. From the computation of the mean length of the random walk, an estimate was made of the number of steps needed to give a specified number of walks. This many random digit cards were selected, a header card was put in with the coordinates of the starting point, and a trailer card on which the answer was to be punched. This deck was then run through the 604, and the sum of the boundary values and the tally of the number of times the boundary was reached was punched on the final card. The quotient of these two numbers gave the value of the function at the starting point.

The tests indicate that the method will give the correct answers, but the speed of convergence is very slow. The smoothing method converges as $1/n$, where n is the number of times the field is smoothed. In this case, we mean that the difference between the approximate solution and the true solution is proportional to $1/n$. But in the Monte Carlo method, the convergence is proportional to $1/\sqrt{n}$. And the convergence here is a statistical convergence; that is, the probable error is proportional to $1/\sqrt{n}$ where n is the number of random walks. With statistical convergence, one can never guarantee that the solution is correct, but one can state that the probability is high that the approximate solution does not deviate from the true one by more than a certain amount.

The great drawback of this statistical method is its slow speed of convergence. This should not cause the method to be discarded, for the ideas of Monte Carlo procedures are still new, and various ways have been and are being found to speed up the convergence of the solution. It is also true that the speed of convergence of the Monte Carlo method is not affected by the dimensionality of the problem, and this may prove to be a very great advantage in problems involving three or more dimensions. Finally, Monte Carlo procedures may be very important in problems where a single number is required as an answer (such as the solution at a single point, or a definite integral involving the solution of the differential equation.) In these cases, the entire solution would have to be found by smoothing techniques. With these limitations and advantages recognized, the simplicity of the procedure certainly makes the Monte Carlo method worth considering for the solution of a partial differential equation.

The best general account of Monte Carlo methods is given by Curtiss [43]. More specialized articles are those of Cutkosky [44], Donsker and Kac [46], Forsythe and Leibler [61], Fortet [63], Kac [95].

APPENDIX D

HOW TO STABILIZE MILNE'S METHOD

The method of Article 28, page 64, using formula 28.1 to predict and 28.2 to correct, is unstable in this form. It can however be made stable by the simple expedient of using formula 19.5, page 48, as an occasional check.

The *Journal of the Association for Computing Machinery* has kindly granted permission to reprint the following two articles, which contain an analysis of the problem and several illustrative computations.

Stability of a Numerical Solution of Differential Equations*

W. E. MILNE AND R. R. REYNOLDS

Oregon State College, Corvallis, Oregon

In 1926 Milne [1] published a numerical method for the solution of ordinary differential equations. This method turns out to be unstable, as shown by Muhin [2], Hildebrand [3], Liniger [4], and others. Instability was not too serious in the day of desk calculators but is fatal in the modern era of high speed computers. The basic cause of the instability in this particular method is the use of Simpson's rule to perform the final integration. Simpson's rule integrates over two intervals, and under certain conditions can produce an error which alternates in sign from step to step and which increases in magnitude exponentially. It is the purpose of this paper to show that the occasional application of Newton's "three eighths" quadrature formula over three intervals can effectively damp out the unwanted oscillation without harm to the desired solution.

Let the given differential equation be

$$\frac{dy}{dx} = f(x, y),$$

and let the step length for the independent variable x be denoted by h. The quantity

$$s = h \frac{\partial f}{\partial y}$$

plays a basic role in the analysis, for it may be shown that when Simpson's rule is used an error E_0 occurring at $x = x_0$ is propagated through subsequent steps according to the second order difference equation

$$\left(1 - \frac{s_{n+1}}{3}\right) E_{n+1} - \left(\frac{4s_n}{3}\right) E_n - \left(1 + \frac{s_{n-1}}{3}\right) E_{n-1} = 0.$$

(See e.g., Hildebrand [3], p. 206, Milne [5], p. 68.)

While in general s is a variable, the special case where s is constant not only permits a simple analysis but also serves to explain the behavior in other cases. Cf. Hildebrand [3], p. 202. Accordingly we treat s as a constant and assume that our differential equation is

$$\frac{dy}{dx} = Gy,$$

the general solution of which, after n steps, is

$$y_n = A e^{ns},$$

where A is an arbitrary constant, $s = hG$, and $x = nh$.

*Reprinted from *Journal of the Association for Computing Machinery*, Vol. 6, No. 2, April 1959. Original pagination is indicated within brackets.

When Simpson's rule is used, the corresponding difference equation is

$$\left(1 - \frac{s}{3}\right) y_{n+1} - \left(\frac{4s}{3}\right) y_n - \left(1 + \frac{s}{3}\right) y_{n-1} = 0, \tag{1}$$

with the general solution

$$y_n = A r_1{}^n + B r_2{}^n, \tag{2}$$

in which r_1 and r_2 are the roots of the quadratic equation

$$\left(1 - \frac{s}{3}\right) r^2 - \left(\frac{4s}{3}\right) r - \left(1 + \frac{s}{3}\right) = 0.$$

From this equation we obtain the derivative

$$\frac{dr}{ds} = (r^2 + 4r + 1)^2 (12r^2 + 12r + 12)^{-1},$$

which is never negative. Hence the roots

$$r_1 = \left[\frac{2s}{3} + \left(1 + \frac{s^2}{3}\right)^{\frac{1}{2}}\right]\left(1 - \frac{s}{3}\right)^{-1},$$

$$r_2 = \left[\frac{2s}{3} - \left(1 + \frac{s^2}{3}\right)^{\frac{1}{2}}\right]\left(1 - \frac{s}{3}\right)^{-1},$$

are monotone increasing functions for all real values of s, except for a discontinuity in r_1 at $s = 3$. Moreover, the roots r_1 and r_2 are analytic within a circle of radius $3^{\frac{1}{2}}$ with center at the origin in the complex s plane. Through terms of degree five in s the power series for r_1 and r_2 are respectively

$$r_1 = 1 + s + \frac{s^2}{2!} + \frac{s^3}{3!} + \frac{s^4}{4!} + \frac{s^5}{72} + \cdots$$
$$= e^s + \frac{s^5}{180} + \cdots, \tag{3}$$

and

$$r_2 = -1 + \frac{s}{3} - \frac{s^2}{18} - \frac{s^3}{54} + \frac{5s^4}{648} + \frac{5s^5}{1944} + \cdots \tag{4}$$

Obviously r_1 is the desired root and r_2 is the unwanted root that produces the oscillation.

Quite apart from questions of stability the process of numerical integration with Simpson's rule requires that the quantity $s/3$ must be numerically less than unity and in practical computation should be considerably less than unity. Cf. Milne [5], p. 67. We shall therefore assume that $|s| < 1$. Table I shows to six decimal places the value of r_1 and r_2 for s ranging from -1 to $+1$ at steps of 0.1. It is evident that in this range r_2 is numerically less than one if s is positive, greater than one if s is negative. Thus the oscillating term increases exponentially with n if G is negative, decreases if G is positive.

APPENDIX D

TABLE I

Values of r_1, r_2, $K(r_1)$, and $K(r_2)$ as functions of s in the interval $-1 \leqq s \leqq 1$

s	r_1	r_2	$K(r_1)$	$K(r_2)$
-1.0	0.366025	-1.366025	0.046575	-0.765324
-0.9	0.405341	-1.328418	0.064930	-0.666147
-0.8	0.448564	-1.290669	0.089191	-0.571332
-0.7	0.496145	-1.252902	0.121503	-0.481257
-0.6	0.548584	-1.215250	0.164758	-0.396239
-0.5	0.606428	-1.177857	0.222862	-0.316521
-0.4	0.670283	-1.140871	0.301085	-0.242261
-0.3	0.740808	-1.104445	0.406537	-0.173536
-0.2	0.818729	-1.068729	0.548807	-0.110330
-0.1	0.904837	-1.033870	0.740818	-0.052544
0	1.	$-1.$	1.	0
0.1	1.105171	-0.967240	1.349859	0.047549
0.2	1.221405	-0.935691	1.822132	0.090404
0.3	1.349877	-0.905432	2.459741	0.128908
0.4	1.491908	-0.876523	3.320868	0.163428
0.5	1.649000	-0.849000	4.484657	0.194343
0.6	1.822876	-0.822876	6.059220	0.222031
0.7	2.015538	-0.798147	8.193055	0.246854
0.8	2.229337	-0.774792	11.091472	0.269159
0.9	2.467061	-0.752775	15.040548	0.289262
1.0	2.732051	-0.732051	20.442545	0.307455

Now suppose that after k steps of the process we recompute y_k from the values already found, using Newton's "three eighths rule", to obtain

$$y_k{}^* = y_{k-3} + \left(\frac{3s}{8}\right)(y_k + 3y_{k-1} + 3y_{k-2} + y_{k-3}).$$

Then we replace the originally computed y_k by the arithmetic mean

$$\bar{y}_k = (y_k + y_k{}^*)/2. \tag{5}$$

From (2) and (5) we find that

$$\bar{y}_k = Ar_1^{k-3}K(r_1) + Br_2^{k-3}K(r_2), \tag{6}$$

in which $K(r)$ is defined by the equation

$$K(r) = \left[r^3 + 1 + \left(\frac{3s}{8}\right)(r + 1)^3\right] \Big/ 2. \tag{7}$$

This function $K(r)$ is the key to the problem. For by means of the series for r_1 and r_2 it can be shown that

$$K(r_1) = r_1^3 + \frac{s^5}{96} + \cdots \tag{8}$$

while

$$K(r_2) = \frac{s}{2} - \frac{s^2}{4} + \cdots .$$

Hence equation (3) becomes

$$\bar{y}_k = Ar_1^{\ k} + \text{terms of degree 5 and higher} \\ + Br_2^{k-3}(s/2) + \text{terms of degree 2 and higher.} \tag{9}$$

Comparing $\bar{y}_k$ with y_k we note that the desired solution is substantially unchanged, and agrees with the true solution e^{ks} through terms of degree 4 in s, while the unwanted solution has been decreased roughly by a factor of magnitude $s/2$.

Table I shows to six decimal places the values of $K(r_1)$ and $K(r_2)$ in the range from $s = -1$ to $s = +1$. It is seen that in this interval the absolute value of $K(r_2)$ is always less than unity.

Consider now the propagation of a single error starting at $n = 0$ and modified after every group of k steps by means of formula (5). Since E_n is a solution of equation (1), in the mth group of k steps the error E_n can be expressed by formula (2) in the form

$$E_n = a_m r_1^{\ j} + b_m r_2^{\ j}, \tag{10}$$

provided $n = mk + j$ and $j < k$. But if $j = k$ the value of E_n can be expressed approximately as

$$E_n = a_m r_1^{\ k} + b_m r_2^{k-3} K(r_2). \tag{11}$$

To obtain this result one must replace $K(r_1)$ by its approximate value $r_1^{\ 3}$, as shown in equation (8).

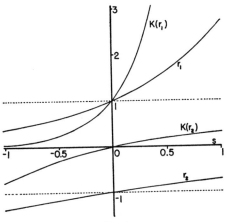

FIG. 1

Similarly in the $(m + 1)$th group of steps we may let

$$E_n = a_{m+1}r_1^{\ j} + b_{m+1}r_2^{\ j},$$

where $n = (m + 1)k + j$ and $j < k$. The coefficients a_{m+1} and b_{m+1} are connected to the coefficients a_m and b_m by the following equations, written in matrix form:

$$\begin{pmatrix} r_1^{-1} & r_2^{-1} \\ 1 & 1 \end{pmatrix}\begin{pmatrix} a_{m+1} \\ b_{m+1} \end{pmatrix} = \begin{pmatrix} r_1^{k-1} & r_2^{k-1} \\ r_1^{k} & r_2^{k-3}K(r_2) \end{pmatrix}\begin{pmatrix} a_m \\ b_m \end{pmatrix}.$$

One may verify the above statement by noting that both members of the first equation are equal to E_{mk+k-1} and both members of the second equation are equal to E_{mk+k}.

Left-hand multiplication by the inverse of the matrix on the left leads to

$$\begin{pmatrix} a_{m+1} \\ b_{m+1} \end{pmatrix} = M \begin{pmatrix} a_m \\ b_m \end{pmatrix} \tag{12}$$

in which

$$M = \begin{pmatrix} u & v \\ 0 & w \end{pmatrix},$$

where

$$u = r_1^{\ k}, \qquad v = r_2^{\ k}P, \qquad w = r_2^{\ k}Q,$$

and

$$P = -\ [r_1 - r_1K(r_2)r_2^{-3}]\ (r_1 - r_2)^{-1},$$
$$Q = [r_1 - r_2K(r_2)r_2^{-3}]\ (r_1 - r_2)^{-1}.$$

The quantities a_m and b_m, and consequently (by equation (10)) the quantity E_n also, will approach zero as m becomes infinite provided both latent roots of M, namely u and w, are less than one in absolute value. For the case under consideration where s lies between -1 and 0 the value of u is always less than one, as we see from table I.

It remains to examine the other latent root $w = r_2^{\ k}Q$. The quantity Q is a function of s alone and its values for s in the interval from -1 to 0 are shown in table II. If we define q by the equation

$$q = -\ \log Q/\log(-r_2) \tag{13}$$

it can be shown that for s between -1 and 0 and for k an integer less than q the latent root w will be less than one in absolute value. Hence to assure that the propagated error E_n will approach zero it is sufficient to choose a value of k which is less than q. For convenience of computers some values of q for s between -1 and 0 are supplied in table II.

It may be noted that the foregoing analysis does not strictly apply if k is less than 3 since formula (6) on which the reasoning depends was derived with the

TABLE II

Values of P, Q, and q as functions of s

s	P	Q	q	s	q
−1.0	0.1479	0.4481	2.57	−0.10	21.3
−0.9	0.1674	0.4515	2.80	−0.09	23.6
−0.8	0.1894	0.4551	3.09	−0.08	26.5
−0.7	0.2143	0.4589	3.45	−0.07	30.2
−0.6	0.2424	0.4631	3.95	−0.06	35.2
−0.5	0.2740	0.4677	4.59	−0.05	42.1
−0.4	0.3097	0.4729	5.68	−0.04	52.5
−0.3	0.3498	0.4786	7.42	−0.03	69.8
−0.2	0.3946	0.4850	10.89	−0.02	104.6
−0.1	0.4445	0.4921	21.29	−0.01	208.5
0	0.5000	0.5000	inf.		

For small values of s the approximate formula

$$q = (\log 2)/\log(-r_2)$$

is adequate for practical estimates of q.

tacit assumption that k is not less than 3. Nevertheless machine tests indicate that the convergence for k less than 3 is just about what might be expected on the basis of the above analysis. However, it is unwise for other reasons than stability to use values of s numerically greater than 0.8, so that the accuracy of table II in this range is unimportant.

To illustrate the foregoing theory several computations were performed on the Alwac III-E at Oregon State College for the system

$$\frac{dy}{dx} = -y, \qquad y(0) = 1.$$

In this case $s < 0, |r_2| > 1$, and Simpson's rule, if uncorrected, produces instability.

Table III shows the difference $E = e^{ns} - y_n$ between the true solution e^{ns} and the computed solution y_n after n steps of the computation. Six values of k are used in table III, $k = \infty$ (that is, no stabilization), $k = 169, 39, 19, 5$, and 3. Four values of s are used, namely $s = -0.10, -0.07, -0.04$, and -0.01. The number n of steps in the computations varies from 300 for larger values of $-s$ to 2000 for the smallest. Not all computations were carried to the full number of steps shown at the left, hence some columns are partially blank.

The number of decimal places is indicated for each division of the table. For example at $s = -0.10, k = 169$, and $n = 300$, the entry -31 means -0.000031, while for $s = -0.04, k = 169, n = 500$, the entry -6 means -0.00000006.

From table II we obtain the integral parts of q corresponding to the given values of s and find that according to theory the solution should be stable for $s = -0.10$ if $k < 21$, for $s = -0.07$ if $k < 30$, for $s = -0.04$ if $k < 52$, and for

APPENDIX D

TABLE III

Values of the error $E = e^{ns} - y_n$

n	$k = \infty$	$k = 169$	$k = 39$	$k = 19$	$k = 5$	$k = 3$
	$s = -0.10$,	$q = 21$	(Entries to be multiplied by 10^{-6})			
100	-1	-1	-1	0	0	0
200	-7	-3	0	0	0	0
300	-71	-31	0		0	0
	$s = -0.07$,	$q = 30$	(Entries to be multiplied by 10^{-7})			
100	0	0	0	0	0	0
200	-5	-2	0	0	0	0
300	-29	-14	0		0	0
	$s = -0.04$,	$q = 52$	(Entries to be multiplied by 10^{-8})			
100	0	0	0	0	0	0
200	-1	0	0	0	0	0
300	-2	-2	0	-1	0	-1
400	-5	-2	0	0	0	0
500	-15	-6	0	0		0
600	-48	-10				
700	-159	-20				
	$s = -0.01$,	$q = 208$	(Entries to be multiplied by 10^{-8})			
100	0	0	0	0	0	0
200	0	0	0	0	0	0
300	1	1	1	0	0	0
400	1	1	1	0	0	0
500	1	1	0	0		0
600	1	0	0	-1		-1
700	1	0	0	-1		-1
800	2	0	0	-1		-1
900	2	0	0			0
1000	3	0	0			0
1200	6	0				
1400	11	0				
1600	21	0				
1800	40	0				
2000	75	-2				

$s = -0.01$ if $k < 208$. Hence in table III the three right-hand columns should be stable for all four values of s, four right-hand columns should be stable for $s = -0.04$, and five right-hand columns for $s = -0.01$. The computations appear to conform to the theory, since the error is negligible in these cases. Occasional errors of one unit in the last place are to be explained by the accidents of roundoff, for if they were due to instability they would increase with n. (The error -2 for $k = 169$ at $n = 2000$ is unexplained, but apparently is not due to instability, as it persisted without increasing through many steps between 1800 and 2000.)

The results shown in table III illustrate the normal situation occurring in practice, where if h is properly chosen and the machine operates correctly, the only source of error is roundoff.

We note that of two consecutive values of k the odd value is likely to give somewhat better results. For if k is even no stabilization occurs at odd values of n, and since Simpson's rule operates over two intervals the effect of stabilization only reaches the odd entries indirectly through the derivative y'.

It is the intent of the authors to treat differential systems of higher order in a future paper.

REFERENCES

1. W. E. MILNE, Numerical integration of ordinary differential equations. *Amer. Math. Month. 33*, 455–460 (1926).
2. I. S. MUHIN, On the accumulation of errors in numerical integration of differential equations. *Akad. Nauk, SSSR Prikl. Mat. Meh. 16*, 753–755 (1952). [Russian]
3. F. B. HILDEBRAND, *Introduction to numerical analysis*, pp. 202–214. McGraw-Hill, New York, 1956.
4. WERNER LINIGER, Zur Stabilität der numerischen Integrationsmethoden für Differentialgleichungen. Thèse présentée à la faculté des sciences de l'Université de Lausanne pour l'obtention du grade de doctor ès sciences. Zurich, 1957.
5. W. E. MILNE, *Numerical solution of differential equations*. John Wiley and Sons, New York, 1953.

Stability of a Numerical Solution of Differential Equations—Part II*

W. E. Milne and R. R. Reynolds

Oregon State College, Corvallis, Oregon

In Part I of this paper [1] the authors have shown that instability in Milne's method of solving differential equations numerically [2] can be avoided by the occasional use of Newton's "three eighths" quadrature formula. Part I dealt with a single differential equation of first order. In Part II the analysis is extended to equations and systems of equations of higher order.

A differential equation of order m or a system of equations of total order m can be expressed by m simultaneous equations of the first order

$$dy_i/dt = F_i(y_1, \cdots, y_m, t), \qquad (i = 1, 2, \cdots, m) \quad (1)$$

in m dependent variables y_i. Let $\mathbf{y}$ denote a vector in m-space which has the components $y_1, \cdots, y_m$. Let $\mathbf{z}$, with components $z_1, z_2, \cdots, z_m$, denote a small variation in the vector $\mathbf{y}$ such as might have been produced by a small error occurring at an earlier value of t. The differential system (1) can be written in vector form as

$$d\mathbf{y}/dt = \mathbf{F}(\mathbf{y}, t) \quad (2)$$

and the varied vector $\mathbf{y} + \mathbf{z}$ satisfies

$$d\mathbf{y}/dt + d\mathbf{z}/dt = \mathbf{F}(\mathbf{y} + \mathbf{z}, t). \quad (3)$$

Let us assume that the functions F_i have continuous partial derivatives $\partial F_i/\partial y_j = a_{ij}$ and let $\mathbf{G}$ denote the $m \times m$ matrix $[a_{ij}]$. From equations (2) and (3) it may be shown that to terms of the first order in the small quantity $\mathbf{z}$ we have

$$d\mathbf{z}/dt = \mathbf{G}\mathbf{z}. \quad (4)$$

The matrix $\mathbf{G}$ is usually a variable depending on the y's and t, but it is possible to forecast the general behavior of the error by treating the simple case where $\mathbf{G}$ is constant. (See [3] of Part I.) Assuming that $\mathbf{G}$ is constant we first of all briefly review some well-known facts about the solution of the differential system (4).

If we introduce a new vector variable $\mathbf{x}$ in place of the variable $\mathbf{z}$ by means of a nonsingular linear transformation $\mathbf{x} = \mathbf{T}\mathbf{z}$, it is seen that the system (4) is transformed into

$$d\mathbf{x}/dt = (\mathbf{T}\mathbf{G}\mathbf{T}^{-1})\mathbf{x}. \quad (5)$$

In particular the transformation $\mathbf{T}$ may be chosen so as to put the matrix $\mathbf{T}\mathbf{G}\mathbf{T}^{-1}$ into the classical canonical form. (Cf. e.g., Turnbull and Aitkin [3].)

*Reprinted from *Journal of the Association for Computing Machinery*, Vol. 7, No. 1, January 1960. Original pagination is indicated within brackets.

Then if the latent roots λ_1, λ_2, $\cdots$, λ_m, of the matrix $\mathbf{G}$ are all distinct, the new matrix $\mathbf{TGT}^{-1}$ has these latent roots in the principal diagonal and zeros elsewhere. In this case the differential system (5) separates into m independent equations

$$dx_i/dt = \lambda_i x_i, \qquad (i = 1, 2, \cdots, m), \quad (6)$$

where x_i is the ith component of $\mathbf{x}$. The solutions are

$$x_i = c_i e^{\lambda_i t},$$

in which the c_i are arbitrary constants.

If the latent roots are not all distinct there may occur ones instead of zeros in the diagonal just above the principal diagonal. Wherever a one occurs the latent root to its left equals the latent root below it. In such a case we have in addition to equations of type (6) certain nonhomogeneous linear equations. For example, if $\lambda_1 = \lambda_2 = \lambda_3$, while the remaining roots are distinct, we would have

$$dx_1/dt = \lambda_1 x_1 + x_2,$$
$$dx_2/dt = \lambda_1 x_2 + x_3, \qquad (7)$$
$$dx_3/dt = \lambda_1 x_3.$$

The solutions of the system (7) are

$$x_3 = c_3 e^{\lambda_1 t}, \qquad x_2 = (c_2 + c_3 t)e^{\lambda_1 t}, \qquad x_1 = (c_1 + c_2 t + c_3 t^2/2)e^{\lambda_1 t}.$$

It should be noted that multiple roots do not always lead to equations of type (7). Such a case is illustrated by example 3 at the end of the paper.

Turning now to numerical integration and the problem of stability we introduce a linear operator S defined by the equation

$$Sf(t) = f(t_{n+1}) - f(t_{n-1}) - (h/3)[f'(t_{n+1}) + 4f'(t_n) + f'(t_{n-1})], \qquad (8)$$

where $f'(t)$ means df/dt and where $t_n = nh$. If $f(t_{n+1})$ has been computed by means of Simpson's rule from $f(t_{n-1})$ and the values of $f'(t)$, it is clear that $Sf(t) = 0$. Since in Milne's method the final values of the variables y_i are found by Simpson's rule we may assume[1] that $Sy_i = 0$ for $i = 1$ to m. Then $Sy = 0$, and since $\mathbf{z}$ represents an error inherited from previous steps we have $S(\mathbf{y} + \mathbf{z}) = 0$, whence $S\mathbf{z} = 0$ also. Moreover, since $\mathbf{x} = \mathbf{Tz}$, it follows that $S\mathbf{x} = S\mathbf{Tz} = \mathbf{T}S\mathbf{z} = 0$. Hence

$$Sx_i = 0, \qquad\qquad i = 1, \cdots, m. \quad (9)$$

From equations (6), (8), and (9) it is clear that in case λ_i is not a multiple root the values x_i satisfy the difference equation

$$\left(1 - \frac{s}{3}\right)x_i(t_{n+1}) - \left(\frac{4s}{3}\right)x_i(t_n) - \left(1 + \frac{s}{3}\right)x_i(t_{n-1}) = 0,$$

[1] In actual calculation this is not exactly true, since the values of y_i' used to compute the final approximation to y_i were themselves computed from the preceding approximation to y_i. For properly chosen step length this error may be ignored.

in which $s = h\lambda_i$. This is identical with equation (1) of Part I. Hence we see by equation (2) of Part I that x_i is expressed in the form

$$x_i(t_n) = Ar_1{}^n + Br_2{}^n,$$

where r_1 and r_2 are roots of the quadratic equation

$$\left(1 - \frac{s}{3}\right) r^2 - \left(\frac{4s}{3}\right) r - \left(1 + \frac{s}{3}\right) = 0. \tag{10}$$

We now consider the following three cases:

Case 1, where λ_i is real and simple. In this case the treatment of Part I applies without change to the stabilization of the particular component x_i.

Case 2, where the root λ_i is complex but not a multiple root.

Consider equation (10) as the defining equation for a complex function r of a complex variable s. This two-valued function has branch points at $s = \pm i\sqrt{3}$. We make branch cuts in the two-sheeted Riemann surface in the s-plane along the axis of imaginaries from $s = + i\sqrt{3}$ up to infinity, and from $s = -i\sqrt{3}$ down to infinity, as shown in figure 1 (here i is the unit of imaginaries).

These branch cuts are mapped in the r-plane on the circle $BCDC'$ with center at $(-2, 0)$ and radius $\sqrt{3}$. Corresponding points in the s- and r-planes are denoted by the same letters. The function which we have called r_2 is mapped conformally on the *interior* of the circle $BCDC'$, the function r_1 on the *exterior*. The interior of the unit circle in the s-plane (which is our only concern in this discussion) is mapped on the shaded regions in figure 2. The segment BD of the imaginary axis in the s-plane goes into the unit circle $BADA'$ with center at the origin in the r-plane.

When λ_i is a complex latent root, but not a multiple root, we proceed exactly as in Part I, and find that the stability of the solution corresponding to $s = h\lambda_i$ depends on the latent roots u and w of the matrix M. These latent roots are approximately expressed by the formulas

$$u = r_1{}^k, \qquad w = r_2{}^k Q,$$

just as in Part I.

(We recall that the above simple forms for u and w were obtained by replacing $K(r_1)$ by $r_1{}^3$. The resulting error in u and w can be shown to be of the order of s^5. In the domain of s that is used in practice we may ignore these errors.)

The value of $|r_1|$ exceeds unity if s is in the right half plane, is less than unity if s is in the left half. This is evident from figure 2, since the section BD of the imaginary axis maps on the unit circle. If s is in the right half plane an error will tend to increase like $r_1{}^n$, but so does the true solution of the differential equation and this nonoscillating error is not regarded as instability.

Our real concern is with the root w. In the complex case it is convenient to write

$$w = r_2{}^{k-2}(r_2{}^2 Q). \tag{11}$$

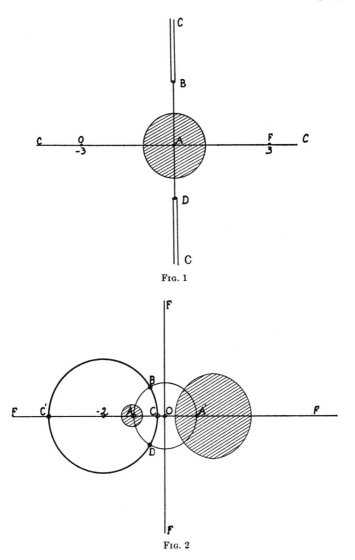

FIG. 1

FIG. 2

For the treatment of stability in the complex case the following theorem is basic:

THEOREM. *For all values of s such that $|s|$ is constant and less than one the value of $|w|$ attains its maximum for s real and negative.*

To show that this is true for all values of k greater than 2 we show that if

$|s|$ is constant and less than one then each of the quantities $|r_2|$ and $|r_2^2 Q|$ attains a maximum value when s is real and negative. These results were first obtained by a rather lengthy analysis, but it seems simpler to display the computed values, as shown in tables I and II.

With the aid of the foregoing theorem we have an easy procedure for the stabilization if λ_i is a nonmultiple complex root. We enter table II of Part I with the real value $-|h\lambda_i|$ playing the role of s in the table and obtain a value q such that $|w| < 1$ if $k < q$. The corresponding solution x_i will then be stable.

Case 3, where λ_i is a multiple root and where one or more equations of type (7) occur.

Suppose in particular that

$$dx_1/dt = \lambda_1 x_1 + x_2 ,$$

$$dx_2/dt = \lambda_1 x_2 .$$

The question of stability arises only if the real part of λ_1 is negative, which is henceforth assumed. Then if h and k are properly chosen the preceding analysis shows that x_2 is stabilized and approaches zero exponentially. The equation $dx_1/dt = \lambda_1 x_1 + x_2$ is then of the form $dx_1/dt = f(x_1, t)$, where $\partial f/\partial x_1 = \lambda_1$, since x_2 is a known function of t. Hence the argument of Case 1 or of Case 2 applies and a proper choice of h and k will assure stability. The same reasoning is easily extended to the case of roots of higher multiplicity.

Since $\mathbf{z} = \mathbf{T}^{-1}\mathbf{x}$, it is clear that if $\mathbf{x}$ is stabilized so also is $\mathbf{z}$. Thus the stabilization process applies to a system of the form (1) provided it has continuous partial derivatives $\partial F_i/\partial y_j$ for all values of i and j from 1 to m.

To test the foregoing analysis numerous computations were carried out on the Alwac III-E at Oregon State College. Of these the following are selected to illustrate several different aspects of the theory.

Example 1 tests the case of pure imaginary latent roots, where, according to theory, instability does not occur. While an error has accumulated at the end of 2000 steps in the computation, it is clearly not an oscillating error, as we note from the steps 2000 to 2003, and hence is not due to instability. Stabilization has no effect on this error.

Example 2 tests the case of equal negative roots. Here instability is to be expected and does in fact occur as we see from steps 1750 to 1753 in the unstabilized case. Stabilization completely removes the oscillation and gives results correct within one unit in the last digit.

Example 3 treats the case of repeated complex roots with negative real parts, where, however, none of the differential equations in x are of the form $dx_i/dt = \lambda_i x_i + x_{i+1}$. This example also agrees with theory, as we note by comparing steps 1300 to 1303 for the unstabilized and stabilized cases.

Example 4 has the same latent roots as example 3 but equations of the form $dx_i/dt = \lambda_i x_i + x_{i+1}$ do occur, and the solution therefore contains terms of the form $te^{-t}\cos t$ and $te^{-t}\sin t$. Here again stabilization wipes out the oscillation.

NOTE. A striking illustration of the stabilization process occurred by acci-

TABLE I
Values of $|r_2|$

Arg s (degrees)	$|s| = 0.25$	$|s| = 0.50$	$|s| = 0.75$	$|s| = 1.00$
0	0.9204	0.8490	0.7863	0.7321
9	9213	8505	7880	7337
18	9240	8550	7932	7387
27	9285	8625	8021	7475
36	9347	8731	8150	7605
45	9425	8870	8324	7788
54	9518	9040	8549	8037
63	9625	9242	8830	8368
72	9743	9473	9170	8802
81	9869	9728	9564	9351
90	1.0000	1.0000	1.0000	1.0000
99	1.0133	1.0279	1.0456	1.0694
108	1.0264	1.0556	1.0906	1.1361
117	1.0390	.1.0820	1.1325	1.1950
126	1.0506	1.1062	1.1697	1.2443
135	1.0610	1.1274	1.2013	1.2840
144	1.0699	1.1453	1.2270	1.3149
153	1.0770	1.1594	1.2468	1.3379
162	1.0822	1.1696	1.2607	1.3537
171	1.0854	1.1758	1.2690	1.3630
180	1.0865	1.1779	1.2718	1.3660

TABLE II
Values of $|r_2^2 Q|$

Arg s (degrees)	$|s| = 0.25$	$|s| = 0.50$	$|s| = 0.75$	$|s| = 1.00$
0	0.4435	0.3980	0.3621	0.3339
9	4440	3988	3627	3343
18	4457	4010	3647	3355
27	4485	4048	3681	3378
36	4524	4103	3734	3415
45	4575	4178	3808	3472
54	4637	4274	3911	3558
63	4709	4394	4050	3683
72	4791	4540	4235	3867
81	4881	4712	4474	4139
90	4978	4908	4775	4544
99	5079	5123	5133	5101
108	5182	5351	5528	5756
117	5282	5579	5929	6407
126	5378	5798	6307	6980
135	5465	5997	6639	7446
144	5541	6169	6913	7802
153	5603	6307	7125	8059
162	5648	6408	7274	8232
171	5676	6469	7362	8330
180	5686	6489	7391	8362

APPENDIX D

dent. In the computation of the unstabilized values for example 4 the switch on the console which turns on the stabilization routine was inadvertently pushed and stabilized values were obtained for a *single* stabilization at the 1000th step. They agree *exactly* with those given above where stabilization occurred at every 50 steps. This of course affects only one line of the computation, and subsequent steps would inherit errors from previous lines. This inadvertent stabilization was typed out separately and had no influence on the actual unstabilized computation.

EXAMPLE 1

Pure Imaginary Roots

$$G = \begin{bmatrix} 0 & 1 \\ -1 & 0 \end{bmatrix}$$

Latent roots: i, $-i$.

Particular solution computed:

$$y_1 = \sin t, \quad y_2 = \cos t$$

By numerical integration (unstabilized) with $h = 0.01$:

Step	t	$10^7 y_1$	Error	$10^7 y_2$	Error
0	0	0	0	10000000	0
500	5.00	−9589243	0	2836623	+1
1000	10.00	−5440213	−2	−8390714	+1
1500	15.00	6502876	−2	−7596882	+3
2000	20.00	9129455	+2	4080816	−5
2001	20.01	9169806	+2	3989319	−5
2002	20.02	9209240	+2	3897423	−5
2003	20.03	9247753	+2	3805138	−4

This computation was repeated with stabilization every 127 steps. The values obtained agree with those shown above for the unstabilized case except for an occasional difference of one unit in the last digit.

EXAMPLE 2

Repeated Real Roots

$$G = \begin{pmatrix} 2 & -1 & -1 & -1 \\ 3 & -2 & -1 & -1 \\ 2 & 0 & -2 & -1 \\ 2 & -1 & 0 & -2 \end{pmatrix}$$

Latent roots: -1, -1, -1, -1.

Particular solution computed:

$$y_1 = [1 + 2t + t^2/2 + t^3/6]e^{-t}, \qquad y_2 = [2t + t^2/2 + t^3/6]e^{-t},$$
$$y_3 = [1 + t + t^2/2 + t^3/6]e^{-t}, \qquad y_4 = [2t + t^3/6]e^{-t}.$$

By numerical integration (unstabilized) with $h = 0.01$:

Step	t	$10^7 y_1$	$10^7 y_2$	$10^7 y_3$	$10^7 y_4$
0	0	10000000	0	10000000	0
500	5.00	2987157	2919778	2650260	2077534
1000	10.00	107901	107449	103362	84745
1500	15.00	2148	2151	2094	1792
1750	17.50	228	242	193	190
1751	17.51	314	300	340	276
1752	17.52	223	237	188	186
1753	17.53	310	296	337	272

By numerical integration (stabilized every 50 steps) with $h = 0.01$:

Step	t	$10^7 y_1$	$10^7 y_2$	$10^7 y_3$	$10^7 y_4$
0	0	10000000	0	10000000	0
500	5.00	2987157	2919778	2650260	2077534
1000	10.00	107901	107447	103361	84747
1500	15.00	2160	2157	2114	1813
1750	17.50	272	272	268	233
1751	17.51	270	270	265	232
1752	17.52	267	267	263	229
1753	17.53	265	265	261	228
1754	17.54	263	263	259	226

By the analytic solution:

t	$10^7 y_1$	$10^7 y_2$	$10^7 y_3$	$10^7 y_4$
0	10000000	0	10000000	0
5.00	2987157	2919777	2650259	2077534
10.00	107901	107447	103361	84747
15.00	2160	2157	2114	1812
17.50	272	272	267	233

EXAMPLE 3

Repeated Complex Roots

$$\mathbf{G} = \begin{pmatrix} -4 & 2 & 0 & 2 \\ -4 & 1 & 1 & 2 \\ -2 & 1 & -1 & 2 \\ -1 & 1 & -1 & 0 \end{pmatrix}$$

Latent roots: $(-1 + i)$, $(-1 - i)$, $(-1 + i)$, $(-1 - i)$.

Particular solution calculated:

$$y_1 = e^{-t}\cos t - 3e^{-t}\sin t, \qquad y_2 = -3e^{-t}\sin t,$$
$$y_3 = e^{-t}\cos t - 2e^{-t}\sin t, \qquad y_4 = -2e^{-t}\sin t.$$

By numerical integration (unstabilized) with $h = 0.01$:

Step	t	$10^7 y_1$	$10^7 y_2$	$10^7 y_3$	$10^7 y_4$
0	0	10000000	0	10000000	0
500	5.00	212948	193836	148337	129224
1000	10.00	360	740	112	493
1300	13.00	−1	−21	6	−17
1301	13.01	−14	−35	−2	−20
1302	13.02	−2	−22	5	−18
1303	13.03	−15	−36	−3	−21

By numerical integration (stabilized every 50 steps) with $h = 0.01$:

Step	t	$10^7 y_1$	$10^7 y_2$	$10^7 y_3$	$10^7 y_4$
0	0	10000000	0	10000000	0
500	5.00	212949	193836	148337	129224
1000	10.00	359	740	112	493
1300	13.00	−7	−27	2	−18
1301	13.01	−8	−28	2	−19
1302	13.02	−8	−28	1	−19
1303	13.03	−9	−28	1	−19

By the analytic solution given above:

Step	t	$10^7 y_1$	$10^7 y_2$	$10^8 y_3$	$10^7 y_4$
0	0	10000000	0	10000000	0
	5.00	212948	193835	148337	129224
	10.00	360	741	113	494
	13.00	−8	−28	2	−19

EXAMPLE 4

Repeated Complex Roots

$$\mathbf{G} = \frac{1}{2}\begin{pmatrix} 1 & -3 & -1 & -1 \\ 3 & -5 & 1 & -3 \\ 1 & 1 & -3 & 1 \\ -1 & 3 & -1 & -1 \end{pmatrix}$$

Latent roots: $(-1 \pm i)$, $(-1 \pm i)$.

Particular solution calculated:

$$y_1 = te^{-t}\cos t + e^{-t}\cos t, \qquad y_2 = te^{-t}\cos t + e^{-t}\sin t,$$
$$y_3 = te^{-t}\sin t + e^{-t}\cos t, \qquad y_4 = te^{-t}\sin t - e^{-t}\sin t.$$

By numerical integration (unstabilized) with $h = 0.01$:

Step	t	$10^7 y_1$	$10^7 y_2$	$10^7 y_3$	$10^7 y_4$
0	0	10000000	0	10000000	0
500	5.00	114678	30954	−303948	−258450
1000	10.00	−4201	−4070	−2846	−2217
1300	13.00	276	259	180	158
1301	13.01	295	290	109	71
1302	13.02	268	252	182	161
1303	13.03	287	283	111	73
1304	13.04	261	245	185	164

Computed values (stabilized every 50 steps):

Step	t	$10^7 y_1$	$10^7 y_2$	$10^7 y_3$	$10^7 y_4$
0	0	10000000	0	10000000	0
500	5.00	114678	30953	−303947	−258448
1000	10.00	−4191	−4057	−2851	−2223
1300	13.00	288	277	144	114
1301	13.01	284	273	145	115
1302	13.02	280	270	146	116
1303	13.03	276	266	147	118
1304	13.04	272	263	148	119

By the analytic solution:

t	$10^7 y_1$	$10^7 y_2$	$10^7 y_3$	$10^7 y_4$
5.00	114678	30953	−303946	−258447
10.00	−4190	−4056	−2851	−2223
13.00	286	275	145	115

Conclusions. It has been shown in these two papers that instability in the Milne method of numerical solution is eradicated by the occasional application of Newton's "three eighths" rule. Specifically this stabilization process consists in replacing the computed **y** by the arithmetic mean of **y** and a new **y** obtained by Newton's rule, as shown in equation (5) of Part I. This operation is performed after every k steps of the computation.

An appropriate value of k can be found from table II of Part I provided we can estimate the magnitude of $h\lambda$, where λ is the largest latent root (in absolute value) of the matrix **G**. In actual problems it is usually out of the question to find the latent roots and a more practical way to estimate k must be devised. The method recommended is similar to that by which the step interval h in the numerical integration is controlled. This is now briefly outlined.

In general it is impractical to compute the theoretical error term for Simpson's rule. But this error is roughly $\frac{1}{29}$ of the difference between the "predicted" and the "corrected" values of the corresponding variable. See [4, p. 66]. The coded routine for numerical solution therefore obtains each difference between a predicted and a corrected value and compares this difference with a constant that does not exceed about 15 times the permissible error. If the above difference exceeds this amount the machine stops and displays the difference. (On a binary machine like the ALWAC these steps may be expeditiously performed by (1) a subtraction, (2) a right shift of a suitable number of bits, and (3) a "nonzero jump" to a stop. The residual difference is automatically displayed on the oscilloscope.) If the difference is only slightly larger than the permitted amount the operator can decide whether to change h or to continue with somewhat reduced accuracy. However, if a large difference suddenly appears it almost certainly indicates faulty coding or a machine failure.

In like manner after each stabilization there should be provision for comparing the unstabilized with the stabilized value. In this case, however, the allowable difference should not exceed the permissible error itself rather than 15

APPENDIX D

times the error as in the former comparison. If the machine stops from this cause the value of k should be decreased.

If instability is not removed for k as small as 10, it may be worthwhile to shorten the interval h. (See [4, p. 67] for a discussion of a third source of error, different from truncation error and instability.) A rough rule of thumb asserts that cutting h in half permits one to double k.

REFERENCES

1. W. E. MILNE AND R. R. REYNOLDS, Stability of a numerical solution of differential equations, *J. Assoc. Comput. Mach. 6*, 196–203 (1959).
2. W. E. MILNE, Numerical integration of ordinary differential equations, *Amer. Math. Month. 33*, 455–460 (1926).
3. TURNBULL AND AITKEN, *An Introduction to the Theory of Canonical Matrices* (Blackie and Son, Ltd., London and Glasgow, 1932).
4. W. E. MILNE, *Numerical Solution of Differential Equations* (John Wiley and Sons, New York, 1953).

APPENDIX E

THE ADAMS-MOULTON METHOD

This method applies to a system of first-order differential equations

(1) $$\bar{y}' = \bar{f}(x,\bar{y}),$$

in which $\bar{y}$ is an m-dimensional vector and $\bar{f}$ is an m-dimensional vector function. If the equation or equations to be solved are not already in this form they must be reduced to it, by the introduction of additional variables if needed. The initial values x_0, $\bar{y}_0$ must also be known, and a suitable step length $\Delta x = h$ has to be chosen.

As its name implies, the Adams-Moulton method uses an open type quadrature formula to integrate ahead, as in Adams' method, and then improves the value so obtained by use of a closed type formula, as in Moulton's method. The first formula is called the "predictor," denoted by P, and the second is the "corrector," denoted by C. The operation of computing $\bar{y}'$ by equation 1 when x and $\bar{y}$ are given is called a "substitution," denoted by S. The difference between the predicted value and the corrected value of $\bar{y}$ is denoted by D, and is used to estimate the error of that step, in much the same way as explained on page 66. Thus if the truncation error of the predictor is $py^{(k)}(s)h^k$, and that of the corrector is $-cy^{(k)}h^k$, we see that

$$D = (p + c)y^{(k)}h^k,$$

whence the error of the corrector is

(2) $$E_c = -cD/(p + c).$$

Since the numerical values of p and c are known for the formulas in use, formula 2 gives a good approximation to the truncation error of that step. The machine should examine the difference D at each step, and should have stored instructions on what to do if D is either too large or too small. Appendix G treats the problem of changing the step length h when the value of D indicates that a change should be made.

The order of operations for one complete step of the regular integration routine (i.e., after the necessary initial steps have been taken) is as follows: 1. predict; 2. shift; 3. substitute to get the new $\bar{y}'$;

4. correct the predicted value; 5. find and store the difference D; 6. substitute to get a corrected $\bar{y}'$; 7. record $\bar{y}$.

Item 7 refers to the disposal to be made of the latest computed $\bar{y}$: whether to store it in memory, or on tape or disk, or to print it out, or to print every pth step, or whatever is desired.

The second item, "shift," deals with the changes needed to go from one step to the next. These can be handled in either one of two ways. The first uses a fixed block of $2m$ addresses in which $\bar{y}_n$ and $\bar{y}_{n-1}'$ are stored, and a fixed block of mk addresses for $\bar{y}_n', \bar{y}_{n-1}', \cdots, \bar{y}_{n-k+1}'$ used in the kth order predictors and correctors. After $\bar{y}_{n+1}$ is predicted we move $\bar{y}_n$ to the $\bar{y}_{n-1}$ spot, blotting out the previous value stored there, and put the new $\bar{y}_{n+1}$ in the $\bar{y}_n$ spot. Similarly, after the substitution gets the new $\bar{y}_{n+1}'$, this value is put in the $\bar{y}_n'$ spot, $\bar{y}_n'$ in the $\bar{y}_{n-1}'$ spot, and so on, the old value of $\bar{y}_{n-k+1}'$ being finally blotted out by the new.

In the other procedure for shifting, the values of $\bar{y}$ and of $\bar{y}'$ are not moved, but each new value is assigned a new address, and the addresses in the predict and correct routines must be changed at each step.

If differences instead of ordinates are used the procedure is somewhat similar, the main difference being that any change in $\bar{y}_n'$ requires a change in the whole string of following differences of the $\bar{y}'$s.

We now consider specific predictors and correctors, first in terms of differences, and then in terms of ordinates.

Predictors and correctors in terms of backward differences are obtained from Table II on page 50. Formula 20.3, with signs changed and fractions reduced to lowest terms, gives the one-step predictor P_1 in Table E–I, and the corrector C is obtained from 20.5. A two-step predictor P_2, which has a slight advantage over P_1 in accuracy and in simplicity of coefficients, is obtained merely by the addition of P_1 and C. The author has used P_2 in preference to P_1 for a number of years.

Still another predictor, which gives slightly better stability than P_1, has been proposed by Krogh (see Appendix H, reference [131]).

In actual computation the quadrature formulas are often expressed and used in terms of ordinates rather than in terms of differences. Predictors of type P_2 and correctors C are given in terms of ordinates in Table E–II for fifth, sixth, and seventh orders.

We now return to the beginning and see how to get the computation started. Quadrature formulas of the kth order, whether in ordinates or in differences, use k values of the derivatives $\bar{y}'$, but at the start only $\bar{y}_0'$ is known. Hence the first $k-1$ steps require a special method, and special formulas. The usual procedure is iteration along the lines illustrated in Article 18, page 44. The necessary starting formulas for the fifth, sixth, and seventh orders are given in differences in Tables E–III, E–IV, and E–V, respectively. For the fifth and sixth orders the same formulas expressed in ordinates instead of differences are found on pages 48 and 49, formulas 19.6 through 19.14. Formulas for the seventh order expressed in ordinates are given here in Table E–VI.

A variation of the foregoing procedure, especially convenient for odd orders, is the use of starting formulas which give values both preceding and following the initial point $(x_0, \bar{y}_0)$. This scheme has the advantage of more rapid convergence, but the disadvantage (in most cases) of giving fewer values that are ultimately useful.

Symmetric starting formulas of this type are given in Table E–VII for the fifth order and in Table E–VIII for the seventh order. The author regularly uses the seventh-order symmetric starting method also to start the Adams-Moulton sixth order, rather than take the trouble to code a special method for the sixth order.

Several references to the extensive literature on predict-correct methods are included in the bibliography in Appendix H. Among the wealth of articles on the important topic of stability in particular one may consult the following references in Appendix H: Crane and Klopfenstein [60], Crane and Lambert [61], Hamming [112], and Krogh [129], [131], [132].

In the tables of these Appendices each entry in a column headed *Div* is the divisor for a given formula. Each entry in a column headed E is the factor p or c of the truncation term $py^{(k)}(s)h^k$ or $-cy^{(k)}h^k$ for a $(k-1)$th order formula. This factor is also the coefficient of $\nabla^{k-1}y_n'$ in the formula when written in terms of backward differences.

Table E–I. Adams-Moulton Predictors and Correctors in Terms of Backward Differences

	y_n'	$\nabla y_n'$	$\nabla^2 y_n'$	$\nabla^3 y_n'$	$\nabla^4 y_n'$	$\nabla^5 y_n'$	$\nabla^6 y_n'$	$\nabla^7 y_n'$	$\nabla^8 y_n'$	$\nabla^9 y_n'$
$P_1\ y_{n+1} = y_n\ + h\big($	1	$\dfrac{1}{2}$	$\dfrac{5}{12}$	$\dfrac{3}{8}$	$\dfrac{251}{720}$	$\dfrac{95}{288}$	$\dfrac{19087}{60480}$	$\dfrac{5257}{17280}$	$\dfrac{1070017}{3628800}$	$\dfrac{25713}{89600}\big)$
$C\ \ y_n\ \ = y_{n-1} + h\big($	1	$-\dfrac{1}{2}$	$-\dfrac{1}{12}$	$-\dfrac{1}{24}$	$-\dfrac{19}{720}$	$-\dfrac{3}{160}$	$-\dfrac{863}{60480}$	$-\dfrac{275}{24192}$	$-\dfrac{33953}{3628800}$	$-\dfrac{57281}{7257600}\big)$
$P_2\ y_{n+1} = y_{n-1} + h\big($	2	0	$\dfrac{1}{3}$	$\dfrac{1}{3}$	$\dfrac{29}{90}$	$\dfrac{14}{45}$	$\dfrac{1139}{3780}$	$\dfrac{41}{140}$	$\dfrac{32377}{113400}$	$\dfrac{3956}{14175}\big)$

Table E–II. Adams-Moulton Predictor P_2 and Corrector C in Terms of Ordinates for Fifth, Sixth, and Seventh Orders

		y_6'	y_5'	y_4'	y_3'	y_2'	y_1'	y_0'	Div	E
FIFTH ORDER	P_2 $y_5 = y_3 + h($			269	-266	294	-146	$29) \div$	90	0.31111
	C $y_4 = y_3 + h($			251	646	-264	106	$-19) \div$	720	-0.01875
SIXTH ORDER	P_2 $y_6 = y_4 + h($		297	-406	574	-426	169	$-28) \div$	90	0.30132
	C $y_5 = y_4 + h($		475	1427	-798	482	-173	$27) \div$	1440	-0.01427
SEVENTH ORDER	P_2 $y_7 = y_5 + h(13613$	-23886	41193	-40672	24183	-8010	$1139) \div$	3780	0.29286
	C $y_6 = y_5 + h(19087$	65112	-46461	37504	-20211	6312	$-863) \div$	60480	-0.01137

Table E–III. Adams-Moulton Five-point, Fifth-Order Start

	y_4'	$\dfrac{\nabla y_4'}{2}$	$\dfrac{\nabla^2 y_4'}{12}$	$\dfrac{\nabla^3 y_4'}{24}$	$\dfrac{\nabla^4 y_4'}{720}$
$y_1 = y_0 + \ \ h(1$	-7	53	-55	$251)$
$y_2 = y_0 + 2h(1$	-6	38	-32	$116)$
$y_3 = y_0 + 3h(1$	-5	27	-21	$81)$
$y_4 = y_0 + 4h(1$	-4	20	-16	$56)$

Table E–IV. Six-point, Sixth-Order Start

	y_5'	$\dfrac{\nabla y_5'}{2}$	$\dfrac{\nabla^2 y_5'}{12}$	$\dfrac{\nabla^3 y_5'}{24}$	$\dfrac{\nabla^4 y_5'}{720}$	$\dfrac{\nabla^5 y_5'}{1440}$
$y_1 = y_0 + \ \ h(1$	-9	95	-161	1901	$-475)$
$y_2 = y_0 + 2h(1$	-8	74	-108	1076	$-224)$
$y_3 = y_0 + 3h(1$	-7	57	-75	711	$-153)$
$y_4 = y_0 + 4h(1$	-6	44	-56	536	$-112)$
$y_5 = y_0 + 5h(1$	-5	35	-45	425	$-95)$

Table E–V. Seven-Point, Seventh-Order Start

	y_6'	$\dfrac{\nabla y_6'}{2}$	$\dfrac{\nabla^2 y_6'}{12}$	$\dfrac{\nabla^3 y_6'}{24}$	$\dfrac{\nabla^4 y_6'}{720}$	$\dfrac{\nabla^5 y_6'}{1440}$	$\dfrac{\nabla^6 y_6'}{60480}$
$y_1 = y_0 + \ \ h(1$	-11	149	-351	6731	-4277	$19087)$
$y_2 = y_0 + 2h(1$	-10	122	-256	4316	-2376	$9112)$
$y_3 = y_0 + 3h(1$	-9	99	-189	2961	-1575	$6165)$
$y_4 = y_0 + 4h(1$	-8	80	-144	2216	-1184	$4576)$
$y_5 = y_0 + 5h(1$	-7	65	-115	1775	-945	$3715)$
$y_6 = y_0 + 6h(1$	-6	54	-96	1476	-792	$2952)$

Table E–VI. Adams-Moulton Iterative Start in Terms of Ordinates

Seventh Order

(For fifth and sixth orders see pages 48–49, III and IV)

	y_0'	y_1'	y_2'	y_3'	y_4'	y_5'	y_6'	Div	E
$y_1 = y_0 + h($	19087	65112	−46461	37504	−20211	6312	−863) ÷	60480	−0.01137
$y_2 = y_0 + h($	1139	5640	33	1328	−807	264	−37) ÷	3780	−0.00847
$y_3 = y_0 + h($	685	3240	1161	2176	−729	216	−29) ÷	2240	−0.01004
$y_4 = y_0 + h($	286	1392	384	1504	174	48	−8) ÷	945	−0.00847
$y_5 = y_0 + h($	3715	17400	6375	16000	11625	5640	−275) ÷	12096	−0.01137
$y_6 = y_0 + h($	41	216	27	272	27	216	41) ÷	140	0

Table E–VII. Adams-Moulton Symmetric Iterative Start
Fifth Order

	y_2'	y_1'	y_0'	y_{-1}'	y_{-2}'	Div	E
$y_2 = y_0 + h($	29	124	24	4	$-1)$	$\div$ 90	0.01111
$y_1 = y_0 + h($	-19	346	456	-74	$11)$	$\div$ 720	0.00764
$y_{-1} = y_0 - h($	11	-74	456	346	$-19)$	$\div$ 720	-0.00764
$y_{-2} = y_0 - h($	-1	4	24	124	$29)$	$\div$ 90	-0.01111

Table E–VIII. Adams-Moulton Symmetric Iterative Start
Seventh Order

	y_3'	y_2'	y_1'	y_0'	$-y_{-1}'$	y_{-2}'	y_{-3}'	Div	E
$y_3 = y_0 + h($	685	3240	1161	2176	-729	216	$-29)$	$\div$ 2240	0.01004
$y_2 = y_0 + h($	-37	1398	4863	1328	33	-30	$5)$	$\div$ 3780	-0.00132
$y_1 = y_0 + h($	271	-2760	30819	37504	-6771	1608	$-191)$	$\div$ 60480	0.00158
$y_{-1} = y_0 - h($	-191	1608	-6771	37504	30819	-2760	$271)$	$\div$ 60480	-0.00158
$y_{-2} = y_0 - h($	5	-30	33	1328	4863	1398	$-37)$	$\div$ 3780	0.00132
$y_{-3} = y_0 - h($	-29	216	-729	2176	1161	3240	$685)$	$\div$ 2240	-0.01004

APPENDIX F

THE COWELL-CROMMELIN METHOD

The process shown in this Appendix is designed for the solution of a system of differential equations in the form

(1) $$\bar{y}'' = \bar{f}(x,\bar{y}),$$

in which the first derivatives $\bar{y}$ do not explicitly appear. Here, as in Appendix E, $\bar{y}$ is a vector with m components, and $\bar{f}$ is a vector function with m components. In addition to equation 1 there must be given the initial values

(2) $$\bar{y} = \bar{y}_0, \qquad \bar{y}' = \bar{y}_0', \qquad \text{when } x = x_0.$$

The predictor P has the form

(3) $$\bar{y}_{n+1} = \bar{y}_{n-1} + 2h\bar{y}_{n-j}' + h^2 \sum_{i=0}^{k-2} a_i \bar{y}_{n-i}''$$

in the kth order case, while the corrector C is

(4) $$\bar{y}_n = \bar{y}_{n-1} + h\bar{y}_{n-j-1}' + h^2 \sum_{i=0}^{k-2} b_i \bar{y}_{n-i}''.$$

There is also a formula Y' to obtain a new value of $\bar{y}'$ at each step:

(5) $$\bar{y}_{n-j}' = \bar{y}_{n-j-1}' + h \sum_{i=0}^{k-2} c_i y_{n-i}''.$$

Formulas 3 and 4 are of kth order, while formula 5 is of order $k - 1$.

The value of j in formulas 3, 4, and 5 is chosen so as to obtain the least truncation error in formula 5. In figure 29 the hatch-marks indicate the intervals of integration in formula 5 in the cases of $k = 5, 6$, and 7 respectively. At first glance it might appear that the predictor and corrector use different values of $\bar{y}'$, but we must recall that the shift operation occurs after the predict and before the correct operations, hence the value of $\bar{y}'$ is actually the same. Formulas 3, 4, and 5 occur in a given step in the order in which they are numbered above. As in the Adams-Moulton method special quadrature formulas are

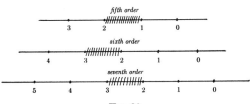

Fig. 29

used iteratively to get the computation under way. For the three cases of fifth, sixth, and seventh orders, the predictors, correctors, "get $\bar{y}'$" formulas, and the starting formulas are given in Tables F–I, F–II, and F–III respectively. For the fifth and seventh orders there is also a special formula to get the value of $\bar{y}'$ needed for the predictor at the first step after the starting values have been obtained. This one extra step is not required in the case of the sixth order with the symmetric starting formulas.

At each step the difference D between the corrected and the predicted values is obtained and used to control the error size in the same manner as in the Adams-Moulton method. Formulas for change of step length h for the present method can be found in Appendix G.

The process described here is stable in the ordinary sense, but is subject to "mathematical instability" and cannot be applied indiscriminately to all equations of type 1. A complete analysis is not available (to the author's knowledge at least), but this much seems almost self-evident: 1. this method can be used for conservative dynamical systems of type 1 ; 2. in the special case of a linear system $\bar{y}'' = \mathbf{A}\bar{y}$ in which $\mathbf{A}$ is an $m \times m$ symmetric matrix, a sufficient condition for stability is that the latent roots of $\mathbf{A}$ be all real, distinct, and negative.

Methods X (page 83), XI (page 88), XII (page 90), and others of this type in Chapter 6 which use the second differences as a replacement for the second derivative turn out to be subject to an error which grows linearly with the number of steps n. This is due to the fact that the associated indicial equation has two roots which approach equality as the step length h approaches zero, and lead to an extraneous solution containing the factor n. All the above-mentioned methods should be discarded in favor of the method treated here, wherever it is applicable.

The term "Cowell–Crommelin method" is perhaps not quite historically accurate as applied to the specific procedure of this Appendix, but obviously the method was inspired by the work of these authors and foreshadowed especially by their method of summation briefly mentioned in Article 40 (page 86).

Table F–I. Fifth-Order Cowell and Crommelin

	y_3''	y_2''	y_1''	y_0''	Div	E
P $y_4 = y_2 + 2hy_2' + h^2($	62	24	6	$-2)$	$\div$ 45	0.05555
C $y_3 = y_2 + hy_1' + h^2($	23	366	159	$-8)$	$\div$ 360	0.00347
Y' $y_2' = y_1' + h($	-1	13	13	$-1)$	$\div$ 24	0.01528

ITERATIVE STARTING FORMULAS

	y_3''	y_2''	y_1''	y_0''	Div	E
$y_1 = y_0 + hy_0' + h^2($	8	-39	114	97)	$\div$ 360	-0.01458
$y_2 = y_0 + 2hy_0' + h^2($	2	-6	66	28)	$\div$ 45	-0.03333
$y_3 = y_0 + 3hy_0' + h^2($	6	27	108	39)	$\div$ 40	-0.05625
SPECIAL						
$y_2' = \qquad y_0' + h($	0	1	4	1)	$\div$ 3	-0.01111

NOTE: The first three starting formulas are iterated until no change occurs. At this point we have values for y_0, y_0'', y_1, y_1'', y_2, y_2', y_3, y_3'' and now use the "special" to get y_2' needed for the predictor. After "predict" all values are shifted, in effect increasing each subscript by 1.

Table F-II. Sixth-Order Cowell and Crommelin

	y_4''	y_3''	y_2''	y_1''	y_0''		Div	E
P $y_5 = y_3 + 2hy_3' + h^2($	129	28	42	−24	5)	÷	90	0.05873
C $y_4 = y_3 + hy_2' + h^2($	97	1444	666	−52	5)	÷	1440	−0.000496
Y' $y_3' = y_2' + h($	−19	346	456	−74	11)	÷	720	0.007639

SYMMETRIC ITERATE STARTING FORMULAS

	y_2''	y_1''	y_0''	y_{-1}''	y_{-2}''		Div	E
$y_2 = y_0 + 2hy_0' + h^2($	5	104	78	−8	1)	÷	90	0.003174
$y_1 = y_0 + hy_0' + h^2($	−17	220	582	−76	11)	÷	1440	0.003670
$y_{-1} = y_0 - hy_0' + h^2($	11	−76	582	220	−17)	÷	1440	0.003670
$y_{-2} = y_0 - 2hy_0' + h^2($	1	−8	78	104	5)	÷	90	0.003174

NOTE: After the symmetric iterate start has converged we get y', then proceed to Predict, Correct, etc.

Table F-III. Seventh-Order Cowell and Crommelin

		y_5''	y_4''	y_3''	y_2''	y_1''	y_0''	Div	E
P	$y_6 = y_4 + 2hy_3' + h^2($	7331	4547	12466	−6110	2299	−373)	÷ 5040	0.06822
C	$y_5 = y_4 + 2hy_2' + h^2($	751	9482	10226	5300	−641	82)	÷ 10080	−0.00472
Y'	$y_3' = y_3'$								
	$+ h ($	11	−93	802	802	−93	11)	÷ 1440	−0.003158

Iterative Starting Formulas

	y_5''	y_4''	y_3''	y_2''	y_1''	y_0''	Div	E
$y_1 = y_0 + hy_0' + h^2($	107	−682	1882	−3044	4315	2462)	÷ 10080	−0.008226
$y_2 = y_0 + 2hy_0' + h^2($	16	−101	272	−370	1088	355)	÷ 630	−0.020106
$y_3 = y_0 + 3hy_0' + h^2($	45	−288	870	−72	3501	984)	÷ 1120	−0.03147
$y_4 = y_0 + 4hy_0' + h^2($	16	−80	608	176	1424	376)	÷ 315	−0.04233
$y_5 = y_0 + 5hy_0' + h^2($	275	1250	6250	2500	11875	3050)	÷ 2016	−0.05684

SPECIAL

	y_5''	y_4''	y_3''	y_2''	y_1''	y_0''	Div	E
$y_3' = \quad y_0' + h ($	3	−21	114	114	219	51)	÷ 160	−0.02946

NOTE: The five starting formulas are iterated until no change occurs. Then the "special" formula gives the value of y_3' needed in the predictor at the first step of the regular routine.

APPENDIX G

CHANGE OF STEP LENGTH

As was indicated in Appendices E and F, it is desirable to modify the step length whenever the difference D between predicted and corrected values becomes too large or too small. Unlike Runge-Kutta methods, where h can be modified at will, the predict-correct procedures using quadrature formulas require some special adjustments in order to change h.

Certainly the simplest, and perhaps also the best way, to change h in the Adams-Moulton or the Milne methods is to replace the value of h stored in the machine by the desired new value, jump to the iterative starting routine, using the latest computed $\bar{y}$ as new initial values, and thus continue the computation with the new h. This process needs but little new coding, and no new formulas to be stored, but probably does use several more substitutions than the regular predict-correct routine for the same number of steps.

Some additional coding is required to compare the difference D with selected upper and lower bounds, to pick a smaller h if D is above the larger bound or a larger h if D is below the lower bound, to insert the new value of h, and to start off the iteration with the latest $\bar{y}$ as initial values. On modern machines the time required for all this is negligible.

For an ardent endorsement of iteration see Rosser [198].*

Substantially the same iteration procedure may be used with the Cowell-Crommelin method, where, however, it is necessary to use one extra quadrature formula to get the proper value of $\bar{y}'$ to use with the iterative starting formulas. This is because the currently stored $\bar{y}'$ is not for the same step as the latest $\bar{y}$.

Some work has been done on doubling or halving the interval for predict-correct methods. Keitel [125] uses a combination of interpolation and quadratures to double or halve the interval for the Milne method.

Haefele and Yungen at Oregon State University Computing Center

---

* Bracketed numbers in this Appendix refer to the bibliography in Appendix H.

used the CDC 3300 computer to obtain coefficients to double or halve the step length for Adams-Moulton fifth, sixth, and seventh orders. However, after some experience with these methods it was decided that to double and halve h is a bit too drastic. As Ceschino [44] puts it, there is the disadvantage of a "variation brusque de l'erreur par pas qui se trouve brutalement multipliée ou divisée par 2^{q+1} pour une formule d'ordre q."

Ceschino, *loc. cit.*, gives general formulas to derive coefficients which enable one to change the step length from h to ωh in the case of Adams' method. This of course requires new coefficients for each new value of ω, but does allow great flexibility in the choice of ω.

To avoid, on the one hand, the task of computing new sets of coefficients for each new value of ω, or on the other hand, of storing many sets of coefficients for prechosen values of ω, a compromise is offered here.

Haefele and Yungen have computed on the CDC 3300 two sets of coefficients, one for $\omega = 1.25$, i.e., a 25% increase, and one for $\omega = 0.80$, a 20% decrease. These were chosen as reciprocals so that in a series of ups and downs the same values of h will appear. Time and funds did not permit exhaustive tests, but a few computations of orbits requiring rather large changes in h indicated that this procedure is superior to the doubling and halving process, and dispelled earlier fears that the process might "chatter," i.e., jump back and forth between "increase h" and "decrease h."

The outstanding virtue of this method of changing h is that it requires no extra computing time, as it simply continues the "predict-correct" process with special formulas which behave exactly like the regular predictors and correctors.

Funds for the above project were supplied in part by the National Science Foundation, and in part by Oregon State University.

These tables appear on the following pages, and supply special coefficients for use in the regular predictors and correctors listed in Appendices E and F and also for the formula to compute $\bar{y}'$, as given in Appendix F.

For a recent contribution to methods for change of step length see Krogh [133].

TABLES FOR CHANGE OF STEP LENGTH IN ADAMS-MOULTON METHOD

The following tables G–I through G–VI provide coefficients for special predictors and correctors to increase the step length by 25%, or to decrease by 20%, for fifth, sixth, and seventh order Adams-Moulton methods.

The coefficients in these tables are numbered in the order in which the independent variable increases. Thus A (1) is the coefficient of the earliest y' used, A (2) of the next, etc.

The value of the step length h stored in the machine *must not* be replaced by the new value until immediately *after* the last predictor has been used.

Table G–I. Adams-Moulton Fifth-Order Coefficients to Increase h by 25%

PREDICTOR No. 1		CORRECTOR No. 1
A (1)	0.6453369141	−0.0554502400
A (2)	−3.2273437500	0.2989157652
A (3)	6.4428222656	−0.6939923322
A (4)	−6.0662109375	1.2855360243
A (5)	4.4553955078	0.4149907826
$E =$	0.642947	$E = \ -0.043615$

PREDICTOR No. 2		CORRECTOR No. 2
A (1)	0.7723141340	−0.0608612669
A (2)	−3.7309695513	0.3035610946
A (3)	6.9806134259	−0.5843098958
A (4)	−5.4904513889	1.1627492877
A (5)	3.9684933803	0.4288607804
$E =$	0.865343	$E = \ -0.058492$

PREDICTOR No. 3		CORRECTOR No. 3
A (1)	0.7651130698	−0.0516699735
A (2)	−3.4618882275	0.2170138889
A (3)	5.5598958333	−0.4803240741
A (4)	−4.1809116809	1.1309523810
A (5)	3.8177910053	0.4340277778
$E =$	1.031664	$E = \ -0.067499$

PREDICTOR NO. 4
A (1)   0.6309175717
A (2)   −2.4305555556
A (3)   4.3518518519
A (4)   −3.8095238095
A (5)   3.7573099415
E =   1.137627

Table G–II. Adams-Moulton Fifth-Order Coefficients to Decrease *h* by 20%

PREDICTOR NO. 1
A (1)   0.1601640000
A (2)   −0.8080560000
A (3)   1.6251840000
A (4)   −1.2868560000
A (5)   2.1095640000
E =   0.152053

CORRECTOR NO. 1
      −0.0126952047
       0.0730209524
      −0.1941807407
       0.6416355556
       0.2922194375
E =   −0.008287

PREDICTOR NO. 2
A (1)   0.1362938012
A (2)   −0.7133866667
A (3)   1.5458607407
A (4)   −1.6287288889
A (5)   2.2599610136
E =   0.115771

CORRECTOR NO. 2
      −0.0113100529
       0.0706005698
      −0.2357333333
       0.6929100529
       0.2835327635
E =   −0.006158

PREDICTOR NO. 3
A (1)   0.1338243386
A (2)   −0.7515168091
A (3)   1.9605333333
A (4)   −2.0808465608
A (5)   2.3380056980
E =   0.095865

CORRECTOR NO. 3
      −0.0130423999
       0.1008888889
      −0.2792592593
       0.7112820513
       0.2801307190
E =   −0.005261

PREDICTOR NO. 4
A (1)   0.1592545668
A (2)   −1.0915555556
A (3)   2.4414814815
A (4)   −2.2851282051
A (5)   2.3759477124
E =   0.085779

Table G–III.   Adams-Moulton Sixth-Order Coefficients to Increase
h by 25%

PREDICTOR No. 1

A (1)	−0.6429473877
A (2)	3.8600738525
A (3)	−9.6568176270
A (4)	12.8722961426
A (5)	−9.2809478760
A (6)	5.0983428955
$E =$	0.640679

CORRECTOR No. 1

0.0415378147
−0.2606959125
0.7015130459
−1.0816786024
1.4599948459
0.3893288084
$E =$ −0.034320

PREDICTOR No. 2

A (1)	−0.8241360780
A (2)	4.8445159314
A (3)	−11.7187500000
A (4)	14.6725501543
A (5)	−8.9518229167
A (6)	4.4776429090
$E =$	0.902012

CORRECTOR No. 2

0.0500466522
−0.3008285479
0.7492146164
−0.9586588542
1.3072942852
0.4029318482
$E =$ −0.048371

PREDICTOR No. 3

A (1)	−0.8827068943
A (2)	4.9975794605
A (3)	−11.3221829530
A (4)	12.1625434028
A (5)	−6.7303502598
A (6)	4.2751172439
$E =$	1.119457

CORRECTOR No. 3

0.0481597851
−0.2682079548
0.5626085069
−0.7638888889
1.2626074735
0.4087210781
$E =$ −0.058597

PREDICTOR No. 4

A (1)	−0.8116851697
A (2)	4.2804493874
A (3)	−8.2552083333
A (4)	9.1310541311
A (5)	−6.0284391534
A (6)	4.1838291381
$E =$	1.282324

CORRECTOR No. 4

0.0382425791
−0.1794704861
0.4444444444
−0.7094494048
1.2448830409
0.4113498264
$E =$ −0.065061

PREDICTOR No. 5

A (1)	−0.6345435348
A (2)	2.8333333333
A (3)	−6.3487654321
A (4)	8.2500000000
A (5)	−5.7412280702
A (6)	4.1412037037
$E =$	1.387369

Table G–IV. Adams-Moulton Sixth-Order Coefficients to Decrease h by 20%

PREDICTOR No. 1

A (1)	-0.1520532000
A (2)	0.9204300000
A (3)	-2.3285880000
A (4)	3.1457160000
A (5)	-2.0471220000
A (6)	2.2616172000

$E = \quad 0.144701$

CORRECTOR No. 1

0.0086328889
-0.0563140117
0.1618163810
-0.2862648889
0.6934328889
0.2786967419

$E = \quad -0.006140$

PREDICTOR No. 2

A (1)	-0.1205949630
A (2)	0.7456157193
A (3)	-1.9537920000
A (4)	2.8322070123
A (5)	-2.3522986667
A (6)	2.4488628980

$E = \quad 0.105287$

CORRECTOR No. 2

0.0070453699
-0.0479627513
0.1495448433
-0.3319466667
0.7531940964
0.2701251084

$E = \quad -0.004362$

PREDICTOR No. 3

A (1)	-0.1096849021
A (2)	0.7044469841
A (3)	-1.9805501994
A (4)	3.4584177778
A (5)	-3.0193706488
A (6)	2.5467409885

$E = \quad 0.083905$

CORRECTOR No. 3

0.0071172679
-0.0527187531
0.2036444444
-0.4015873016
0.7771509972
0.2663933452

$E = \quad -0.003553$

PREDICTOR No. 4

A (1)	-0.1160434824
A (2)	0.8061576001
A (3)	-2.7669333333
A (4)	4.4359788360
A (5)	-3.3590883191
A (6)	2.5999286988

$E = \quad 0.072192$

CORRECTOR No. 4

0.0088256841
-0.0811111111
0.2511111111
-0.4317948718
0.7883660131
0.2646031746

$E = \quad -0.003156$

PREDICTOR No. 5

A (1)	-0.1464409810
A (2)	1.2533333333
A (3)	-3.5101234568
A (4)	4.9107692308
A (5)	-3.5356862745
A (6)	2.6281481481

$E = \quad 0.065910$

Table G–V.   Adams-Moulton Seventh-Order Coefficients to Increase
h by 25%

PREDICTOR NO. 1

A (1)	0.6406785148
A (2)	−4.4870184762
A (3)	13.4702515738
A (4)	−22.4703879220
A (5)	22.4824738639
A (6)	−13.1250189645
A (7)	5.7390214103
$E =$	0.638845

CORRECTOR NO. 1

	−0.0329476008
	0.2376544863
	−0.7452194542
	1.3351207542
	−1.5392841696
	1.6247328501
	0.3699431339
$E =$	−0.028127

PREDICTOR NO. 2

A (1)	0.8659319196
A (2)	−5.9784927426
A (3)	17.5788088673
A (4)	−28.3712869162
A (5)	26.6993823716
A (6)	−13.2814825149
A (7)	4.9871390152
$E =$	0.933788

CORRECTOR NO. 2

	−0.0425241487
	0.2983694884
	−0.8961666295
	1.4862998603
	−1.4230225578
	1.4439051974
	0.3831387900
$E =$	−0.041299

PREDICTOR NO. 3

A (1)	0.9841382195
A (2)	−6.6296530745
A (3)	18.7755145336
A (4)	−28.3805787578
A (5)	22.9093327598
A (6)	−9.8919425320
A (7)	4.7331888515
$E =$	1.199562

CORRECTOR NO. 3

	−0.0445655441
	0.2990086502
	−0.8321463808
	1.1626389922
	−1.1114178577
	1.3872891328
	0.3891930073
$E =$	−0.052091

PREDICTOR NO. 4

A (1)	0.9752669130
A (2)	−6.3012310380
A (3)	16.6216088958
A (4)	−21.3862020503
A (5)	16.7363355599
A (6)	−8.7569573284
A (7)	4.6111790480
$E =$	1.420244

CORRECTOR NO. 4

	−0.0397885643
	0.2469580307
	−0.5792023190
	0.8817579026
	−1.0140069917
	1.3619669645
	0.3923149772
$E =$	−0.060146

PREDICTOR No. 5

A	(1)	0.8484626867
A	(2)	-5.0852513124
A	(3)	11.3573288690
A	(4)	-15.6741622575
A	(5)	14.7444727891
A	(6)	-8.2379590280
A	(7)	4.5471082530
$E =$		1.586322

CORRECTOR No. 5

-0.0301070478
0.1572110615
-0.4474481922
0.8006129535
-0.9743368839
1.3501932457
0.3938748632

$E = \quad -0.065325$

PREDICTOR No. 6

A	(1)	0.6357715395
A	(2)	-3.2137896825
A	(3)	8.6247795414
A	(4)	-13.9878117914
A	(5)	13.9193817878
A	(6)	-7.9929728836
A	(7)	4.5146414888
$E =$		1.694250

Table G–VI. Adams-Moulton Seventh-Order Coefficients to Decrease h by 20%

PREDICTOR No. 1

A	(1)	0.1447010229
A	(2)	-1.0202593371
A	(3)	3.0909453429
A	(4)	-5.2226084571
A	(5)	5.3162313429
A	(6)	-2.9153281371
A	(7)	2.4063182229
$E =$		0.138279

CORRECTOR No. 1

-0.0063512661
0.0470051217
-0.1532543893
0.2933783220
-0.3885908430
0.7394795683
0.2683334865

$E = \quad -0.004782$

PREDICTOR No. 2

A	(1)	0.1089172808
A	(2)	-0.7786368677
A	(3)	2.4080373734
A	(4)	-4.2099356735
A	(5)	4.5869854250
A	(6)	-3.1419489524
A	(7)	2.6265814143
$E =$		0.097001

CORRECTOR No. 2

-0.0048681844
0.0369897172
-0.1258537022
0.2613882599
-0.4341785397
0.8065733116
0.2599491376

$E = \quad -0.003276$

PREDICTOR NO. 3

A (1)   0.0936439486
A (2)   −0.6856916157
A (3)   2.2027501618
A (4)   −4.1319598904
A (5)   5.4249406984
A (6)   −4.0461683309
A (7)   2.7424850282
  $E =$ 0.074857

CORRECTOR NO. 3

        −0.0045171126
        0.0359578589
        −0.1331070066
        0.3424397884
        −0.5320340534
        0.8351758065
        0.2560847190
$E =$ −0.002568

PREDICTOR NO. 4

A (1)   0.0917775723
A (2)   −0.7020194434
A (3)   2.4394661610
A (4)   −5.5869426455
A (5)   7.0863635279
A (6)   −4.5380219780
A (7)   2.8093768057
  $E =$ 0.062329

CORRECTOR NO. 4

        −0.0049267981
        0.0428302512
        −0.1966984127
        0.4345830184
        −0.5799837200
        0.8501774043
        0.2540182569
$E =$ −0.002192

PREDICTOR NO. 5

A (1)   0.1028822606
A (2)   −0.8565303090
A (3)   3.6670476190
A (4)   −7.3414159738
A (5)   8.0052747253
A (6)   −4.8264425770
A (7)   2.8491842549
  $E =$ 0.055016

CORRECTOR NO. 5

        −0.0064478860
        0.0697936508
        −0.2483156966
        0.4785225885
        −0.6044911298
        0.8580083144
        0.2529301587
$E =$ −0.001993

PREDICTOR NO. 6

A (1)   0.1361602264
A (2)   −1.4059682540
A (3)   4.7163315697
A (4)   −8.2369719170
A (5)   8.5053968254
A (6)   −4.9863643235
A (7)   2.8714158730
  $E =$ 0.050939

TABLES FOR CHANGE OF STEP LENGTH IN COWELL-CROMMELIN METHOD

Tables G–VII through G–XII supply predictors, correctors, and formulas for y' to be used for change of step length in the fifth, sixth, and seventh order Cowell-Crommelin methods.

The coefficients in these tables are numbered in the order in which the independent variable increases. Thus A (1) is the coefficient of the earliest y' used, A (2) of the next, etc.

The value of the step length h stored in the machine must be replaced by the new value immediately *before* the first predictor is to be used.

Table G–VII. **Cowell-Crommelin Fifth-Order Coefficients to Increase h by 25%**

PREDICTOR No. 1	CORRECTOR No. 1	y' No. 1
A (1) −0.1050468750	−0.0058493590	−0.0358974359
A (2) 0.3910781250	0.2997685185	0.4444444444
A (3) −0.0620156250	0.9289583333	0.4133333333
A (4) 1.3959843750	0.0771225071	−0.0218803419
$E =$ 0.045425	−0.002613	0.005860
PREDICTOR No. 2	CORRECTOR No. 2	y' No. 2
A (1) −0.0801282051	−0.0330687831	−0.0620039683
A (2) 0.2314814815	0.4583333333	0.5729166667
A (3) 0.4583333333	1.0092592593	0.5277777778
A (4) 1.3903133903	0.0654761905	−0.0386904762
$E =$ 0.048889	0.002361	0.01319
PREDICTOR No. 3		
A (1) −0.0661375661		
A (2) 0.1666666667		
A (3) 0.5185185185		
A (4) 1.3809523810		
$E =$ 0.053333		

Table G–VIII. **Cowell-Crommelin Fifth-Order Coefficients to Decrease h by 20%**

PREDICTOR No. 1	CORRECTOR No. 1	y' No. 1
A (1) 0.0070875000	−0.0357916667	−0.0483630952
A (2) −0.0972000000	0.5909814815	0.6597222222
A (3) 1.1856375000	1.1601458333	0.7161458333
A (4) 1.4357250000	0.0346643519	−0.0775049603
$E =$ 0.032506	0.023513	0.040266

PREDICTOR No. 2		CORRECTOR No. 2	y' No. 2
A (1)	−0.0243809524	−0.0145868946	−0.0273504274
A (2)	0.0758518519	0.4283333333	0.5166666667
A (3)	0.5813333333	1.0240740741	0.5555555556
A (4)	1.3671957672	0.0621794872	−0.0448717949
$E =$	0.063889	0.004861	0.017882

PREDICTOR No. 3	
A (1)	−0.0291737892
A (2)	0.1066666667
A (3)	0.5481481481
A (4)	1.3743589744
$E =$	0.058333

Table G–IX.   Cowell-Crommelin Sixth-Order Coefficients to Increase h by 25%

PREDICTOR No. 1		CORRECTOR No. 1	y' No. 1
A (1)	0.1109003906	−0.0060033701	0.0134640523
A (2)	−0.5486484375	0.0177023237	−0.0887179487
A (3)	1.0564804688	0.2657494213	0.5207407407
A (4)	−0.5056171875	0.9493697917	0.3675555556
A (5)	1.5068847656	0.0731818334	−0.0130423999
$E =$	0.036563	−0.002487	0.003484

PREDICTOR No. 2		CORRECTOR No. 2	y' No. 2
A (1)	0.1123366013	0.0047297899	0.0264311788
A (2)	−0.5208333333	−0.0506365741	−0.1601769180
A (3)	0.8680555555	0.4804687500	0.6966145833
A (4)	0.0763888889	0.9971509972	0.4601139601
A (5)	1.4640522876	0.0682870370	−0.0229828042
$E =$	0.044597	−0.001046	0.005556

PREDICTOR No. 3		CORRECTOR No. 3	y' No. 3
A (1)	0.1068376068	0.0054389446	0.0239313562
A (2)	−0.4629629630	−0.0395833333	−0.1180555556
A (3)	0.6666666667	0.4648148148	0.6435185185
A (4)	0.2450142450	1.0017857143	0.4761904762
A (5)	1.4444444444	0.0675438596	−0.0255847953
$E =$	0.051797	−0.000635	0.007028

PREDICTOR No. 4	
A (1)	0.0870231133
A (2)	−0.3222222222
A (3)	0.5037037037
A (4)	0.2952380952
A (5)	1.4362573099
$E =$	0.056508

Table G–X. Cowell-Crommelin Sixth-Order Coefficients to Decrease
h by 20%

PREDICTOR No. 1		CORRECTOR No. 1	y' No. 1
A (1)	0.0133143750	0.0101377193	0.0173611111
A (2)	−0.0461700000	−0.0770666667	−0.1190476190
A (3)	−0.0173137500	0.6551870370	0.7696759259
A (4)	1.1323800000	1.1119916667	0.6336805556
A (5)	1.4490393750	0.0497502437	−0.0516699735
$E =$	0.065338	0.011080	0.025113

PREDICTOR No. 2		CORRECTOR No. 2	y' No. 2
A (1)	0.0275461988	0.0023703704	0.0087195767
A (2)	−0.1365333333	−0.0247977208	−0.0649116809
A (3)	0.2503111111	0.4470000000	0.5853333333
A (4)	0.4504888889	1.0092592593	0.5010582011
A (5)	1.4081871345	0.0661680912	−0.0301994302
$E =$	0.079147	0.000441	0.010712

PREDICTOR No. 3		CORRECTOR No. 3	y' No. 3
A (1)	0.0284444444	0.0021451316	0.0094385789
A (2)	−0.1517037037	−0.0333333333	−0.0905555556
A (3)	0.3306666667	0.4601851852	0.6231481481
A (4)	0.3703703704	1.0038461538	0.4852564103
A (5)	1.4222222222	0.0671568627	−0.0272875817
$E =$	0.067897	−0.000322	0.008403

PREDICTOR No. 4	
A (1)	0.0343221049
A (2)	−0.2222222222
A (3)	0.4296296296
A (4)	0.3282051282
A (5)	1.4300653595
$E =$	0.061508

Table G–XI. Cowell-Crommelin Seventh-Order Coefficients to Increase h by 25%

PREDICTOR No. 1		CORRECTOR No. 1	y' No. 1
A	(1)	−0.1115840541	0.0072269906	−0.0068253968
A	(2)	0.6688206613	−0.0417132061	0.0471895425
A	(3)	−1.6644889787	0.0877485405	−0.1548717949
A	(4)	2.1723210100	0.1982975088	0.5844444444
A	(5)	−1.0635374582	0.9797231523	0.3388888889
A	(6)	1.6184688197	0.0687170139	−0.0088256841
	$E =$	0.029026	−0.001852	0.001104

PREDICTOR No. 2		CORRECTOR No. 2	y' No. 2
A	(1)	−0.1296178193	0.0027313245	−0.0145062760
A	(2)	0.7528011204	−0.0083665611	0.0959869124
A	(3)	−1.7771291209	−0.0263147794	−0.2893518519
A	(4)	2.0778218696	0.4600384425	0.8051215278
A	(5)	−0.4680059524	1.0050396227	0.4182168594
A	(6)	1.5441299027	0.0668719508	−0.0154671717
	$E =$	0.038349	−0.001536	0.002711

PREDICTOR No. 3		CORRECTOR No. 3	y' No. 3
A	(1)	−0.1352482281	0.0013824742	−0.0153022614
A	(2)	0.7553355210	−0.0007769921	0.0927340052
A	(3)	−1.6673162321	−0.0296626984	−0.2278645833
A	(4)	1.6783234127	0.4566748067	0.7336182336
A	(5)	−0.1456103809	1.0055650038	0.4343584656
A	(6)	1.5145159074	0.0668174058	−0.0175438596
	$E =$	0.046784	−0.001527	0.003656

PREDICTOR No. 4		CORRECTOR No. 4	y' No. 4
A	(1)	−0.1230402046	0.0008093668	−0.0124642480
A	(2)	0.6402414767	0.0003720238	0.0630208333
A	(3)	−1.2051587302	−0.0305996473	−0.1876543210
A	(4)	1.2281644282	0.4571853741	0.7151785714
A	(5)	−0.0411186697	1.0053884712	0.4403508772
A	(6)	1.5009116995	0.0668444114	−0.0184317130
	$E =$	0.053164	−0.001546	0.004226

PREDICTOR No. 5	
A	(1)	−0.0958290236
A	(2)	0.4226190476
A	(3)	−0.9192239859
A	(4)	1.0959183673
A	(5)	0.0020050125
A	(6)	1.4945105820
	$E =$	0.057302

Table G–XII. Cowell-Crommelin Seventh-Order Coefficients to Decrease h by 20%

	PREDICTOR No. 1	CORRECTOR No. 1	y' No. 1
A (1)	−0.0214100357	−0.0037819957	−0.0085720486
A (2)	0.1203645536	0.0292467502	0.0606725146
A (3)	−0.2602703571	−0.1159671939	−0.2072172619
A (4)	0.1967866071	0.6955283245	0.8611111111
A (5)	1.0253298214	1.0892996925	0.5822482639
A (6)	1.4704494107	0.0556744224	−0.0382425791
$E =$	0.095445	0.005644	0.018386

	PREDICTOR No. 2	CORRECTOR No. 2	y' No. 2
A (1)	−0.0270154497	−0.0001655152	−0.0040160692
A (2)	0.1640453133	0.0032314437	0.0296126984
A (3)	−0.4144065306	−0.0266523403	−0.1099122507
A (4)	0.5384759083	0.4492603175	0.6401777778
A (5)	0.2883961905	1.0078430202	0.4666945141
A (6)	1.4505045683	0.0664830741	−0.0225566704
$E =$	0.094668	−0.001300	0.007608

	PREDICTOR No. 3	CORRECTOR No. 3	y' No. 3
A (1)	−0.0254558716	0.0001429259	−0.0037248677
A (2)	0.1608755858	0.0013483684	0.0302034523
A (3)	−0.4369400081	−0.0312698413	−0.1443333333
A (4)	0.6782984127	0.4577286470	0.6871693122
A (5)	0.1525559932	1.0051689052	0.4507834758
A (6)	1.4706658881	0.0668809948	−0.0200980392
$E =$	0.077449	−0.001581	0.005627

	PREDICTOR No. 4	CORRECTOR No. 4	y' No. 4
A (1)	−0.0272658558	0.0002918546	−0.0044945614
A (2)	0.1863199981	0.0014880952	0.0458333333
A (3)	−0.6158730159	−0.0317019400	−0.1706790123
A (4)	0.8982615268	0.4579212454	0.7038461538
A (5)	0.0758648759	1.0051120448	0.4446078431
A (6)	1.4826924709	0.0668886999	−0.0191137566
$E =$	0.067295	−0.001583	0.004799

	PREDICTOR No. 5
A (1)	−0.0345555886
A (2)	0.2904761905
A (3)	−0.7887125220
A (4)	1.0087912088
A (5)	0.0347338936
A (6)	1.4892668178
$E =$	0.061706

APPENDIX H

BIBLIOGRAPHY (1969)*

Reviews known to the compilers are in brackets with the volume and page (or review number preceded by #) of the reviewing journal. CR = *Computing Reviews*, MR = *Mathematical Reviews*. Volume numbers are in boldface.

1. Adachi, R., Approximate formulas for definite integrals and differential coefficients, *Kumamoto J. Sci. Ser. A*, **2**, 196–209 (1955) [*MR* **18** 73].
2. Adachi, R., A method on numerical solution of some differential equation, *Kumamoto J. Sci. Ser. A*, **2**, 40–46 (1954) [*MR* **17** 539].
3. Adachi, R., A method on the numerical solutions of some differential equations, *Kumamoto J. Sci. Ser. A*, **2**, 244–252 (1955) [*MR* **19** 323].
4. Adachi, R., On the numerical solution of the second order differential equation under some conditions, *Kumamoto J. Sci. Ser. A*, **1**, no. 3, 14–33 (1954) [*MR* **16** 631].
5. Adachi, R., On the numerical solution of the simultaneous differential equations under some conditions. I, *Kumamoto J. Sci. Ser. A*, **1**, no. 4, 28–30 (1954) [*MR* **16** 631].
6. Adachi, R., On the numerical solution of the simultaneous differential equations under some conditions. II, *Kumamoto J. Ser. A*, **2**, 24–39 (1954) [*MR* **17** 539].
7. Albrecht, J., Beiträge zum Runge-Kutta Verfahren, *Z. Angew. Math. Mech.*, **35**, 100–110 (1955) (English summary) [*MR* **17** 90].
8. Albrecht., J., Zum Differenzenverfahren bei parabolischen Differentialgleichungen, *Z. Angew. Math. Mech.*, **37**, 202–212 (1957) (English summary) [*MR* **19** 462].
9. Alonso, R., A starting method for the three-point Adams predictor-corrector method, *J. ACM*, **7**, 176–180 (1960) [*MR* **26** #3195].
10. Anderson, W. H., The solution of simultaneous ordinary differential equations using a general purpose digital computer, *Comm. ACM*, **3**, 355–360 (1960) [*MR* **23** #B583].
11. Anderson, W. H., Ball, R. B., and Voss, J. R., A numerical method for solving control differential equations on digital computers, *J. ACM*, **7**, 61–68 (1960) [*MR* **22** #1082].
12. Ardouin, P. G., and Lapierre, G., A compiler for solving systems of linear differential equations. In *The Computing and Data Processing Society of Canada, Proceedings*, Univ. Toronto, **2**, 276–298 (1960) [*CR* **2** #867] =[*MR* **22** #10215].
13. Artemov, G. A., On a modification of Čaplygin's method for systems of ordinary differential equations of first order, *Dokl. Akad. Nauk SSSR*, **101**, 197–200 (1955) (Russian) [*MR* **17** 90].

* Compiled with the aid of R. R. Reynolds and F. T. Krogh.

303

14. Azbelev, N. V., On an approximate solution of ordinary differential equations of the nth order based upon S. A. Čaplygin's method, *Dokl. Akad. Nauk SSSR*, **83**, 517–519 (1952) (Russian) [*MR* **13** 992].

15. Azbelev, N. V., On extension of Čaplygin's method beyond the limit of applicability of the theorem on differential inequalities, *Dokl. Akad. Nauk SSSR*, **102**, 429–430 (1955) (Russian) [*MR* **17** 90].

16. Babkin, B. N., Approximate integration of systems of ordinary differential equations of the first order by the method of S. A. Čaplygin, *Izv. Akad. Nauk SSSR Ser. Mat.*, **18**, 477–484 (1954) (Russian) [*MR* **16** 631].

17. Bahvalov, N. S., On estimation of the error in the numerical integration of differential equations by the Adams extrapolation method, *Dokl. Akad. Nauk SSSR*, **104**, 683–686 (1955) (Russian) [*MR* **17** 412].

18. Bahvalov, N. S., Some remarks concerning numerical integration of differential equations by the method of finite differences, *Dokl. Akad. Nauk SSSR*, **104**, 805–808 (1955) (Russian) [*MR* **17** 667].

19. Bajcsay, P., and Lovass-Nagy, V., Ein Iterationsverfahren zur näherungsweisen Lösung von Matrizendifferentialgleichungen, *Z. Angew. Math. Mech.*, **39**, 8–13 (1959) (English summary) [*MR* **20** #7390].

Ball, R. B., see Anderson, W. H.

20. Ballester, C., and Pereyra, V., On the construction of discrete approximations to linear differential expressions, *Math. Comp.*, **21**, 297–302 (1967) [*MR* **37** #3751].

Bennett, M. M., see Riley, J. D.

21. Bernstein, B., and Truesdell, C., The solution of linear differential systems by quadratures, *J. Reine Angew. Math.*, **197**, 104–111 (1957) [*MR* **18** 654].

22. Blanc, C., Sur les formules d'intégration approchée d'équations différentielles, *Arch. Math. (Basel)*, **5**, 301–308 (1954) [*MR* **16** 290].

23. Blanch, G., On the numerical solution of parabolic partial differential equations, *J. Res. Nat. Bur. Standards*, **50**, 343–356 (1953) [*MR* **15** 474].

24. Blanch, G., and Rhodes, I., Seven-point Lagrangian integration formulas, *J. Math. and Phys.*, **22**, 204–207 (1943) [*MR* **5** 159].

25. Boček, L. V., On numerical integration of equations in a complex region, *Vyčisl. Mat. Vyčisl. Tehn.*, **2**, 94–96 (1955) (Russian) [*MR* **17** 90].

Bradshaw, C. L., see Mann, W. R.

26. Bridgland, T. F., Jr., A note on numerical integrating operators, *J. SIAM*, **6**, 240–256 (1958) [*MR* **20** #2842].

27. Brock, P., and Murray, F. J., The use of exponential sums in step by step integration, *MTAC*, **6**, 63–78 (1952) [*MR* **13** 873].

28. Brock, P., and Murray, F. J., The use of exponential sums in step by step integration. II, *MTAC*, **6**, 138–150 (1952) [*MR* **14** 413].

29. Brodskiĭ, M. L., Asymptotic estimates of the errors in numerical integration of systems of ordinary differential equations by difference methods, *Dokl. Akad. Nauk SSSR*, **93**, 599–602 (1953) (Russian) [*MR* **15** 651].

Brousseau, R., see Evans, G. W.

Browne, H. N., see Lotkin, M.

30. Brush, D. G., Kohfeld, J. J., and Thompson, G. T., Solution of ordinary differential equations using two "off-step" points, *J. ACM*, **14**, 769–784 (1967) [*CR* **9** #14089].

31. Budak, B. M., and Gorbunov, A. D., Stability of calculation processes involved in the solution of the Cauchy problem for the equation $dy/dx = f(x,y)$ by multipoint difference methods, *Dokl. Akad. Nauk SSSR*, **124**, 1191–1194 (1959) (Russian) [*MR* **21** #434].

32. Budden, K. G., The numerical solution of differential equations governing reflexion of long radio waves from the ionosphere, *Proc. Roy. Soc. Ser. A*, **227**, 516–537 (1955) [*MR* **16** 752].

33. Bukovics, E., Beiträge zur numerischen Integration. I. Der Fehler beim Blaess'schen Verfahren zur numerischen Integration gewöhnlicher Differentialgleichungen n-ter Ordnung, *Monatsh. Math.*, **57**, 217–245 (1953) [*MR* **15** 561].

34. Bukovics, E., Beiträge zur numerischen Integration. II. Der Fehler beim Runge-Kutta-Verfahren zur numerischen Integration gewöhnlicher Differentialgleichungen n-ter Ordnung, *Monatsh. Math.*, **57**, 333–350 (1954) [*MR* **15** 561].

35. Bukovics, E., Beiträge zur numerischen Integration. III. Nachträge, *Monatsh. Math.*, **58**, 258–265 (1954) [*MR* **16** 631].

36. Burgerhout, T. J., On the numerical solution of partial differential equations of the elliptic type. I, *Appl. Sci. Res. Sect. B*, **4**, 161–172 (1954) [*MR* **16** 406].

37. Butcher, J. C., A modified multistep method for the numerical integration of ordinary differential equations, *J. ACM*, **12**, 124–135 (1965) [*CR* **6** #7685], [*MR* **31** #2830].

38. Butcher, J. C., A multistep generalization of Runge-Kutta methods with four or five stages, *J. ACM*, **14**, 84–99 (1967) [*CR* **8** #12022], [*MR* **35** #3896].

39. Call, D. H., and Reeves, R. F., Error estimates in Runge-Kutta procedures, *Comm. ACM*, **1**, 79–88 (1958).

40. Carr, J. W., IIIrd, Error bounds for the Runge-Kutta single-step integration process, *J. ACM*, **5**, 39–44 (1958) [*MR* **20** #1417].

41. Čermák, J., Über lineare Systeme von Differenzengleichungen mit periodischen Koeffizienten, *Časopis Pěst. Mat.*, **79**, 141–150 (1954) (Czech) (German summary) [*MR* **17** 43].

42. Ceschino, F., Critère d'utilisation du procédé Runge-Kutta, *C. R. Acad. Sci. Paris.*, **238**, 1553–1555 (1954) [*MR* **16** 290].

43. Ceschino, F., L'intégration approchée des équations différentielles, *C. R. Acad. Sci. Paris*, **243**, 1478–1479 (1956) [*MR* **19** 771].

44. Ceschino, F., Modification de la longueur du pas dans l'intégration numérique par les méthodes à pas liés, *Chiffres*, **2**, 101–106 (1961).

45. Ceschino, F., and Kuntzmann, J., Impossibilité d'un certain type de formule d'intégration approchée à pas liés, *Chiffres*, **1**, 95–101 (1958) [*MR* **21** #432].

46. *Ceschino, F., and Kuntzmann, J., *Numerical Solution of Initial Value Problems*, Prentice-Hall, 318 pp. (1966) [*CR* **5** #6011 (review of 1963 French ed.)], [*CR* **7** #10484], [*MR* **28** #2639 (review of 1963 French ed.)].

47. Četaev, N. G., On estimates of approximate integrations, *Prikl. Mat. Meh.*, **21**, 419–421 (1957) (Russian) [*MR* **20** #1416].

* Refers to books.

48. Chase, P. E., Stability properties of predictor-corrector methods for ordinary differential equations, *J. ACM*, **9**, 457–468 (1962) [*CR* **4** #4243], [*MR* **29** #738].

49. Clenshaw, C. W., and Olver, F. W. J., Solution of differential equations by recurrence relations, *MTAC*, **5**, 34–39 (1951).

50. Collatz, L., Einige funktionalanalytische Methoden bei der numerischen Behandlung von Differentialgleichungen, *Z. Angew. Math. Mech.*, **38**, 264–267 (1958) [*MR* **20** #6792].

51. Collatz, L., Über die Instabilität beim Verfahren der zentralen Differenzen für Differentialgleichungen zweiter Ordnung, *Z. Angew. Phys.*, **4**, 153–154 (1953) (English summary) [*MR* **14** 907].

52. Collatz, L., Zur Stabilität des Differenzenverfahrens bei der Stabsschwingungsgleichung, *Z. Angew. Math. Mech.*, **31**, 392–393 (1951) [*MR* **13** 693].

53. Conte, S. D., and Reeves, R. F., A Kutta third-order procedure for solving differential equations requiring minimum storage, *J. ACM*, **3**, 22–25 (1956) [*MR* **17** 667].

54. Conte, S. D., and Royster, W. C., A study of finite difference approximations to a fourth order parabolic differential equation, *Ballistic Research Laboratories*, Aberdeen Proving Ground, Md., Rep. no. **959**, 22 pp. (1955) [*MR* **19** 773].

55. Cooper, G. J., Interpolation and quadrature methods for ordinary differential equations, *Math. Comp.*, **22**, 69–76 (1968) [*MR* **36** #7333].

Cox, J. G., *see* Mann, W. R.

Craggs, J. W., *see* Mitchell, A. R.

56. Crandall, S. H., Implicit vs. explicit recurrence formulas for the linear diffusion equation, *J. ACM*, **2**, 42–49 (1955).

57. Crandall, S. H., Numerical treatment of a fourth order parabolic partial differential equation, *J. ACM*, **1**, 111–118 (1954) [*MR* **16** 525].

58. Crandall, S. H., On a stability criterion for partial difference equations, *J. Math. and Phys.*, **32**, 80–81 (1953) [*MR* **14** 908].

59. Crandall, S. H., An optimum implicit recurrence formula for the heat conduction equation, *Quart. Appl. Math.*, **13**, 318–320 (1955) [*MR* **17** 413].

60. Crane, R. L., and Klopfenstein, R. W., A predictor-corrector algorithm with an increased range of absolute stability, *J. ACM*, **12**, 227–241 (1965) [*CR* **6** #8323], [*MR* **31** #6378].

61. Crane, R. L., and Lambert, R. J., Stability of a generalized corrector formula, *J. ACM*, **9**, 104–117 (1962) [*MR* **24** #B1283].

Creemer, A. L., *see* Hull, T. E.

62. Dahlquist, G., Convergence and stability in the numerical integration of ordinary differential equations, *Math. Scand.*, **4**, 33–53 (1956) [*MR* **18** 338].

63. *Dahlquist, G., *Stability and Error Bounds in the Numerical Integration of Ordinary Differential Equations*, Thesis, Univ. Stockholm, 87 pp. (1958) (published in [64]).

64. Dahlquist, G., Stability and error bounds in the numerical integration of ordinary differential equations, *Kungl. Tekn. Hogsk. Handl. Stockholm*, no. **130**, 87 pp. (1959) [MR **21** #1706].

* Refers to books.

65. Davis, P. J., On the numerical integration of periodic analytic functions. In *Langer, R. E., ed., *On Numerical Approximation*, Univ. Wisconsin, 45–59 (1959) [*MR* 20 #6787].

66. Day, J. T., A one-step method for the numerical solution of second order linear ordinary differential equations, *Math. Comp.*, 18, 664–668 (1964) [*CR* 6 #7331], [*MR* 29 #5385].

67. Dennis, S. C. R., and Poots, G., The solution of linear differential equations, *Proc. Cambridge Philos. Soc.*, 51, 422–432 (1955) [*MR* 17 412].

68. Dieulefait, C. E., New principles in the problem of mechanical quadrature, *An. Soc. Ci. Argentina*, 166, 23–25 (1958) (Spanish) [*MR* 21 #969].

69. Douglas, J., Jr., On the numerical integration of $\partial^2 u/\partial x^2 + \partial^2 u/\partial y^2 = \partial u/\partial t$ by implicit methods, *J. SIAM*, 3, 42–65 (1955) [*MR* 17 196].

70. Douglas, J., Jr., and Gallie, T. M., Jr., On the numerical integration of a parabolic differential equation subject to a moving boundary condition, *Duke Math. J.*, 22, 557–571 (1955) [*MR* 17 1241].

71. Douglas, J., Jr., and Gallie, T. M., Jr., Variable time steps in the solution of the heat flow equation by a difference equation, *Proc. Amer. Math. Soc.*, 6, 787–793 (1955) [*MR* 17 1241].

72. Douglas, J., Jr., On the relation between stability and convergence in the numerical solution of linear parabolic and hyperbolic differential equations, *J. SIAM*, 4, 20–37 (1956) [*MR* 18 236].

73. Douglas, J., Jr., and Rachford, H. H., Jr., On the numerical solution of heat conduction problems in two and three space variables, *Trans. Amer. Math. Soc.*, 82, 421–439 (1956) [*MR* 18 827].

74. Douglas, J., Jr., A note on the alternating direction implicit method for the numerical solution of heat flow problems, *Proc. Amer. Math. Soc.*, 8, 409–412 (1957) [*MR* 19 884].

75. Du Fort, E. C., and Frankel, S. P., Stability conditions in the numerical treatment of parabolic differential equations, *MTAC*, 7, 135–152 (1953) [*MR* 15 474].

76. van den Dungen, F. H., Sur le contrôle des intégrations numériques. In *Studies in Mathematics and Mechanics Presented to Richard von Mises*, Academic Press, 103–110 (1954) [*MR* 16 751].

77. Èl'sgol'c, L. È., Stability of solutions of differential-difference equations, *Uspehi Mat, Nauk*, 9, no. 4, 95–112 (1954) (Russian) [*MR* 17 44].

78. Emanuel, G., The Wilf stability criterion for numerical integration, *J. ACM*, 10, 557–561 (1963) [*CR* 5 #6374], [*MR* 29 #4192].

79. Evangelisti, G., Il calcolo delle differenze finite nella matematica applicata, *Rend. Sem. Mat. Fis. Milano*, 26, 69–87 (1957) [*MR* 19, 288].

80. Evans, G. W., Brousseau, R., and Keirstead, R., Stability considerations for various difference equations derived for the linear heat conduction equation, *J. Math. and Phys.*, 34, 267–285 (1956) [*MR* 19 773].

81. Favard, J., Sur les quadratures mécaniques, *Enseignement Math.*, 3, 263–275 (1957) [*MR* 19 983].

82. Fehlberg, E., Bemerkungen zur numerischen Lösung von Randwertaufgaben für nichtlineare gewöhnliche Differentialgleichungen nach der Picardschen Iterationsmethode, *Z. Angew. Math. Mech.*, 32, 23–26 (1952) [*MR* 13 692].

* Refers to books.

83. Fehlberg, E., Eine Methode zur Fehlerverkleinerung beim Runge-Kutta-Verfahren, *Z. Angew. Math. Mech.*, **38**, 421–426 (1958) (English summary) [*MR* **20** #6791].

84. Fehlberg, E., Neue genauere Runge-Kutta-Formeln für Differential-gleichungen zweiter Ordnung, *Z. Angew. Math. Mech.*, **40**, 252–259 (1960) (English summary) [*MR* **22** #4118].

85. Fehlberg, E., Numerisch stabile Interpolationsformeln mit günstiger Fehlerfortpflanzung für Differentialgleichungen erster und zweiter Ordnung, *Z. Angew. Math. Mech.*, **41**, 101–110 (1961) (English summary) [*MR* **26** #3193].

86. Filippov, A. F., On stability of difference equations, *Dokl. Akad. Nauk SSSR*, **100**, 1045–1048 (1955) (Russian) [*MR* **16** 829].

Filippov, A. F., *see* Rjaben'kiĭ, V. S.

87. Forrington, C. V. D., Extensions of the predictor-corrector method for the solution of systems of ordinary differential equations, *Comput. J.*, **4**, 80–84 (1961) [*CR* **3** #1459].

88. Fox, L., A note on the numerical integration of first-order differential equations, *Quart. J. Mech. Appl. Math.*, **7**, 367–378 (1954) [*MR* **16** 1055].

89. *Fox, L., The Numerical Solution of Two-point Boundary Problems in Ordinary Differential Equations*, Oxford, 371 pp. (1957) [*MR* **21** #972].

90. Fox, L., and Goodwin, E. T., Some new methods for the numerical integration of ordinary differential equations, *Proc. Cambridge Philos. Soc.*, **45**, 373–388 (1949) [*MR* **10** 744].

Frankel, S. P., *see* Du Fort, E. C.

91. Froese, C., An evaluation of Runge-Kutta methods for higher order differential equations, *J. ACM*, **8**, 637–644 (1961) [*CR* **3** #2241], [*MR* **23** #B3155].

92. Gaier, D., Über die Konvergenz des Adamsschen Extrapolationsverfahrens, *Z. Angew. Math. Mech.*, **36**, 230 (1956) [*MR* **18** 154].

93. Galler, B. A., and Rozenberg, D. P., A generalization of a theorem of Carr on error bounds for Runge-Kutta procedures, *J. ACM*, **7**, 57–60 (1960) [*CR* **1** #50], [*MR* **26** #3202].

Gallie, T. M., Jr., *see* Douglas, J., Jr.

94. Garwick, J. V., The solution of boundary-value problems by step by step methods. *Arch. Math. Naturvid.*, **52**, 95–161 (1955) [*MR* **17** 412].

95. Gautschi, W., Über den Fehler des Runge-Kutta-Verfahrens für die numerische Integration gewöhnlicher Differentialgleichungen n-ter Ordnung, *Z. Angew. Math. Phys.*, **6**, 456–461 (1955) [*MR* **17** 1010].

96. Gear, C. W., The numerical integration of ordinary differential equations, *Math. Comp.*, **21**, 146–156 (1967) [*MR* **37** #1087].

97. Ginzburg, B. L., Formulas for numerical quadrature most convenient for application, *Uspehi Mat. Nauk*, **9**, no. 2, 137–142 (1954) (Russian) [*MR* **15** 832].

98. Glinskaja, N. N., and Mysovskih, I. P., On numerical solution of a boundary problem for a nonlinear ordinary differential equation, *Vestnik Leningrad. Univ.*, **9**, no. 8, 49–54 (1954) (Russian) [*MR* **17** 412].

* Refers to books.

99. Goheen, H., On the remainder in formulas of numerical interpolation, differentiation, and quadrature, *Amer. Math. Monthly*, **61**, 44–46 (1954).

100. Gol'cov, N. A., The use of a certain functional series in deducing formulas involved in various numerical methods of solving ordinary differential equations, *Dokl. Akad. Nauk SSSR*, **120**, 450–453 (1958) (Russian) [*MR* **20** #4348].

101. Goodman, T. R., and Lance, G. N., The numerical integration of two-point boundary value problems, *MTAC*, **10**, 82–86 (1956) [*MR* **18** 420].
Goodwin, E. T., *see* Fox, L.
Gorbunov, A. D., *see* Budak, B. M.

102. Gragg, W. B., and Stetter, H. J., Generalized multistep predictor-corrector methods, *J. ACM*, **11**, 188–209 (1964) [*CR* **5** #6399], [*MR* **28** #4680].

103. Grosswald, E., Transformations useful in numerical integration methods, *J. SIAM*, **7**, 76–84 (1959) [*MR* **21** #1705].

104. Grunsky, H., Eine Methode zur Lösung von Anfangswertproblemen bei gewöhnlichen und partiellen linearen Differentialgleichungen zweiter Ordnung, *Z. Angew. Math. Mech.*, **34**, 291–292 (1954) [*MR* **19** 1082].

105. Gurk, H. M., The use of stability charts in the synthesis of numerical quadrature formulae, *Quart. Appl. Math.*, **13**, 73–78 (1955) [*MR* **16** 1055].

106. Hahn, S. G., Stability criteria for difference schemes, *Comm. Pure Appl. Math.*, **11**, 243–255 (1958) [MR **20** #4350].

107. Hammer, P. C., The midpoint method of numerical integration, *Math. Mag.*, **31**, 193–195 (1957–58) [*MR* **20** #6191].

108. Hammer, P. C., Numerical evaluation of multiple integrals. In *Langer, R. E., ed., On Numerical Approximation*, Univ. Wisconsin, 99–115 (1959) [*MR* **20** #6788].

109. Hammer, P. C., and Hollingsworth, J. W., Trapezoidal methods of approximating solutions of differential equations, *MTAC*, **9**, 92–96 (1955) [*MR* **17** 302].

110. Hammer, P. C., and Stroud, A. H., Numerical evaluation of multiple integrals. II, *MTAC*, **12**, 272–280 (1958) [*MR* **21** #970].

111. *Hamming, R. W., *Numerical Methods for Scientists and Engineers*, McGraw-Hill, 411 pp. (1962) [*CR* **3** #3367], [*MR* **25** #735].

112. Hamming, R. W., Stable predictor-corrector methods for ordinary differential equations, *J. ACM*, **6**, 37–47 (1959) [*MR* **21** #973].

113. Hartree, D. R., A method for the numerical integration of the linear diffusion equation, *Proc. Cambridge Philos. Soc.*, **54**, 207–213 (1958) [*MR* **20** #3637].

114. *Henrici, P., *Discrete Variable Methods in Ordinary Differential Equations*, Wiley, 407 pp. (1962) [*CR* **4** #3733], [*MR* **24** #B1772].

115. *Henrici, P., *Error Propagation for Difference Methods*, Wiley, 73 pp. (1963) [*CR* **5** #5083], [*MR* **27** #4365].

116. Hersch, J., Contribution à la méthode des équations aux différences, *Z. Angew. Math. Phys.*, **9a**, 129–180 (1958) [*MR* **21** #1708].

* Refers to books.

117. *Hildebrand, F. B., *Introduction to Numerical Analysis*, McGraw-Hill, 511 pp. (1956) [*MR* **17** 788].

Hollingsworth, J. W., *see* Hammer, P. C.

118. Hopkin, H. R., Routine computing methods for stability and response investigations on linear systems, *Ministry of Supply, Aeronautical Research Council*, Great Britain, R. & M. no. **2392**, 50 pp. (1950) [*MR* **12** 639].

119. Householder, A. S., Bibliography on numerical analysis, *Oak Ridge National Laboratory*, Rep. **ORNL 1897**, 32 pp. (1955) [*MR* **16** 1053].

120. Hull, T. E., and Creemer, A. L., Efficiency of predictor-corrector procedures, *J. ACM*, **10**, 291–301 (1963) [*CR* **5** #5059], [*MR* **27** #4367].

121. Hull, T. E., and Luxemburg, W. A. J., Numerical methods and existence theorems for ordinary differential equations, *Numer. Math.*, **2**, 30–41 (1960) [*MR* **22** #4847].

122. Hull, T. E., and Newbery, A. C. R., Error bounds for a family of three-point integration procedures, *J. SIAM*, **7**, 402–412 (1959) [*MR* **24** #B2118].

123. Hull, T. E., and Newbery, A. C. R., Integration procedures which minimize propagated errors, *J. SIAM*, **9**, 31–47 (1961) [*MR* **22** #11519].

Hyman, M. A., *see* O'Brien, G. G.

124. Juncosa, M. L., and Young, D., On the Crank-Nicolson procedure for solving parabolic partial differential equations, *Proc. Cambridge Philos. Soc.*, **53**, 448–461 (1957) [*MR* **19** 583].

Kaplan, S., *see* O.Brien, G. G.

Keirstead, R., *see* Evans, G. W.

125. Keitel, G. H., An extension of Milne's three-point method, *J. ACM*, **3**, 212–222 (1956) [*MR* **18** 338].

126. *Keller, H. B., *Numerical Methods for Two-point Boundary-value Problems*, Blaisdell, 184 pp. (1968) [*MR* **37** 6038].

Klopfenstein, R. W., *see* Crane, R. L.

127. Kohfeld, J. J., and Thompson, G. T., Multistep methods with modified predictors and correctors, *J. ACM*, **14**, 155–166 (1967) [*CR* **8** #12023].

Kohfeld, J. J., *see* Brush, D. G.

128. Krogh, F. T., Integration coefficients for the numerical solution of ordinary differential equations, *TRW Systems*, Redondo Beach, Calif. (Nov. 1967).

129. Krogh, F. T., A note on the effect of conditionally stable correctors, *Math. Comp.*, **21**, 717–719 (1967) [*MR* **36** #7335].

130. Krogh, F. T., The numerical integration of stiff differential equations, *TRW Systems*, Redondo Beach, Calif. (Mar. 1968).

131. Krogh, F. T., Predictor-corrector methods of high order with improved stability characteristics, *J. ACM*, **13**, 374–385 (1966) [*CR* **7** #10917], [*MR* **33** #5127].

132. Krogh, F. T., A test for instability in the numerical solution of differential equations, *J. ACM*, **14**, 351–354 (1967) [*CR* **8** #12820].

133. Krogh, F. T., A variable step, variable order multistep method for the numerical solution of ordinary differential equations (May 1967). An addendum to the above (Oct. 1967), *TRW Systems*, Redondo Beach,

---

* Refers to books.

Calif. (Presented at the IFIP Congress 68 in Edinburgh, Scotland, Aug. 1968).

134. Kuntzmann, J., Deux formules optimales du type de Runge-Kutta, *Chiffres*, **2**, 21–26 (1959) [*CR* 1 #268], [*MR* **24** #B2119].

135. Kuntzmann, J., Évaluation de l'erreur sur un pas dans les méthodes à pas séparés, *Chiffres*, **2**, 97–102 (1959) (English summary) [*CR* 1 #11, #261], [*MR* **21** #4547].

136. Kuntzmann, J., Remarques sur la méthode de Runge-Kutta, *C. R. Acad. Sci. Paris*, **242**, 2221–2223 (1956) [*MR* **17** 1010].

Kuntzmann, J., *see* Ceschino, F.

137. Kunz, K. S., High accuracy quadrature formulas from divided differences with repeated arguments, *MTAC*, **10**, 87–90 (1956) [*MR* **18** 419].

Lambert, R. J., *see* Crane, R. L.

Lance, G. N., *see* Goodman, T. R.

138. Lanczos, C., Trigonometric interpolation of empirical and analytical functions, *J. Math. and Phys.*, **17**, 123–199 (1938).

Lapierre, G., *see* Ardouin, P. G.

139. *Liniger, W., Zur Stabilität der numerischen Integrationsmethoden für Differentialgleichungen*, Thesis, Univ. Lausanne, 95 pp. (1957).

140. Lotkin, M., On the accuracy of Runge-Kutta's method, *MTAC*, **5**, 128–133 (1951) [MR **13** 286].

141. Lotkin, M., A new integrating procedure of high accuracy, *J. Math. and Phys.*, **31**, 29–34 (1952) [*MR* **13** 782].

142. Lotkin, M., The propagation of error in numerical integrations, *Ballistic Research Laboratories*, Aberdeen Proving Ground, Md., Rep. no. **875**, 30 pp. (1953) [*MR* **16** 405].

143. Lotkin, M., The improvement of accuracy in integration, *Ballistic Research Laboratories*, Aberdeen Proving Ground, Md., Rep. no. **912**, 25 pp. (1954) [*MR* **16** 865].

144. Lotkin, M., The propagation of error in numerical integrations, *Proc. Amer. Math. Soc.*, **5**, 869–887 (1954).

145. Lotkin, M., On the improvement of accuracy in integration, *Quart. Appl. Math.*, **13**, 47–54 (1955) [*MR* **16** 865].

146. Lotkin, M., and Browne, H. N., On the accuracy of the adjoint method of differential corrections, *Amer. Math. Monthly*, **63**, 97–105 (1956) [*MR* **17** 792].

147. Lotkin, M., A note on the midpoint method of integration, *J. ACM*, **3**, 208–211 (1956) [*MR* **18** 338].

148. Lotkin, M., The numerical integration of heat conduction equations, *J. Math. and Phys.*, **37**, 178–187 (1958) [*MR* **20** #5562].

149. Loud, W. S., On the long-run error in the numerical solution of certain differential equations, *J. Math. and Phys.*, **28**, 45–49 (1949) [*MR* **11** 57].

Lovass-Nagy, V., *see* Bajcsay, P.

150. Löwdin, P.-O., On the numerical integration of ordinary differential equations of the first order, *Quart. Appl. Math.*, **10**, 97–111 (1952) [*MR* **14** 413].

Luxemburg, W. A. J., *see* Hull, T. E.

151. Mann, W. R., Bradshaw, C. L., and Cox, J. G., Improved approximations

* Refers to books.

to differential equations by difference equations, *J. Math. and Phys.*, **35**, 408–415 (1957) [*MR* **19** 179].

152. Martin, D. W., Runge-Kutta methods for integrating differential equations on high speed digital computers, *Comput. J.*, **1**, 118–123 (1958) [*MR* **20** #3634].

153. Matthieu, P., Über die Fehlerabschätzung beim Extrapolationsverfahren von Adams. I. Gleichungen 1. Ordnung, *Z. Angew. Math. Mech.*, **31**, 356-370 (1951) (English summary) [*MR* **13** 691].

154. Matthieu, P., Über die Fehlerabschätzung beim Extrapolationsverfahren von Adams. II. Gleichungen zweiter und höherer Ordnung, *Z. Angew. Math. Mech.*, **33**, 26–41 (1953) (English summary) [*MR* **14** 800].

McCormick, E., *see* Riley, J. D.

155. Meriam, J. L., Procedure for the machine or numerical solution of ordinary linear differential equations for two-point linear boundary values, *MTAC*, **3**, 532–539 (1949) [*MR* **11** 744].

156. Mikeladze, Š. E., On the approximate integration of linear differential equations with discontinuous coefficients, *Soobšč. Akad. Nauk Gruzin. SSR*, **3**, 633–639 (1942) (Russian) [*MR* **5** 246].

157. Mikeladze, Š. E., New formulas for the numerical integration of differential equations, *Soobšč. Akad. Nauk Gruzin. SSR*, **4**, 215–218 (1943) (Russian) [*MR* **6** 133].

158. Mikeladze, Š. E., On the numerical integration of a function depending on a parameter, *Soobšč. Akad. Nauk Gruzin. SSR*, **5**, 575–583 (1944) (Georgian) (Russian summary) [*MR* **7** 338].

159. Mikeladze, Š. E., On numerical integration, *Dokl. Akad. Nauk SSSR*, **49**, 166–167 (1945) (English) [*MR* **8** 56].

160. Mikeladze, Š. E., New quadrature formulas and their application to the integration of differential equations, *Dokl. Akad. Nauk SSSR*, **61**, 613–615 (1948) (Russian) [*MR* **10** 331].

161. Mikeladze, Š. E., New formulas for the numerical integration of differential equations, *Dokl. Akad. Nauk SSSR*, **61**, 789–790 (1948) (Russian) [*MR* **10** 485].

162. Mikeladze, Š. E., Numerical integration, *Uspehi Mat. Nauk*, **3**, no. 6, 3–88 (1948) (Russian) [*MR* **10** 575].

163. Mikeladze, Š. E., Numerical integration of differential equations by means of summation formulas, *Dokl. Akad. Nauk SSSR*, **65**,125–128 (1949) (Russian) [*MR* **10** 576].

164. Mikeladze, Š. E., Approximate formulas for multiple integrals, *Soobšč. Akad. Nauk Gruzin. SSR*, **13**, 193–200 (1952) (Russian) [*MR* **14** 907].

165. Mikeladze, Š. E., Numerical solution of a system of differential equations. Application of the method to the computation of rotating shells, *Prikl. Mat. Meh.*, **17**, 382–386 (1953) (Russian) [*MR* **15** 165].

166. Mikeladze, Š. E., Numerical solution of boundary problems for nonlinear ordinary differential equations, *Soobšč. Akad. Nauk Gruzin. SSR*, **14**, 133–137 (1953) (Russian) [*MR* **16** 631].

167. Mikeladze, Š. E., Numerical solution of boundary problems for nonlinear ordinary differential equations, 1–6 (1954) (English translation of [166]) [*MR* **16** 631].

168. Mikeladze, Š. E., Numerical integration of differential equations in the complex plane, *Soobšč. Akad. Nauk Gruzin. SSR*, **17**, 97–102 (1956) (Russian) [*MR* **18** 73].

169. Mikeladze, Š. E., Numerical solution of the inhomogeneous polyharmonic equation, *Inžen. Sb.*, **23**, 190–202 (1956) (Russian) [*MR* **18** 938].

170. Mikeladze, Š. E., Quadrature formulas for multiple integrals with the greatest possible degree of accuracy, *Soobšč. Akad. Nauk Gruzin. SSR*, **18**, no. 1, 3–10 (1957) (Russian) [*MR* **21** #971].

171. Miller, J. C. P., and Mursi, Z., Notes on the solution of the equation $y'' - xy = f(x)$, *Quart. J. Mech. Appl. Math.*, **3**, 113–118 (1950) [*MR* **12** 288].

172. Mitchell, A. R., and Craggs, J. W., Stability of difference relations in the solution of ordinary differential equations, *MTAC*, **7**, 127–129 (1953) [*MR* **14** 908].

173. Mohr, E., Über das Verfahren von Adams zur Integration gewöhnlicher Differentialgleichungen, *Math. Nachr.*, **5**, 209–218 (1951) [*MR* **13** 286].

174. Morel, H., Évaluation de l'erreur sur un pas dans la méthode de Runge-Kutta, *C. R. Acad. Sci. Paris*, **243**, 1999–2002 (1956) [*MR* **18** 603].

Morrison, D., *see* Stoller, L.

175. Muhin, I. S., Application of the Markov-Hermite interpolation polynomials for numerical integration of ordinary differential equations, *Prikl. Mat. Meh.*, **16**, 231–238 (1952) (Russian) [*MR* **13** 783].

176. Muhin, I. S., On the accumulation of errors in numerical integration of differential equations, *Prikl. Mat. Meh.*, **16**, 753–755 (1952) (Russian) [*MR* **14** 587].

Murray, F. J., *see* Brock, P.

Mursi, Z., *see* Miller, J. C. P.

Mysovskih, I. P., *see* Glinskaja, N. N.

Newbery, A. C. R., *see* Hull, T. E.

Nise, S., *see* Urabe, M.

177. Nishimura, T., On a new method of finite differences for solving differential equations. In **Proceedings of the Second Japan National Congress for Applied Mechanics, 1952*, Science Council of Japan, 303–304 (1953) [*MR* **17** 668].

178. Nyström, E. J., Zur praktischen Integration von linearen Differentialgleichungen, *Soc. Sci. Fenn. Comment. Phys.-Math.*, **11**, no. 14, 1–14 (1943) [*MR* **7** 339].

179. O'Beirne, T. H., Can numerical integration be exact? *Math. Gaz.*, **41**, 59–60 (1957) [*MR* **18** 937].

180. Obrechkoff, N., On mechanical quadratures, *Spisanie Bulgar. Akad. Nauk*, **65**, 191–289 (1942) (Bulgarian) (French summary) [*MR* **10** 70].

181. O'Brien, G. G., Hyman, M. A., and Kaplan, S., A study of the numerical solution of partial differential equations, *J. Math. and Phys.*, **29**, 223–251 (1951) [*MR* **12** 751].

Olver, F. W. J., *see* Clenshaw, C. W.

182. Osborne, M. R., A method of finite-difference approximation to ordinary differential equations, *Comput. J.*, **7**, 58–65 (1964) [*CR* **5** #6743], [*MR* **31** #5338].

183. Osborne, M. R., Minimising truncation error in finite difference approximations to ordinary differential equations, *Math. Comp.*, **21**, 133–145 (1967) [*MR* **36** #6156].

184. Papoulis, A., On the accumulation of errors in the numerical solution of differential equations, *J. Appl. Phys.*, **23**, 173–176 (1952) [*MR* **13** 691].

* Refers to books.

185. Peaceman, D. W., and Rachford, H. H., Jr., The numerical solution of parabolic and elliptic differential equations, *J. SIAM*, **3**, 28–41 (1955) [*MR* **17** 196].

Pereyra, V., *see* Ballester, C.

186. Pickard, W., Tables for the step-by-step integration of ordinary differential equations of first order. *J. ACM*, **11**, 229–233 (1964) [*CR* **5** #6744], [*MR* **31** #1773].

Poots, G., *see* Dennis, S. C. R.

187. Quade, W., Über die Stabilität numerischer Methoden zur Integration gewöhnlicher Differentialgleichungen erster Ordnung, *Z. Angew. Math. Mech.*, **39**, 117–134 (1959) (English summary) [*MR* **21** #4546].

Rachford, H. H., Jr., *see* Douglas, J., Jr.

Rachford, H. H., Jr., *see* Peaceman, D. W.

188. Ralston, A. Relative stability in the numerical solution of ordinary differential equations, *SIAM Rev.*, **7**, 114–125 (1965) [*CR* **6** #7969], [*MR* **31** #2831].

189. Ralston, A., Some theoretical and computational matters relating to predictor-corrector methods of numerical integration, *Comput. J.*, **4**, 64–67 (1961) [*CR* **2** #1295].

190. Reed, H. L., Jr., Numerical integration of oscillatory systems, *Ballistic Research Laboratories*, Aberdeen Proving Ground, Md., Rep. no. **957**, 15 pp. (1955) [*MR* **17** 667].

Reeves, R. F., *see* Call, D. H.

Reeves, R. F., *see* Conte, S. D.

Rhodes, I., *see* Blanch, G.

191. Rice, J. R., Split Runge-Kutta method for simultaneous equations, *J. Res. Nat. Bur. Standards*, **64B**, 151–170 (1960) [*CR* **2** #68], [*MR* **23** #B2173].

192. *Richter, W., Estimation de l'erreur commise dans la méthode de M. W. E. Milne pour l'intégration d'un système de n équations différentielles du premier ordre*, Thesis, Univ. Neuchâtel, 43 pp. (1952) [*MR* **16** 865].

193. *Richtmyer, R. D., Difference Methods for Initial-value Problems*, Interscience, 238 pp. (1957) [*MR* **20** #438].

194. Ridley, E. C., A numerical method of solving second-order linear differential equations with two-point boundary conditions, *Proc. Cambridge Philos. Soc.*, **53**, 442–447 (1957) [*MR* **19** 178].

195. Riley, J. D., Bennett, M. M., and McCormick, E., Numerical integration of variational equations, *Math. Comp.*, **21**, 12–17 (1967) [*CR* **8** #12830].

196. *Rjaben'kiĭ, V. S., and Filippov, A. F., On Stability of Difference Equations*, Gosudarstv. Izdat. Tehn.-Teor. Lit., 171 pp. (1956) (Russian) [*MR* **19** 865].

197. Rose, M. E., Finite difference schemes for differential equations, *Math. Comp.*, **18**, 179–195 (1964) [*CR* **6** #6994], [*MR* **32** #605].

198. Rosser, J. B., A Runge-Kutta for all seasons, *SIAM Rev.*, **9**, 417–442 (1967) [*CR* **9** #14321], [*MR* **36** #2325].

Royster, W. C., *see* Conte, S. D.

Rozenberg, D. P., *see* Galler, B. A.

* Refers to books.

199. Rutishauser, H., Bemerkungen zur numerischen Integration gewöhnlicher Differentialgleichungen n-ter Ordnung, *Z. Angew. Math. Phys.*, **6**, 497–498 (1955) (English summary) [*MR* **17** 667].

200. Rutihauser, H., On the instability of methods for the integration of ordinary differential equations, *NACA Tech. Memo.*, no. **1403**, 15 pp. (1956) (English translation of [201]).

201. Rutihauser, H., Über die Instabilität von Methoden zur Integration gewöhnlicher Differentialgleichungen, *Z. Angew. Math. Phys.*, **3**, 65–74 (1952) (English summary) [*MR* **13** 692].

202. Salzer, H. E., Coefficients for facilitating the use of the Gaussian quadrature formula, *J. Math. and Phys.*, **25**, 244–246 (1946) [*MR* **8** 172].

203. Salzer, H. E., Coefficients for repeated integration with central differences, *J. Math. and Phys.*, **28**, 54–61 (1949) [*MR* **10** 576].

204. Salzer, H. E., Equally weighted quadrature formulas over semi-infinite and infinite intervals, *J. Math. and Phys.*, **34**, 54–63 (1955) [*MR* **16** 1055].

205. Salzer, H. E., Numerical integration of $y'' = \phi(x,y,y')$ using osculatory interpolation, *J. Franklin Inst.*, **263**, 401–409 (1957) [*MR* **19** 65].

206. Salzer, H. E., Osculatory extrapolation and a new method for the numerical integration of differential equations, *J. Franklin Inst.*, **262**, 111–119 (1956) [*MR* **18** 419].

207. Salzer, H. E., Osculatory quadrature formulas, *J. Math. and Phys.*, **34**, 103–112 (1955) [*MR* **17** 538].

208. Salzer, H. E., Table of coefficients for double quadrature without differences for integrating second order differential equations, *J. Math. and Phys.*, **24**, 135–140 (1945) [*MR* **8** 172].

209. Salzer, H. E., Tables for facilitating the use of Chebyshev's quadrature formula, *J. Math. and Phys.*, **26**, 191–194 (1947) [*MR* **9** 251].

210. Schröder, J., Verbesserung einer Fehlerabschätzung für gewöhnliche Differentialgleichungen erster Ordnung, *Numer. Math.*, **3**, 125–130 (1961) [*CR* **3** #2017], [*MR* **26** #3192].

211. Sefton, P. and Vaillancourt, R., A simple technique for coding differential equations, *Comm. ACM*, **3**, 616–617 (1960) [*CR* **1** #684].

212. Seidel, W., Bibliography of numerical methods in conformal mapping. In *Beckenbach, E. F., ed., Construction and Applications of Conformal Maps*, National Bureau of Standards Applied Mathematics Series, **18**, U.S. Government Printing Office, 269–280 (1952).

213. Selmer, E. S., Numerical integration by nonequidistant ordinates, *Nordisk Mat. Tidskr.*, **6**, 97–108, 136 (1958) [*MR* **20** #4918].

214. Servais, F., Sur l'estimation des erreurs dans l'intégration numérique des équations différentielles linéaires du second ordre, *Ann. Soc. Sci. Bruxelles Sér. I*, **70**, 5–8 (1956) [*MR* **18** 73].

215. Sharma, A., On Gołąb's contribution to Simpson's formula, *Ann. Polon. Math.*, **3**, 240–246 (1957) [*MR* **19** 178].

216. Sheldon, J. W., On the numerical solution of elliptic difference equations, *MTAC*, **9**, 101–112 (1955) [*MR* **17** 668].
    Sheldon, J. W., *see* Zondek, B.

217. Skalkina, M. A., On preservation of asymptotic stability in the passage from differential equations to the corresponding difference equations, *Dokl. Akad. Nauk SSSR*, **104**, 505–508 (1955) (Russian) [*MR* **17** 631].

* Refers to books.

218. Smith, E. S., Men vs. machines on quadrature in weapons analysis, *Ballistic Research Laboratories*, Aberdeen Proving Ground, Md., Ordnance Computer Research Report, **3**, no. 3, 6–14 (1956) [*MR* **18** 338].

219. Sokolov, Ju. D., On a method of approximate solution of linear integral and differential equations, *Dopovĭdĭ Akad. Nauk Ukraïn. RSR*, 107–111 (1955) (Ukrainian) (Russian summary) [*MR* **17** 196].

220. Squire, W., Approximate solution of linear second order differential equations, *J. Roy. Aeronaut. Soc.*, **63**, 368–369 (1959).

221. Stancu, D. D., Sur certaines formules générales d'intégration numérique, *Acad. R. P. Romîne. Stud. Cerc. Mat.*, **9**, 209–216 (1958) (Romanian) (French summary) [*MR* **20** #4917].

222. Stein, P., A note on numerical integration, *Math. Gaz.*, **40**, 268–270 (1956) [*MR* **18** 937].

223. Sterne, T. E., The accuracy of numerical solutions of ordinary differential equations, *MTAC*, **7**, 159–164 (1953) [*MR* **15** 256].

224. Stetter, H. J., Stabilizing predictors for weakly unstable correctors, *Math. Comp.*, **19**, 84–89 (1965) [*MR* **31** #2833].
    Stetter, H. J., *see* Gragg, W. B.

225. Stoller, L., and Morrison, D., A method for the numerical integration of ordinary differential equations, *MTAC*, **12**, 269–272 (1958) [*MR* **21** #974].

226. Stroud, A. H., Remarks on the disposition of points in numerical integration formulas, *MTAC*, **11**, 257–261 (1957) [*MR* **20** #431].
    Stroud, A. H., *see* Hammer, P. C.

227. Thacher, H. C., Jr., Optimum quadrature formulas in *s* dimensions, *MTAC*, **11**, 189–194 (1957) [*MR* **19** 883].

228. Thompson, G. T., On Bateman's method for solving linear integral equations, *J. ACM*, **4**, 314–328 (1957) [*MR* **20** #1423].
    Thompson, G. T., *see* Brush, D. G.
    Thompson, G. T., *see* Kohfeld, J. J.

229. Tlegenov, K. B., On mechanical solution of certain systems of linear differential equations with constant coefficients, *Izv. Akad. Nauk Kazah. SSR. Ser. Mat. Meh.*, no. 6 (**10**), 87–96 (1957) (Russian) [*MR* **20** #6790].

230. Todd, J., A direct approach to the problem of stability in the numerical solution of partial differential equations, *Comm. Pure Appl. Math.*, **9**, 597–612 (1956) [*MR* **18** 338].

231. Tollmien, W., Bemerkung zur Fehlerabschätzung beim Adamsschen Interpolationsverfahren, *Z. Angew. Math. Mech.*, **33**, 151–155 (1953) [*MR* **14** 1129].
    Truesdell, C., *see* Bernstein, B.

232. Urabe, M., and Nise, S., A method of numerical integration of analytic differential equations, *J. Sci. Hiroshima Univ. Ser. A*, **19**, 307–320 (1955) [*MR* **17** 1138].

233. Urabe, M., and Yanagihara, H., On numerical integration of the differential equation $y^{(n)} = f(x,y)$, *J. Sci. Hiroshima Univ. Ser. A*, **18**, 55–76 (1954) [*MR* **16** 865].
    Vaillancourt, R., *see* Sefton, P.

234. Vernotte, P., Généralisation d'un procédé d'intégration pratique des équations aux dérivées partielles. Application à la diffusion de la

matière ou de la chaleur, *C. R. Acad. Sci. Paris*, **241**, 1699–1700 (1955).

235. Vietoris, L., Der Richtungsfehler einer durch das Adamssche Interpolationsverfahren gewonnenen Näherungslösung eines Systems von Gleichungen, $y_k' = f_k(x, y_1, y_2, \cdots, y_m)$, *Österreich. Akad. Wiss. Math.-Natur. Kl. S.-B. IIa.*, **162**, 293–299 (1953) [*MR* **16** 78].

236. de Vogelaere, R., A method for the numerical integration of differential equations of second order without explicit first derivatives, *J. Res. Nat. Bur. Standards*, **54**, 119–125 (1955) [*MR* **16** 962].

Voss, J. R., *see* Anderson, W. H.

237. Wall, D. D., Note on predictor-corrector formulas, *MTAC*, **10**, 167 (1956) [*MR* **18** 336].

238. Warga, J., On a class of iterative procedures for solving normal systems of ordinary differential equations, *J. Math. and Phys.*, **31**, 223–243 (1953) [*MR* **14** 587].

239. Weissinger, J., Eine Fehlerabschätzung für die Verfahren von Adams und Störmer, *Z. Angew. Math. Mech.*, **32**, 62–67 (1952) (English summary) [*MR* **13** 873].

240. Weissinger, J., Numerische Integration impliziter Differentialgleichungen, *Z. Angew. Math. Mech.*, **33**, 63–65 (1953) [*MR* **14** 800].

241. Wilf, H. S., An open formula for the numerical integration of first order differential equations, *MTAC*, **11**, 201–203 (1957) [*MR* **19** 884].

242. Wilf, H. S., An open formula for the numerical integration of first order differential equations. II, *MTAC*, **12**, 55–58 (1958) [*MR* **21** #435].

243. Wilf, H. S., A stability criterion for numerical integration, *J. ACM*, **6**, 363–365 (1959) [*MR* **21** #6095 (German)].

244. Winn, E. A., A matrix method for the numerical solution of linear differential equations with variable coefficients, *J. Roy. Aeronaut. Soc.*, **61**, 133–134 (1957) [*MR* **18** 937].

245. Witting, H., Über die numerische Lösung parabolischer Differentialgleichungen. In **Aktuelle Probleme der Rechentechnik, Bericht über das Internationale Mathematiker-Kolloquium, Dresden, Nov. 1955*, VEB Deutscher Verlag der Wissenschaften, 127–132 (1957) [*MR* **19** 462].

Yanagihara, H., *see* Urabe, M.

Young, D., *see* Juncosa, M. L.

246. Young, R. L., Report on experiments in approximating the solution of a differential equation, *J. ACM*, **3**, 26–28 (1956) [*MR* **17** 791].

247. Zondek, B., and Sheldon, J. W., On the error propagation in Adams' extrapolation method, *MTAC*, **13**, 52–55 (1959) [*MR* **21** #4557].

248. Zurmühl, R., Runge-Kutta-Verfahren unter Verwendung höherer Ableitungen, *Z. Angew. Math. Mech.*, **32**, 153–154 (1952) [*MR* **14** 413].

249. Zurmühl, R., Runge-Kutta-Verfahren zur numerischen Integration von Differentialgleichungen n-ter Ordnung, *Z. Angew. Math. Mech.*, **28**, 173–182 (1948) [*MR* **10** 212].

* Refers to books.

APPENDIX I

A SHORT GUIDE TO THE NUMERICAL SOLUTION OF PARTIAL DIFFERENTIAL EQUATIONS

Robert R. Reynolds

THE BOEING COMPANY

One of the most fundamental papers on the theory of finite differences for solving partial differential equations is Courant, Friedrichs, and Lewy [52].* Recent assessments of this work—forty years later—are found in Lax [145], Widlund [234], and Parter [174].

The book by Kantorovich and Krylov [125] describes finite difference and variational methods for the solution of certain elliptic equations, as well as some techniques of conformal mapping. The excellent treatises by Collatz [47] and Crandall [54] contain the description of a number of iteration procedures, the development of finite difference methods for solving several types of initial, boundary, and mixed problems, the description of variational methods, the use of characteristics, the calculation of eigenvalues and eigenfunctions, theories of error estimation, and numerous worked out examples from physics and engineering. These books however lack certain important recent developments.

The books of Southwell [210], [211], Allen [3], Shaw [205], and Grinter [107] describe in great detail the method of *relaxation* and its application to boundary value problems of technology during the two decades after 1935. The method applies Gauss-Seidel iteration (see [225]) to groups of linear algebraic equations (which approximate a self-adjoint elliptic equation at a number of grid points) and accelerates the convergence of the process by ingenious use of an *over-relaxation factor* ω.

Since the efficient reduction of certain residuals generally requires a chaotic ordering of the equations, a complete theoretical analysis of relaxation is not possible; however, Frankel and Young discovered in 1950 that, if the equations are ordered in a particular way and then

---

* Bracketed numbers refer to the bibliography at the end of this Appendix.

used cyclically, there is a formula for calculating an optimum ω which maximizes the rate of convergence in an asymptotic sense. Several methods of estimating this ω have been devised: see Varga [225], Wachspress [230], Carré [28], Hageman and Kellogg [110], Kulsrud [135], Randall [190], Reid [191], Rigler [197], Evans [67], Concus [49], and Wrigley [236]. In the simplest form of this process of *successive overrelaxation* (SOR) each equation involves values of the approximate solution at five points. All but one of these have already been calculated during the present or some previous iteration cycle, while the remaining one is unknown but easily determined. In a more sophisticated form, values at all the points of a line (or two lines) are unknown, and a system must be solved for the group of values on each line (or pair of lines). Cuthill and Varga [57] and Varga [223] describe some very efficient techniques for handling these cases of *successive line overrelaxation* (SLOR), which require fewer iterations than SOR.

The *cyclic Chebyshev semiiterative* (CCSI) method maximizes the average rate of convergence rather than the asymptotic rate and requires a smaller total number of iterations than SOR. Effectively, the formulas ((5.60) of Varga [225]) resemble those of SOR with ω varying at each iteration. There is a related method of *cyclic reduction* (CR) in which the algebraic system is modified to accelerate the convergence (see § 5.4 of Varga [225]). Both these procedures have block forms where values at a group of points are computed by solving a system (see Hageman and Varga [111], Hodgkins [118], and Rigler [196])—these are better than SLOR.

The *alternating direction implicit* (ADI) method (developed by Peaceman, Rachford, and Douglas in 1955–56) is superior to the SOR, CCSI, and CR methods. Proof of its effectiveness is limited to special situations, but considerable experience has demonstrated that it is useful for a much broader class of problems (see Widlund [235]). The equations are written in groups so that each group is a system for all the points on a horizontal line. Each system is solved independently to produce tentative values for this cycle. Final values for the cycle are obtained by carrying through the same steps for groups of points on vertical lines. Great efficiency is attained by using a particular sequence of parameters (see Varga [225] and Wachspress [230] for a remarkable development of the theories of selecting these sequences by the methods of Wachspress and Jordan). The article by Birkhoff, Varga, and Young [20] is a most illuminating exposition of the ADI method. The paper by Young and Ehrlich [241] compares SOR and ADI methods for several domains (square, square with square hole removed, Greek cross, triangle, L-shaped region) with a variety of mesh spacings. Spanier [212] describes in great practical detail a computer program for solving a self-adjoint elliptic problem; this includes cogent remarks on the treatment of the boundary. Since 1964 Mitchell, Fairweather, and Gourlay have published a number of new ADI formulas for elliptic, parabolic, and hyperbolic equations. Other papers are Douglas *et al.* [60], [61], Evans [67], Guittet [108], [109], Hubbard [120], Lynch *et al.* [154], [156], Widlund [234], and Young and Wheeler [243].

Varga [225] is an excellently written text on *linear elliptic* problems (and contains a comprehensive chapter on parabolic equations). The properties of the matrices of the finite difference equations are described, and the SOR, ADI, CCSI, and CR methods are developed. Finite difference equations are derived by integration over sub-regions; in particular, this produces a symmetric positive definite matrix for the first (Dirichlet), second (Neumann), or third (Robin) boundary value problem when the equation is self-adjoint, even with variable mesh spacing. Wachspress [230] covers much of the same material in somewhat lesser detail. He describes a discretization process when coefficients are discontinuous along mesh lines. Also, he derives a very efficient method by combining the ADI method with Lanczos's method of minimized iterations.

A new method of solving elliptic equations has been developed in Stone [214a] and Dupont, Kendall, and Rachford [63a]. It is faster than the ADI method and requires at each iteration the solution of a system with a matrix which factors into the product of lower and upper triangular matrices with no more than three nonzero elements in each row.

There are monographs by Young [237], [238], [242] which are very clear introductions to methods for solving elliptic and parabolic problems. Parter [178] compares convergence rates for several methods. Greenspan [99] is a carefully written book containing material about boundary conditions and SOR. Sheldon [206] is a very good introduction. Friedrichs and Keller [82] discuss systems of equations obtained by application of variational principles. Other works on *variational* methods are Mikhlin and Smolitskiy [163], Vorobyëv [229], and Greenspan [102]. The booklet by Engeli, Ginsburg, Rutishauser, and Stiefel [66] includes a treatment of SOR and some examples using *gradient* methods. Babuška, Práger, and Vitásek [8] and Chow [33] are general works on elliptic problems.

Material on *nuclear reactors* is in Cuthill [56] (which is very comprehensive), Sangren [200], Birkhoff [16], [17], H. Greenspan [105], and

Clark and Hansen [39], as well as in reference [254], Wachspress [230], and Fox [79]. The *biharmonic* equation is treated in Bramble [24], Cannon and Cecchi [26], [27], and Parter [180]. *Elasticity* problems are treated in Richtmyer and Morton [195], Griffin and Varga [106], Soare [209], and Kaplan [126].

The most comprehensive and scholarly book on *finite difference* methods for all types of linear partial differential equations is Forsythe and Wasow [78]. This contains an excellent discussion of the general theory of convergence, develops Young's theory with Bernard Friedman's elegant proofs, and covers many other topics (see the reviews). Fox [79] is a collection of well written reports presented in a summer school session at Oxford in 1961. Basic methods are clearly expounded, and several physical applications are given. Shorter presentation of finite difference methods are in *Modern Computing Methods* [249], Smith [208], Walsh [231], and Fox [80]. Other general works are Mikhlin and Smolitskiy [163], Ralston and Wilf [189], Langer [138]–[142], Berezin and Zhidkov [11], Gram, Naur, and Poulsen [98], Isaacson and Keller [121], Wendroff [233], Ladyženskaya [136], Lieberstein [151], von Rosenberg [198], Godunov and Ryabenki [86], Rektorys [192], and Trifonov, Roslyakov, and Žogolev [222].

The most thorough book on *initial value problems* is Richtmyer and Morton [195], which contains the general theory developed by Lax, Richtmyer, Wendroff, Kreiss, Buchanan, and others; there are also applications to diffusion, sound waves, elasticity, and fluid dynamics. The *transport* equation is considered here, as well as in Douglis [62], [63] and Fox [79]. There are treatments of hyperbolic equations and characteristics in [258], Abbott [1], Lister [152], and P. Fox [81]. Douglas [59] is the basic work on *parabolic* equations, and the books by Saul'yev [202], Dusinberre [64], and Crank [55] are very informative. An *alternating direction explicit* method has been developed for these equations by Saul'yev, extended by Larkin [144], and used by Holton [119] and others with great success. Lowan [153] and Todd [221] offer clear explanations of *stability*, the phenomenon of uncontrolled error growth which masks the solution.

Papers on *fluid flow* are in Alder, Fernbach, and Rotenberg [2], Armitage [7], Birkhoff [18], Courant and Friedrichs [51], Courant, Isaacson, and Rees [53], Emery [65], Flügge-Lotz *et al.* [76], [22], [58], Giese [84], Harlow [112]–[114], von Neumann [169], Noh [170], Richtmyer [194], Roslyakov and Chudov [199], Sauer [201], and   Pearson

[181a].   Methods for handling the Navier-Stokes equations are in Bye [25] and Chorin [32].

Papers on *motion of the atmosphere* are in Carroll and Wetherald [30], Charney [31], Holton [119], Kolsky [132], Leith [150], Phillips [183], Richardson [193], Smagorinsky [207], Thompson [220], reference [257], and issues of the *Monthly Weather Review*, published by the Environmental Science Service Administration (ESSA) of the U.S. Department of Commerce.

Ames [6] is a general treatise on *nonlinear* methods. Nonlinear extensions of SOR are in Greenspan [99], Ortega and Rheinboldt [171], and Concus [49]. The process of quasilinearization, which generates a sequence of linear problems by use of Newton's method and has a wide variety of applications to functional equations, is described in Kalaba [124], Bellman and Kalaba [9], Ames [6], Lee [148], and Radbill and McCue [187]. Other articles are in Ames [5], Gourlay and Morris [95]–[97], Greenspan [100]–[104], McAllister [159]–[161], Parter [175], [176], [179], Rall [187a], Schröder [203], and several works by Collatz.

Functional analysis is used by Collatz in a large number of books and papers (see especially [44]), Schröder [203], and Shampine [204] to estimate errors. See also the references in Rall [188] 1.

Richardson's *deferred approach to the limit* (also called, under certain circumstances, "h^2-extrapolation" and "Romberg's method") and Fox's *difference correction* (or "deferred correction") method are effective procedures for obtaining better estimates of the solution of the differential equation from the computed solution of the discretized system. See Fox [79], Walsh [231], Gourlay and Morris [95], and Pereyra [181b].

A number of techniques have evolved for using a *direct* method instead of iteration to solve the huge number of algebraic equations. The equation and region are restricted however. Various kinds of partitioning are used. The long paper by Bickley and McNamee [15] offers an excellent, easy-to-read introduction to this field. The articles by Lynch, Rice, and Thomas [155]–[157] develop some general theory and give applications. Hockney [117] uses an ingenious folding technique to reduce the number of equations—this is aided by the fast Fourier transform of Cooley and Tukey [50]. A kindred procedure is Polozhii's method of *summary representation* [184]. Kron's method of *tearing* [133] is expounded in Steward [214] and Spillers and Hickerson [213]. Other direct methods are those of Osborne [172] and Merzrath [162].

The classical *Schwarz alternating procedure* is an iterative method for solving an equation successively in a number of overlapping regions. Its numerical aspects are described in Miller [164], Kantorovich and Krylov [125], and Polozhii [184].

The *integral operator* method of Bergman has numerous applications to compressible fluid flow. See Bergman and Schiffer [13], v. Mises and Schiffer [167], Bergman [12], and v. Krzywoblocki [134]. Milnes, Potts, and Chow [165], [166], [34]–[38] have developed the *boundary contraction* method, Synge [216] the *hypercircle* method, and Thom and Apelt [219] the method of *squares*.

Eigenvalues are treated in Collatz [43] (which unfortunately has never been translated into English), and Engeli *et al.* [66]. The most comprehensive treatise is Gould [87] (the reviews by Diaz and Reid are wonderful; also see Weinstein [232]). Articles by Birkhoff, de Boor, Swartz, and Wendroff [19] and Swartz [215] are also interesting, as well as a chapter in Kantorovich and Krylov [125].

Description of the preparation of partial differential equations for *analog* and *hybrid* equipment is in Karplus [127], Fifer [75], Volynskii and Bukhman [228], Kafka [123], and Vichnevetsky [227].

Most books on the subject are in the list below, as well as numerous articles in collections. While some attempt was made to gather important original papers since the publication of Varga [225], it is obvious that many have been missed. There is a considerable mine of theoretical and applied material in the journal *U.S.S.R. Computational Mathematics and Mathematical Physics*. With volume **9**, page iv (1968), *Computing Reviews* began to publish abstracts of graduate theses in computing science; the early issues of 1968 absorbed a considerable backlog of these. Voigt [227a] is a computer-prepared author index and KWIC index covering the period 1960–1966; the usefulness of this important work has been unnecessarily impaired by the exasperating reduction of machine printout toward the vanishing point.

The author wishes to thank G. E. McCormick and K. D. Wiegand of the Boeing Company for their encouragement during the preparation of this Guide.

BIBLIOGRAPHY

Reviews known to the compiler are in brackets with the volume and page (or review number preceded by #) of the reviewing journal followed by the reviewer's name. *CJ = Computer Journal*, *CR = Computing Reviews*, *MC = Mathematics of Computation*, *MR = Mathematical Reviews*, *Q = Quarterly of*

Applied Mathematics, SR = SIAM Review. Volume numbers are in boldface. Reviews marked with a dagger (†) are especially comprehensive. Certain collections are placed at the end.

1. *Abbott, M. B., *An Introduction to the Method of Characteristics*, American Elsevier, 243 pp. (1966) [*CR* **7** #10482 Fox], [*MR* **34** #985 Shu].

2. *Alder, B., Fernbach, S., and Rotenberg, M., eds., *Methods in Computational Physics*, Academic Press, **3**, 386 pp. (1964) [*CR* **6** #7501–7511 Anderson]†, [*MC* **20** 190 Roberts], [*MR* **34** #8706]†; **4**, 385 pp. (1965) [*MR* **34** #8707]†.

3. *Allen, D. N. de G., *Relaxation Methods in Engineering and Science*, McGraw-Hill, 257 pp. (1954) [*MR* **15** 831 Fox]†.

4. *Alt, F. L., *Electronic Digital Computers, Their Use in Science and Engineering*, Academic Press, 336 pp. (1958) [*CR* **2** #930 Corbató], [*MR* **21** #4564 Rutishauser].

5. *Ames, W. F., ed., *Nonlinear Partial Differential Equations*, Academic Press, 316 pp. (1967) [*MR* **36** #510].

6. *Ames, W. F., *Nonlinear Partial Differential Equations in Engineering*, Academic Press, 511 pp. (1965) [*MR* **35** #1235], [*Q* **24** 396 Wasow]. Apelt, C. J., *see* Thom, A.

7. Armitage, J. V., The Lax-Wendroff method applied to axial symmetric swirl flow. In *Mond, B., ed., *Blanch Anniversary Volume*, U.S. Air Force Aerospace Research Laboratories (1967) [*MC* **22** 895 Gautschi], [*MR* **35** #1423].

8. *Babuška, I., Práger, M., and Vitásek, E., *Numerical Processes in Differential Equations*, Interscience, 351 pp. (1966) [*CJ* **10** 52 Curtis], [*CR* **8** #13227 Goldstine], [*MC* **22** 222 Keller]†, [*MR* **36** #6150 Hamming]. Bateman, H., *see* Bennett, A. A.

9. *Bellman, R. E., and Kalaba, R. E., *Quasilinearization and Nonlinear Boundary-value Problems*, American Elsevier, 206 pp. (1965) [*CJ* **9** 395 Tee], [*CR* **7** #9062 Gargantini]†, [*MC* **21** 121 Moser], [*MR* **31** #2828 (German) Collatz], [*SR* **8** 401 Rall].

10. *Bennett, A. A., Milne, W. E., and Bateman, H., *Numerical Integration of Differential Equations*, Dover, 108 pp. (1956, reprint of 1933 ed.) [*MR* **18** 826 Forsythe].

11. *Berezin, I. S., and Zhidkov, N. P., *Computing Methods*, Pergamon, **2**, 679 pp. (1965) [*CR* **4** #4238 (review of 1959 Russian ed.) Bečvář], [*CR* **9** #15210], [*MR* **30** #4372 Householder].

12. Bergman, S., Application of integral operators to singular differential equations and to computations of compressible fluid flows. In [23] 257–287 [*CR* **8** #12422 Herriot].

13. *Bergman, S., and Schiffer, M., *Kernel Functions and Elliptic Differential Equations in Mathematical Physics*, Academic Press, 432 pp. (1953) [*MR* **14** 876 Garabedian].

14. *Bernstein, D. L., *Existence Theorems in Partial Differential Equations*, Princeton Univ., 228 pp. (1950) [*MR* **12** 262 (French) Janet].

15. Bickley, W. G., and McNamee, J., Matrix and other direct methods for

* Refers to books.

the solution of systems of linear difference equations, *Philos. Trans. Roy. Soc. London Ser. A*, **252**, 69–131 (1960) [*MR* **22** #4897 Todd].

16. Birkhoff, G., Numerical solution of the reactor kinetics equations. In [101] 3–20 [*MR* **35** #3996 Coveyou].

17. Birkhoff, G., Some mathematical problems of nuclear reactor theory. In [139] 23–44 [*MR* **26** #2279 Varga].

18. Birkhoff, G., Synthetic materials for hydrodynamical computations. In **Proceedings of a Harvard Symposium on Digital Computers and Their Applications, 1961*, Harvard Univ. Comput. Lab., 23–31 (1962) [*CR* **5** #5082 Hammidi].

19. Birkhoff, G., de Boor, C., Swartz, B., and Wendroff, B., Rayleigh-Ritz approximation by piecewise cubic polynomials, *SIAM J. Numer. Anal.*, **3**, 188–203 (1966) [*CR* **8** #12043 Boehm], [*MR* **34** #3773 Clenshaw].

20. Birkhoff, G., Varga, R. S., and Young, D., Alternating direction implicit methods. In **Alt, F. L., and Rubinoff, M., eds., Advances in Computers*, Academic Press, **3**, 189–273 (1962) [*MC* **19** 351 Cuthill], [*MR* **29** #5395 Douglas].

21. Birkhoff, G., Young, D. M., and Zarantonello, E. H., Numerical methods in conformal mapping. In [252] 117–140 [*MR* **15** 258 Saltzer].

22. Blottner, F. G., and Flügge-Lotz, I., Finite-difference computation of the boundary layer with displacement thickness interaction, *J. Mécanique*, **2**, 397–423 (1963) [*MR* **29** #3075 Van Dyke].

23. **Bramble, J. H., ed., Numerical Solution of Partial Differential Equations*, Academic Press, 373 pp. (1966) [*CR* **8** #11281–11284], [*MR* **33** #7220].

24. Bramble, J. H., A second order finite difference analog of the first biharmonic boundary value problem, *Numer. Math.*, **9**, 236–249 (1966) [*CR* **8** #12041 Takata], [*MR* **34** #5305 Garder].

Bukhman, V. Ye., *see* Volynskii, B. A.

25. Bye, J. A. T., Obtaining solutions of the Navier-Stokes equations by relaxation processes, *Comput. J.*, **8**, 53–56 (1965) [*CR* **6** #8621 Meinardus], [*MR* **31** #5344].

26. Cannon, J. R., and Cecchi, M. M., Numerical experiments on the solution of some biharmonic problems by mathematical programming techniques, *SIAM J. Numer. Anal.*, **4**, 147–154 (1967) [*CR* **9** #14075 Pomentale], [*MR* **35** #3918 Coveyou].

27. Cannon, J. R., and Cecchi, M. M., The numerical solution of some biharmonic problems by mathematical programming techniques, *SIAM J. Numer. Anal.*, **3**, 451–466 (1966) [*CR* **9** #15226 Rudin].

28. Carré, B. A., The determination of the optimum accelerating factor for successive over-relaxation, *Comput. J.*, **4**, 73–78 (1961) [*CR* **2** #1286 Forsythe].

29. Carré, B. A., The partitioning of network equations for block iteration, *Comput. J.*, **9**, 84–97 (1966) [*CR* **7** #10910 Frank], [*MR* **33** #3445].

30. Carroll, A. B., and Wetherald, R. T., Application of parallel processing to numerical weather prediction, *J. ACM*, **14**, 591–614 (1967).

Cecchi, M. M., *see* Cannon, J. R.

31. Charney, J. G., Numerical experiments in atmospheric hydrodynamics. In [255] 289–310 [*CR* **5** #5523 Frenzel].

* Refers to books.

32. Chorin, A. J., Numerical solution of the Navier-Stokes equations, *Math. Comp.*, **22**, 745–762 (1968).

33. *Chow, T. S., *A Survey of Computational Methods for the Solution of Elliptic Differential Equations by Difference Approximations*, Boeing Company, Commercial Airplane Division, **D6-29502**, 161 pp. (1968).

34. Chow, T. S., and Milnes, H. W., Boundary contraction method for numerical solution of partial differential equations: convergence and boundary conditions, *Quart. Appl. Math.*, **20**, 209–230 (1962) [*MR* **25** #5593 Wasow].

35. Chow, T. S., and Milnes, H. W., Boundary contraction solution of Laplace's differential equation. II, *J. ACM*, **7**, 37–45 (1960) [*MR* **23** #B592 Wasow].

36. Chow, T. S., and Milnes, H. W., Numerical solution of a class of hyperbolic-parabolic partial differential equations by boundary contraction, *J. SIAM*, **10**, 124–148 (1962) [*MR* **24** #B2556 Wasow].

37. Chow, T. S., and Milnes, H. W., Numerical solution of the Neumann and mixed boundary value problems by boundary contraction, *J. ACM*, **8**, 336–358 (1961) [*CR* **3** #2270 Sackman]†, [*MR* **23** #B1130 Potts].

38. Chow, T. S., and Milnes, H. W., Solution of Laplace's equation by boundary contraction over regions of irregular shape, *Numer. Math.*, **4**, 209–225 (1962) [*MR* **26** #4492 Wasow].

Chudov, L. A., *see* Roslyakov, G. S.

39. *Clark, M., and Hansen, K. F., *Numerical Methods of Reactor Analysis*, Academic Press, 340 pp. (1964) [*CR* **6** #7500 Anderson].

40. Clark, M. M., Finite-difference representation of partial differential equations, *Bul. Inst. Politehn. Iaşi* (*N.S.*), **12** (**16**), no. 1–2, 31–37 (1966) [*MR* **35** #3903 Fox].

41. Collatz, L., Applications of functional analysis to error estimation. In [188] **2**, 253–269 [*CR* **8** #11532 Bareiss], [*MR* **32** #3295 Householder].

42. Collatz, L., Approximation in partial differential equations. In [141] 413–422 [*MR* **21** #2361 Hyman].

43. *Collatz, L., *Eigenwertaufgaben mit technischen Anwendungen*, Akademische Verlagsgesellschaft, 466 pp. (1949) [*MR* **8** 514 (review of 1945 ed.) Milne], [*MR* **11** 137 Milne].

44. *Collatz, L., *Functional Analysis and Numerical Mathematics*, Academic Press, 473 pp. (1966) [*CR* **6** #6955 (review of 1964 German ed.) Newhouse], [*MC* **22** 213 Ortega], [*MR* **29** #2931 (review of 1964 German ed.) Householder], [*SR* **8** 396 (review of 1964 German ed.) Gram].

45. Collatz, L., Functional analysis as an aid in numerical mathematics. In [247] **1**, 145–153 [*CR* **6** #8580 Hamming].

46. Collatz, L., Monotonicity and related methods in non-linear differential equations problems. In [101] 65–87 [*CR* **8** #12057 Kolsky], [*MR* **35** #1223 Parter].

47. *Collatz, L., *The Numerical Treatment of Differential Equations*, 3rd ed., Springer, 568 pp. (1960) [*MC* **15** 426 Roberts], [*MR* **13** 285 (review of 1951 German ed.) Milne]†, [*MR* **16** 962 (review of 1955 German ed.) Milne], [*MR* **22** #322].

* Refers to books.

48. *Collatz, L., Meinardus, G., and Unger, H., eds., *Funktionalanalysis, Approximationstheorie, numerische Mathematik*, Birkhäuser, 232 pp. (1967) [*MR* **36** #4772]†.

49. Concus, P., Numerical solution of the minimal surface equation by block nonlinear successive overrelaxation. In [248] *A*, 76–80 [*CR* **9** #15610].

50. Cooley, J. W., and Tukey, J. W., An algorithm for the machine calculation of complex Fourier series, *Math. Comp.*, **19**, 297–301 (1965) [*CR* **6** #8275 Hamming].

51. *Courant, R., and Friedrichs, K. O., *Supersonic Flow and Shock Waves*, Interscience, 464 pp. (1948) [*MR* **10** 637 Bers]†.

52. Courant, R., Friedrichs, K., and Lewy, H., On the partial difference equations of mathematical physics, *IBM J. Res. Develop.*, **11**, 215–234 (1967, translation of 1928 German original) [*MR* **35** #4621].

53. Courant, R., Isaacson, E., and Rees, M., On the solution of nonlinear hyperbolic differential equations by finite differences, *Comm. Pure Appl. Math.*, **5**, 243–255 (1952) [*MR* **14** 756 Polachek].

54. *Crandall, S. H., *Engineering Analysis, a Survey of Numerical Procedures*, McGraw-Hill, 417 pp. (1956) [*MR* **18** 674 Forsythe]†.

55. *Crank, J., *The Mathematics of Diffusion*, Oxford, 347 pp. (1956) [*MR* **18** 616 (German) Bhagwandin].

56. Cuthill, E., Digital computers in nuclear reactor design. In *Alt, F. L., and Rubinoff, M., eds., *Advances in Computers*, Academic Press, **5**, 289–348 (1964) [*CR* **6** #7514 Sangren].

57. Cuthill, E. H., and Varga, R. S., A method of normalized block iteration, *J. ACM*, **6**, 236–244 (1959) [*MR* **22** #8651 Young].

Dauwalder, J. H., *see* Young, D. M.

58. Davis, R. T., and Flügge-Lotz, I., Second-order boundary-layer effects in hypersonic flow past axisymmetric blunt bodies, *J. Fluid Mech.*, **20**, 593–623 (1964) [*MR* **30** #3666 Ray].

de Boor, C., *see* Birkhoff, G.

59. Douglas, J., A survey of numerical methods for parabolic differential equations, In *Alt, F. L., ed., *Advances in Computers*, Academic Press, **2**, 1–54 (1961) [*MR* **25** #5604 Laasonen].

60. Douglas, J., Garder, A. O., and Pearcy, C., Multistage alternating direction methods, *SIAM J. Numer. Anal.*, **3**, 570–581 (1966) [*CR* **8** #13232 Stetter], [*MR* **35** #3904 Fischer].

61. Douglas, J., Kellogg, R. B., and Varga, R. S., Alternating direction iteration methods for *n* space variables, *Math. Comp.*, **17**, 279–282 (1963) [*MR* **28** #3545 Lees].

62. Douglis, A., A finite-difference method for generalized radial transport equations, *J. Differential Equations*, **3**, 451–481 (1967) [*MR* **36** #3521 Anselone].

63. Douglis, A., The solutions of multidimensional generalized transport equations and their calculation by difference methods. In [23] 197–256 [*CR* **8** #11284 Bareiss], [*MR* **34** #4861 Anselone].

Downing, J. A., *see* Young, D. M.

63a. Dupont, T., Kendall, R. P., and Rachford, H. H., An approximate factorization procedure for solving self-adjoint elliptic difference equations, *SIAM J. Numer. Anal.*, **5**, 559–573 (1968).

* Refers to books.

328 APPENDIX I

64. *Dusinberre, G. M., *Heat-transfer Calculations by Finite Differences*, International Textbook, 293 pp. (1961) [MR 23 #B2817 Stanton]. Ehrlich, L., *see* Young, D.

65. Emery, A. F., An evaluation of several differencing methods for inviscid fluid flow problems, *J. Computational Phys.*, **2**, 306–331 (1968) [MR 37 #2477].

66. *Engeli, M., Ginsburg, T., Rutishauser, H., and Stiefel, E., *Refined Iterative Methods for Computation of the Solution and the Eigenvalues of Self-adjoint Boundary Value Problems*, Birkhäuser, 107 pp. (1959) [MR 26 #3218 Henrici]†.

67. Evans, D. J., Estimation of the line over-relaxation factor and convergence rates of an alternating direction line over-relaxation technique, *Comput. J.*, **7**, 318–321 (1965) [CR 6 #7974 Luke], [MR 30 #5503 Young].

68. Fairweather, G., and Gourlay, A. R., Some stable difference approximations to a fourth-order parabolic partial differential equation, *Math. Comp.*, **21**, 1–11 (1967) [CR 8 #12829 Luke], [MR 36 #4837 Strang].

69. Fairweather, G., Gourlay, A. R., and Mitchell, A. R., Some high accuracy difference schemes with a splitting operator for equations of parabolic and elliptic type, *Numer. Math.*, **10**, 56–66 (1967) [CR 9 #13494 Karplus], [MR 35 #3907].

70. Fairweather, G., and Mitchell, A. R., A generalized alternating direction method of Douglas-Rachford type for solving the biharmonic equation, *Comput. J.*, **7**, 242–245 (1964) [CR 8 #11985 Varga], [MR 31 #6385 Lieberstein].

71. Fairweather, G., and Mitchell, A. R., A high accuracy alternating direction method for the wave equation, *J. Inst. Math. Appl.*, **1**, 309–316 (1965) [MR 35 #1224 Laasonen]†.

72. Fairweather, G., and Mitchell, A. R., A new alternating direction method for parabolic equations in three space variables, *J. SIAM*, **13**, 957–965 (1965) [CR 7 #9764 Conte]†, [MR 32 #4863].

73. Fairweather, G., and Mitchell, A. R., A new computational procedure for A.D.I. methods, *SIAM J. Numer. Anal.*, **4**, 163–170 (1967) [CR 9 #13492 Wachspress], [MR 36 #1116 Laasonen].

74. Fairweather, G., and Mitchell, A. R., Some computational results of an improved A.D.I. method for the Dirichlet problem, *Comput. J.*, **9**, 298–303 (1966) [CR 8 #13237 Wouk].
   Fairweather, G., *see* Mitchell, A. R.
   Fernbach, S., *see* Alder, B.

75. *Fifer, S., *Analogue Computation: Theory, Techniques, and Applications*, McGraw-Hill, **3**, 965 pp. (1961) [CR 2 #1122 Brock].

76. Flügge-Lotz, I., The computation of compressible boundary-layer flow. In [5] 109–124.
   Flügge-Lotz, I., *see* Blottner, F. G.
   Flügge-Lotz, I., *see* Davis, R. T.

77. Forsythe, G. E., What are relaxation methods? In *Beckenbach, E. F., ed., *Modern Mathematics for the Engineer*, McGraw-Hill, 428–447 (1956).

* Refers to books.

78. *Forsythe, G. E., and Wasow, W. R., *Finite-difference Methods for Partial Differential Equations*, Wiley, 444 pp. (1960) [*CJ* 4 78 Fox]†, [*MC* 16 379 Polachek], [*MR* 23 #B3156 Young]†.

79. *Fox, L., ed., *Numerical Solution of Ordinary and Partial Differential Equations*, Pergamon, 509 pp. (1962) [*CR* 4 #4639 Hamming], [*MC* 18 528 Polachek], [*MR* 26 #4488 Greenspan]†.

80. Fox, L., Partial differential equations, *Comput. J.*, 6, 69–74 (1963) [*CR* 4 #4899 Herriot], [*MR* 27 #5377].

80a. *Fox, L., and Parker, I. B., *Chebyshev Polynomials in Numerical Analysis*, Oxford, 205 pp. (1968) [*CJ* 11 310 Clenshaw]†, [*MR* 37 #3733 Lance].

81. Fox, P., The solution of hyperbolic partial differential equations by difference methods. In [189] 180–188 [*MR* 22 #8696 Young].

Frank, T. G., *see* Young, D. M.

82. Friedrichs, K. O. and Keller, H. B., A finite difference scheme for generalized Neumann problems. In [23] 1–19 [*MR* 34 #3803 Thomée].

Friedrichs, K., *see* Courant, R.

Fromm, J. E., *see* Harlow, F. H.

83. Gaier, D., and Todd, J., On the rate of convergence of optimal ADI processes, *Numer. Math.*, 9, 452–459 (1967) [*MR* 35 #7603 Young]†.

Garder, A. O., *see* Douglas, J.

84. Giese, J. H., Computers in fluid mechanics. In *Freiberger, W. F., and Prager, W., eds., *Applications of Digital Computers*, Ginn, 97–137 (1963) [*CR* 5 #5489 Alt].

Ginsburg, T., *see* Engeli, M.

85. *Girerd, J., and Karplus, W. J., *Traitement des équations différentielles sur calculateurs électroniques*, Gauthier-Villars, 559 pp. (1968) [*CR* 9 #15616 Butcher], [*MR* 37 #2459 Booth].

86. *Godunov, S. K., and Ryabenki, V. S., *Theory of Difference Schemes*, Interscience, 289 pp. (1964) [*CR* 5 #5408 (review of 1962 Russian ed.) Kotz]†, [*MR* 29 #724 (review of 1962 Russian ed.) Johnson], [*MR* 31 #5346 Lohwater], [*SR* 11 92 Aronson]†.

87. *Gould, S. H., *Variational Methods for Eigenvalue Problems*, 2d ed., Univ. Toronto, 275 pp. (1966) [*MR* 19 287 (review of 1957 ed.) Diaz]†, [*MR* 35 #559 Reid]†.

88. Gourlay, A. R., The acceleration of the Peaceman-Rachford method by Chebyshev polynomials, *Comput. J.*, 10, 378–382 (1968) [*CR* 9 #14331 Wachspress]†, [*MR* 36 #7345 Styś].

89. Gourlay, A. R., and Mitchell, A. R., Alternating direction methods for hyperbolic systems, *Numer. Math.*, 8, 137–149 (1966) [*MR* 34 #971 Parter].

90. Gourlay, A. R.. and Mitchell, A. R., High accuracy A.D.I. methods for parabolic equations with variable coefficients, *Numer. Math.*, 12, 180–185 (1968).

91. Gourlay, A. R., and Mitchell, A. R., Intermediate boundary corrections for split operator methods in three dimensions, *BIT*, 7, 31–38 (1967) [*CR* 8 #12838 Conte]†, [*MR* 36 #1117 Garder].

92. Gourlay, A. R., and Mitchell, A. R., Split operator methods for hyper-

* Refers to books.

bolic systems in p space variables, *Math. Comp.*, **21**, 351–354 (1967) [*MR* **37** #2471 Evans].

93. Gourlay, A. R., and Mitchell, A. R., A stable implicit difference method for hyperbolic systems in two space variables, *Numer. Math.*, **8**, 367–375 (1966) [*CR* **8** #11606 Dames], [*MR* **33** #6850 Froese].

94. Gourlay, A. R., and Mitchell, A. R., Two-level difference schemes for hyperbolic systems, *SIAM J. Numer. Anal.*, **3**, 474–485 (1966) [*CR* **9** #13465], [*MR* **34** #7038 Kopáček].

95. Gourlay, A. R., and Morris, J. L., Deferred approach to the limit in nonlinear hyperbolic systems, *Comput. J.*, **11**, 95–101 (1968) [*CR* **9** #15442 Concus] [*MR* **37** #1101].

96. Gourlay, A. R., and Morris, J. L., Finite-difference methods for nonlinear hyperbolic systems, *Math. Comp.*, **22**, 28–39 (1968) [*MR* **36** #6163 Strang]; II, *Math. Comp.*, **22**, 549–556 (1968) [*MR* **37** #3785].

97. Gourlay, A. R., and Morris, J. L., A multistep formulation of the optimized Lax-Wendroff method for nonlinear hyperbolic systems in two space variables, *Math. Comp.*, **22**, 715–719 (1968).

Gourlay, A. R., *see* Fairweather, G.

98. Gram, C., Naur, P., and Poulsen, E. T., Partial differential equations. In *Gram, C., ed., Selected Numerical Methods for Linear Equations, Polynomial Equations, Partial Differential Equations, Conformal Mapping*, Regencentralen, 29–113 (1962) [*CR* **4** #4863 Conte], [*MR* **28** #1727 Douglas].

99. *Greenspan, D., Introductory Numerical Analysis of Elliptic Boundary Value Problems*, Harper and Row, 164 pp. (1965) [*CJ* **9** 44 Osborne], [*CR* **6** #8611 Willoughby]†, [*CR* **7** #10696 Gallaher], [*MR* **31** #4193 Isaacson].

100. *Greenspan, D., Lectures on the Numerical Solution of Linear, Singular, and Nonlinear Differential Equations*, Prentice-Hall, 185 pp. (1968).

101. *Greenspan, D., ed., Numerical Solutions of Nonlinear Differential Equations*, Wiley, 347 pp. (1966) [*CR* **8** #11277], [*MC* **22** 896 Gautschi]†, [*MR* **34** #3749].

102. Greenspan, D., On approximating extremals of functionals. I, *ICC Bull.*, **4**, 99–120 (1965) [*CR* **7** #10104 Varga], [*MR* **32** #8526 Madan]; II, *Internat. J. Engrg. Sci.*, **5**, 571–588 (1967) [*MR* **36** #3508 Madan].

103. Greenspan, D., and Parter, S. V., Mildly nonlinear elliptic partial differential equations and their numerical solution. II, *Numer. Math.*, **7**, 129–146 (1965) [*CR* **7** #9058 Herriot], [*MR* **31** #1777 Thomée].

104. Greenspan, D., and Yohe, M., On the approximate solution of $\Delta u = F(u)$, *Comm. ACM*, **6**, 564–568 (1963) [*CR* **5** #5079 Froese].

105. *Greenspan, H., et al., eds., Computing Methods in Reactor Physics*, Gordon and Breach (1968).

106. Griffin, D. S., and Varga, R. S., Numerical solution of plane elasticity problems, *J. SIAM*, **11**, 1046–1062 (1963) [*CR* **5** #6371], [*MR* **28** #3544 Fox].

107. *Grinter, L. E., ed., Numerical Methods of Analysis in Engineering*, Macmillan, 207 pp. (1949).

108. Guittet, J., Méthodes de directions alternées, *Chiffres*, **9**, 95–107 (1966) [*CR* **8** #12394 Wouk], [*MR* **35** #3867 Young].

* Refers to books.

109. Guittet, J., Une nouvelle méthode de directions alternées à q variables, *J. Math. Anal. Appl.*, **17**, 199–213 (1967) [*MR* **36** #4792 Young].

110. Hageman, L. A., and Kellogg, R. B., Estimating optimum overrelaxation parameters, *Math. Comp.*, **22**, 60–68 (1968) [*MR* **37** #4945 (French) Gastinel].

111. Hageman, L. A., and Varga, R. S., Block iterative methods for cyclically reduced matrix equations, *Numer. Math.*, **6**, 106–119 (1964) [*CR* **5** #6722 Fox], [*MR* **29** #4185 (German) Schröder].

Hansen, K. F., *see* Clark, M.

112. Harlow, F. H., Numerical fluid dynamics, *Amer. Math. Monthly*, **72**, no. 2, pt. II, 84–91 (1965).

113. Harlow, F. H., The particle-in-cell method for numerical solution of problems in fluid dynamics. In [255] 269–288.

114. Harlow, F. H., and Fromm, J. E., Computer experiments in fluid dynamics, *Sci. Amer.*, **212**, 104–110 (Mar. 1965) [*CR* **7** #8808 Wood].

Hickerson, N., *see* Spillers, W. R.

115. *Hildebrand, F. B., *Finite-difference Equations and Simulations*, Prentice-Hall, 338 pp. (1968) [*CR* **10** #16007 Breslau], [*MR* **37** #3769 Kreiss].

116. *Hildebrand, F. B., *Methods of Applied Mathematics*, Prentice-Hall, 523 pp. (1952) [*MR* **15** 204]; 2nd ed., 362 pp. (1965).

117. Hockney, R. W., A fast direct solution of Poisson's equation using Fourier analysis, *J. ACM*, **12**, 95–113 (1965) [*CR* **6** #7975 Kennedy]†, [*MR* **35** #3913 Varga].

118. Hodgkins, W. R., On the relation between dynamic relaxation and semi-iterative matrix methods, *Numer. Math.*, **9**, 446–451 (1967) [*CR* **9** #13472 Householder], [*MR* **35** #3908].

119. Holton, J. R., A stable finite difference scheme for the linearized vorticity and divergence equation system, *J. Appl. Meteorology*, **6**, 519–522 (1967).

120. Hubbard, B. E., Alternating direction schemes for the heat equation in a general domain, *J. SIAM Ser. B Numer. Anal.*, **2**, 448–463 (1965) [*CR* **7** #10533 Peaceman], [*MR* **33** #5136 Laasonen].

121. *Issaacson, E., and Keller, H. B., *Analysis of Numerical Methods*, Wiley, 541 pp. (1966) [*CJ* **10** 270 Albasiny], [*CR* **8** #11966 Froese], [*MR* **34** #924 Householder], [*SR* **11** 100 Moore].

Isaacson, E., *see* Courant, R.

122. *John, F., *Lectures on Advanced Numerical Analysis*, Gordon and Breach, 179 pp. (1967) [*CJ* **11** 346 Gilles], [*MC* **22** 683 Isaacson], [*MR* **36** #4773 (German) Collatz].

123. Kafka, J., Accuracy of analogue computing networks designed for the solution of the diffusion equation, *Information Processing Machines*, no. **13**, 181–213 (1967) [*CR* **9** #15424 Kovach].

124. Kalaba, R., On nonlinear differential equations, the maximum operation, and monotone convergence, *J. Math. Mech.*, **8**, 519–574 (1959) [*MR* **21** #6453 (German) Schröder].

Kalaba, R. E., *see* Bellman, R. E.

125. *Kantorovich, L. V., and Krylov, V. I., *Approximate Methods of Higher Analysis*, Interscience, 681 pp. (1958) [*MC* **14** 90 Householder], [*MR* **21** #5268 Freiberger], [*SR* **2** 299 Rivlin].

* Refers to books.

126. Kaplan, W., Numerical methods in the solution of problems of non-linear elasticity. In [251] 194–196 [*MR* **10** 759 Levy].

127. *Karplus, W. J., *Analog Simulation: Solution of Field Problems*, McGraw-Hill, 434 pp. (1958) [*MC* **15** 109 Keen]†, [*MR* **20** #3647 Michel].

128. Karplus, W. J., and Vemuri, V., Numerical solution of partial differential equations. In *Klerer, M., and Korn, G. A., eds., *Digital Computer User's Handbook*, McGraw-Hill (1967) [*MC* **22** 893 Ortega and Schweppe].

Karplus, W. J., *see* Girerd, J.

129. Keast, P., and Mitchell, A. R., Finite difference solution of the third boundary problem in elliptic and parabolic equations, *Numer. Math.*, **10**, 67–75 (1967) [*MR* **35** #3909].

130. Keller, H. B., The numerical solution of parabolic differential equations. In [189] 135–143 [*MR* **22** #8692 Douglas].

Keller, H. B., *see* Friedrichs, K. O.

Keller, H. B., *see* Isaacson, E.

131. Kellogg, R. B., Difference equations on a mesh arising from a general triangulation, *Math. Comp.*, **18**, 203–210 (1964) [*CR* **5** #6752 Peaceman]†, [*MR* **31** #1780 Rall].

Kellogg, R. B., *see* Douglas, J.

Kellogg, R. B., *see* Hageman, L. A.

Kendall, R. P., *see* Dupont, T.

132. Kolsky, H. G., Some computer aspects of meteorology, *IBM J. Res. Develop.*, **11**, 584–600 (1967) [*CR* **9** #15112 Rosenthal].

133. Kron, G., *Diakoptics*, Macdonald (1963).

Krylov, V. I., *see* Kantorovich, L. V.

134. *v. Krzywoblocki, M. Z., *Bergman's Linear Integral Operator Method in the Theory of Compressible Fluid Flow*, Springer, 188 pp. (1960) [*MR* **22** #6289 Cherry].

135. Kulsrud, H. E., A practical technique for the determination of the optimum relaxation factor of the successive over-relation method, *Comm. ACM*, **4**, 184–187 (1961) [*MR* **26** #895 Varga].

136. Ladyženskaya, O. A., The method of finite differences in the theory of partial differential equations. In *American Mathematical Society Translations Ser. 2*, **20**, 77–104 (1962) [*MR* **20** #3395 (review of 1957 Russian original) (French) Lions].

137. *Lance, G. N., *Numerical Methods for High Speed Computers*, Iliffe, 166 pp. (1960) [*CR* **2** #670 Acton]†, [*MR* **23** #B1096 Herriot]†.

138. *Langer, R. E., ed., *Boundary Problems in Differential Equations*, Univ. Wisconsin, 324 pp. (1960) [*MC* **14** 395 Taub]†.

139. *Langer, R. E., ed., *Frontiers of Numerical Mathematics*, Univ. Wisconsin, 132 pp. (1960) [*MC* **15** 106 Taub].

140. *Langer, R. E., ed., *Nonlinear Problems*, Univ. Wisconsin, 334 pp. (1962) [*CR* **6** #6956 Perry]†.

141. *Langer, R. E., ed., *On Numerical Approximation*, Univ. Wisconsin, 462 pp. (1959).

142. *Langer, R. E., ed., *Partial Differential Equations and Continuum Mechanics*, Univ. Wisconsin, 397 pp. (1961) [*MC* **16** 399 Truesdell], [*MR* **22** #8183].

* Refers to books.

143. *Lapidus, L., *Digital Computation for Chemical Engineers*, McGraw-Hill, 407 pp. (1962) [*CR* **5** #5536 Naphtali].

144. Larkin, B. K., Some stable explicit difference approximations to the diffusion equation, *Math. Comp.*, **18**, 196–202 (1964) [*CR* **6** #6752 Peaceman]†, [*MR* **29** #1747 Garder].

145. Lax, P. D., Hyperbolic difference equations: a review of the Courant-Friedrichs-Lewy paper in the light of recent developments, *IBM J. Res. Develop.*, **11**, 235–238 (1967) [*MR* **36** #2330 Laasonen].

146. Lax, P. D., Numerical solution of partial differential equations, *Amer. Math. Monthly*, **72**, no. 2, pt. II, 74–84 (1965) [*CR* **8** #51180] = [*MR* **31** #5349 Kreiss].

147. Lax, P. D., Survey of stability of different schemes for solving initial value problems for hyperbolic equations. In [255] 251–258 [*CR* **5** #5733 Conte], [*MR* **28** #3549 Kreiss].

148. *Lee, E. S., *Quasilinearization and Invariant Imbedding, with Applications to Chemical Engineering and Adaptive Control*, Academic Press, 329 pp. (1968).

149. Lees, M., Approximate solutions of parabolic equations, *J. SIAM*, **7**, 167–183 (1959) [*MR* **22** #1092 Douglas].

150. Leith, C. E., Numerical hydrodynamics of the atmosphere. In [256] 125–137 [*CR* **10** #16670 Duquet].

Lewy, H., *see* Courant, R.

151. *Lieberstein, H. M., *A Course in Numerical Analysis*, Harper and Row, 258 pp. (1968) [*CR* **9** #15421 Hammer], [*SR* **11** 294 Parter]†.

152. Lister, M., The numerical solution of hyperbolic partial differential equations by the method of characteristics. In [189] 165–179 [*MR* **22** #8695 Fox].

153. *Lowan, A. N., *The Operator Approach to Problems of Stability and Convergence of Solutions of Difference Equations and the Convergence of Various Iteration Procedures*, Scripta Mathematica, 104 pp. (1957).

154. Lynch, R. E., and Rice, J. R., Convergence rates of ADI methods with smooth initial error, *Math. Comp.*, **22**, 311–335 (1968).

155. Lynch, R. E., Rice, J. R., and Thomas, D. H., Direct solution of partial difference equations by tensor product methods, *Numer. Math.*, **6**, 185–199 (1964) [*MR* **35** #3912 Varga].

156. Lynch, R. E., Rice, J. R., and Thomas, D. H., Tensor product analysis of alternating direction implicit methods, *J. SIAM*, **13**, 995–1006 (1965) [*CR* **7** #9765 Wachspress] = [*MR* **33** #6854].

157. Lynch, R. E., Rice, J. R., and Thomas, D. H., Tensor product analysis of partial difference equations, *Bull. Amer. Math. Soc.*, **70**, 378–384 (1964) [*MR* **29** #6640 Varga].

158. Mason, J. C., Chebyshev methods for separable partial differential equations. In [248] *A*, 60–64 [*CR* **9** #15607].

159. McAllister, G. T., Difference methods for a nonlinear elliptic system of partial differential equations, *Quart. Appl. Math.*, **23**, 355–359 (1966) [*CR* **7** #10542 Perry], [*MR* **32** #6698].

160. McAllister, G. T., Quasilinear uniformly elliptic partial differential equations and difference equations, *SIAM J. Numer. Anal.*, **3**, 13–33 (1966) [*MR* **34** #2213 Parter].

* Refers to books.

161. McAllister, G. T., Some nonlinear elliptic partial differential equations and difference equations, *J. SIAM*, **12**, 772–777 (1964) [*MR* **31** #4195 Isaacson].

McCue, G. A., *see* Radbill, J. R.

McNamee, J., *see* Bickley, W. G.

Meinardus, G., *see* Collatz, L.

162. Merzrath, E., Direct solution of partial difference equations, *Numer. Math.*, **9**, 431–436 (1967) [*MR* **35** #3914 Elliott].

163. *Mikhlin, S. G., and Smolitskiy, K. L., *Approximate Methods for Solution of Differential and Integral Equations*, American Elsevier, 308 pp. (1967) [*CR* **9** #13670], [*MR* **33** #855 (review of 1965 Russian ed.) Gautschi], [*SR* **10** 391 Wouk].

164. Miller, K., Numerical analogs to the Schwarz alternating procedure, *Numer. Math.*, **7**, 91–103 (1965) [*MR* **31** #1783 Hitotumatu].

Milne, W. E., *see* Bennett, A. A.

165. Milnes, H. W., and Potts, R. B., Boundary contraction solution of Laplace's differential equation, *J. ACM*, **6**, 226–235 (1959) [*MR* **21** #4554 Forsythe].

166. Milnes, H. W., and Potts, R. B., Numerical solution of partial differential equations by boundary contraction, *Quart. Appl. Math.*, **18**, 1–13 (1960) [*MR* **22** #5129 Wasow].

Milnes, H. W., *see* Chow, T. S.

167. v. Mises, R., and Schiffer, M., On Bergman's integration method in two-dimensional compressible fluid flow. In *von Mises, R., and von Kármán, T., eds., *Advances in Applied Mechanics*, Academic Press, **1**, 249–285 (1948) [*MR* **10** 642 Bers].

168. Mitchell, A. R., and Fairweather, G., Improved forms of the alternating direction methods of Douglas, Peaceman, and Rachford for solving parabolic and elliptic equations, *Numer. Math.*, **6**, 285–292 (1964) [*CR* **6** #8618 Lowell], [*MR* **30** #4391 Lees].

Mitchell, A. R., *see* Fairweather, G.

Mitchell, A. R., *see* Gourlay, A. R.

Mitchell, A. R., *see* Keast, P.

Morris, J. L., *see* Gourlay, A. R.

Morton, K. W., *see* Richtmyer, R. D.

Naur, P., *see* Gram, C.

169. *von Neumann, J., *Collected Works*, Macmillan, **5**, 784 pp. (1963) [*MR* **28** #1104]; **6**, 538 pp. (1963) [*CR* **5** #4929 Hamming], [*MC* **19** 346 Polachek], [*MR* **28** #1105].

170. Noh, W. F., A general theory for the numerical solution of the equations of hydrodynamics. In [101] 181–211 [*CR* **8** #11278 Garder], [*MR* **35** #6377 Polachek].

171. Ortega, J. M., and Rheinboldt, W. C., Monotone iterations for nonlinear equations with application to Gauss-Seidel methods, *SIAM J. Numer. Anal.*, **4**, 171–190 (1967) [*CR* **9** #13477 Gelder].

172. Osborne, M. R., Direct methods for the solution of finite-difference approximations to separable partial differential equations, *Comput. J.*, **8**, 150–156 (1965) [*CR* **7** #9396 Willoughby], [*MR* **32** #6699].

* Refers to books.

173. *Panov, D. J., *Formulas for the Numerical Solution of Partial Differential Equations by the Method of Differences*, Ungar, 133 pp. (1963) [*MR* **14** 93 (review of 1950 Russian ed.) Milne].

Parker, I. B., *see* Fox, L.

174. Parter, S. V., Elliptic equations, *IBM J. Res. Develop.*, **11**, 244–247 (1967) [*MR* **37** #1097].

175. Parter, S. V., Maximal solutions of mildly non-linear elliptic equations. In [101] 213–238 [*CR* **8** #11572 Wigley], [*MR* **36** #1118 Young].

176. Parter, S. V., Mildly nonlinear elliptic partial differential equations and their numerical solution. I, *Numer. Math.*, **7**, 113–128 (1965) [*CR* **7** #9057 Herriot], [*MR* **30** #5499 Thomée].

177. Parter, S. V., Numerical methods for generalized axially symmetric potentials, *J. SIAM Ser. B Numer. Anal.*, **2**, 500–516 (1965) [*MR* **32** #8522 Bramble].

178. Parter, S. V., On estimating the "rates of convergence" of iterative methods for elliptic difference equations, *Trans. Amer. Math. Soc.*, **114**, 320–354 (1965) [*CR* **8** #11600] = [*MR* **31** #5350 Keller].

179. Parter, S. V., Remarks on the numerical computation of solutions of $\Delta u = f(P, u)$. In [23] 73–82 [*CR* **8** #11617 Greenspan].

180. Parter, S. V., Some computational results on "two-line" iterative methods for the biharmonic difference equation, *J. ACM*, **8**, 359–365 (1961) [*CR* **3** #1809 Rodriguez], [*MR* **26** 907 Varga].

181. Parter, S. V., Some unusual problems in numerical analysis. In [247] **1**, 183–186 [*CR* **7** #10134 Kolsky].

Parter, S. V., *see* Greenspan, D.

Pearcy, C., *see* Douglas, J.

181a. Pearson, C. E., A numerical method for incompressible viscous flow problems in a spherical geometry. In *Studies in Numerical Analysis. I*, Society for Industrial and Applied Mathematics, 65–78 (1969).

181b. Pereyra, V., On improving an approximate solution of a functional equation by deferred corrections, *Numer. Math.*, **8**, 376–391 (1966) [*CR* **8** #11571 Froberg], [*MR* **34** #3814 Stetter].

182. *Petrovskii, I. G., *Partial Differential Equations*, Iliffe, 410 pp. (1967) [*MR* **13** 241 (review of 1950 Russian ed.) Diaz], [*MR* **16** 133 (review of 1953 Russian ed.)], [*MR* **25** #2309 (review of 1961 Russian ed.) Gould].

183. Phillips, N. A., Numerical weather prediction. In *Alt, F. L., ed., *Advances in Computers*, Academic Press, **1**, 43–90 (1960) [*MR* **22** #5461 Rogers].

184. *Polozhii, G. N., *The Method of Summary Representation for Numerical Solution of Problems of Mathematical Physics*, Pergamon, 283 pp. (1965) [*CJ* **10** 59 Price]†, [*CR* **8** #11276 Kennedy], [*MC* **21** 123 Solomon]†.

Potts, R. B., *see* Milnes, H. W.

Poulsen, E. T., *see* Gram, C.

Práger, M., *see* Babuška, I.

185. Price, H. S., Monotone and oscillation matrices applied to finite difference approximations, *Math. Comp.*, **22**, 489–516 (1968) [*MR* **38** #875 Albrecht].

* Refers to books.

186. Price, H. S., A practical application of block diagonally dominant matrices, *Math. Comp.*, **19**, 307–313 (1965) [*MR* **33** #5110 (French) Gastinel].

Price, H. S., *see* Varga, R. S.

Rachford, H. H., *see* Dupont, T.

187. *Radbill, J. R., and McCue, G. A., *Quasilinearization and Nonlinear Problems in Fluid and Orbital Mechanics*, American Elsevier (1969).

187a. *Rall, L. B., *Computational Solution of Nonlinear Operator Equations*, Wiley, 225 pp. (1969) [*CR* **10** #17311 Gargantini].

188. *Rall, L. B., ed., *Error in Digital Computation*, Wiley, **1**, 324 pp. (1965) [*CR* **6** #8266 Gregory], [*MR* **30** #5510]; **2**, 288 pp. (1965) [*CR* **8** #11224 Gregory].

189. *Ralston, A., and Wilf, H. S., eds., *Mathematical Methods for Digital Computers*, Wiley, 293 pp. (1960) [*CR* **2** #640] = [*SR* **3** 181 Hull], [*MC* **15** 86 Gregory], [*MR* **22** #8680].

190. Randall, T. J., A note on the estimation of the optimum successive over-relaxation parameter for Laplace's equation, *Comput. J.*, **10**, 400–401 (1968) [*MR* **36** #4825].

Rees, M., *see* Courant, R.

191. Reid, J. K., A method for finding the optimum successive overrelaxation parameter, *Comput. J.*, **9**, 200–204 (1966) [*MR* **33** #3475].

192. *Rektorys, K., *et al.*, eds., *A Survey of Applied Mathematics*, Státní Naklad. Tech. Lit., 1137 pp. (1963) (Czech) [*MR* **28** #1799 Nádeník].

Rheinboldt, W. C., *see* Ortega, J. M.

Rice, J. R., *see* Lynch, R. E.

193. *Richardson, L. F., *Weather Prediction by Numerical Process*, Dover, 236 pp. (1965, reprint of 1922 ed.) [*CR* **7** #9952 Chapman], [*MC* **20** 633 Phillips].

194. *Richtmyer, R. D., *A Survey of Difference Methods for Non-steady Fluid Dynamics*, National Center of Atmospheric Research, Tech. Note **63–2**, (1962).

195. *Richtmyer, R. D., and Morton, K. W., *Difference Methods for Initial-value Problems*, 2d ed., Interscience, 405 pp. (1967) [*CR* **9** #15229 Herriot]†, [*MC* **22** 465 Isaacson], [*MR* **20** #438 (review of 1957 ed.) Davis], [*MR* **36** #3515 Sjöberg], [*SR* **10** 381 Wendroff]†.

196. Rigler, A. K., An application of cyclic reduction to Ritz type difference equations, *Math. Comp.*, **18**, 292–296 (1964) [*MR* **30** #728 (German) Collatz].

197. Rigler, A. K., Estimation of the successive overrelaxation factor, *Math. Comp.*, **19**, 302–307 (1965) [*CR* **6** #8616 Kovach], [*MR* **31** #5351 Delves].

198. *von Rosenberg, D. V., *Methods for the Numerical Solution of Partial Differential Equations*, American Elsevier (1969).

199. *Roslyakov, G. S., and Chudov, L. A., eds., *Numerical Methods in Gas Dynamics*, Israel Program for Scientific Translations, 166 pp. (1966).

Roslyakov, G. S., *see* Trifonov, N. P.

Rotenberg, M., *see* Alder, B.

Rutishauser, H., *see* Engeli, M.

Ryabenki, V. S., *see* Godunov, S. K.

* Refers to books.

200. *Sangren, W. C., *Digital Computers and Nuclear Reactor Calculations*, Wiley, 208 pp. (1960) [*MC* **15** 216] = [*MR* **22** #11520 Varga]†, [*SR* **3** 78 Hull].

201. *Sauer, R., *Introduction to Theoretical Gas Dynamics*, Edwards, 222 pp. (1947) [*MR* **7** 92 (review of 1945 German ed.) Bers]†, [*MR* **14** 107 (review of 1951 German ed.) Gilbarg].

202. *Saul'yev, V. K., *Integration of Equations of Parabolic Type by the Method of Nets*, Pergamon, 346 pp. (1964) [*CJ* **8** 156 Clenshaw], [*MC* **20** 462 Lees], [*MR* **23** #A428 (review of 1960 Russian ed.) Householder], [*SR* **7** 576 Perry].

Schiffer, M., *see* Bergman, S.

Schiffer, M., *see* v. Mises, R.

203. Schröder, J., Estimations in nonlinear equations. In [247] **1**, 187–194 [*CR* **7** #10135 Perry].

204. Shampine, L. F., Monotone iterations and two-sided convergence, *SIAM J. Numer. Anal.*, **3**, 607–615 [*CR* **9** #14092], [*MR* **35** #6329 (German) Collatz].

205. *Shaw, F. S., *An Introduction to Relaxation Methods*, Dover, 396 pp. (1953) [*MR* **15** 353 Gilles].

206. Sheldon, J. W., Iterative methods for the solution of elliptic partial differential equations. In [189] 144–156 [*MR* **22** #8693 Young]†.

207. Smagorinsky, J., On the application of numerical methods to the solution of systems of partial differential equations arising in meteorology. In [139] 107–125.

208. *Smith, G. D., *Numerical Solution of Partial Differential Equations*, Oxford, 179 pp. (1965) [*CJ* **9** 204 Johnston], [*SR* **10** 396 Perry].

Smolitskiy, K. L., *see* Mikhlin, S. G.

209. *Soare, M., *Applications of Finite Difference Equations to Shell Analysis*, Pergamon, 439 pp. (1967).

210. *Southwell, R. V., *Relaxation Methods in Engineering Science*, Oxford, 252 pp. (1940) [*MR* **3** 152 Poritsky].

211. *Southwell, R. V., *Relaxation Methods in Theoretical Physics*, Oxford, **1**, 1–248 (1946) [*MR* **8** 355 Milne]†; **2**, 249–522 (1956) [*MR* **18** 677 Milne].

212. Spanier, J., Alternating direction methods applied to heat conduction problems. In *Ralston, A., and Wilf, H. S., *Mathematical Methods for Digital Computers*, Wiley, **2**, 215–245 (1967) [*CR* **8** #12421 Erdelyi].

213. Spillers, W. R., and Hickerson, N., Optimal elimination for sparse symmetric systems as a graph problem, *Quart. Appl. Math.*, **26**, 425–432 (1968).

214. Steward, D. V., Partitioning and tearing systems of equations, *J. SIAM Ser. B Numer. Anal.*, **2**, 345–365 (1965) [*CR* **7** #10120 Harman] = [*MR* **36** #2307].

Stiefel, E., *see* Engeli, M.

214a. Stone, H. L., Iterative solution of implicit approximations of multi-dimensional partial differential equations, *SIAM J. Numer. Anal.*, **5**, 530–558 (1968).

215. Swartz, B. K., Explicit $O(h^2)$ bounds on the eigenvalues of the half-*L*, *Math. Comp.*, **22**, 40–59 (1968) [*MR* **36** #6161].

Swartz, B., *see* Birkhoff, G.

* Refers to books.

216. *Synge, J. L., *The Hypercircle in Mathematical Physics*, Cambridge Univ., 424 pp. (1957) [*MR* **20** #4073 Diaz]†.

217. Tee, G. J., An application of p-cyclic matrices, for solving periodic parabolic problems, *Numer. Math.*, **6**, 142–159 (1964) [*CR* **6** #7288] = [*MR* **30** #1619 Varga].

218. Tee, G. J., Eigenvectors of the successive overrelaxation process and its combination with Chebyshev semi-iteration, *Comput. J.*, **6**, 250–263 (1963).

219. *Thom, A., and Apelt, C. J., *Field Computations in Engineering and Physics*, Van Nostrand, 165 pp. (1961) [*MR* **24** #B529 Forsythe]†.

Thomas, D. H., *see* Lynch, R. E.

220. *Thompson, P. D., *Numerical Weather Analysis and Prediction*, Macmillan, 170 pp. (1961) [*MC* **16** 503 Smagorinsky]†, [*SR* **5** 87 Saltzer].

221. Todd, J., A direct approach to the problem of stability in the numerical solution of partial differential equations, *Comm. Pure Appl. Math.*, **9**, 597–612 (1956) [*MR* **18** 338 Hale].

Todd, J., *see* Gaier, D.

222. *Trifonov, N. P., Roslyakov, G. S., and Zogolev, E. A., eds., *Numerical Methods and Programming*, Izdat. Moskov. Univ., **1**, 350 pp. (1962) (Russian) [*MR* **29** #4172 Juncosa]†.

Tukey, J. W., *see* Cooley, J. W.

Unger, H., *see* Collatz, L.

223. Varga, R. S., Factorization and normalized iterative methods. In [138] 121–142 [*MR* **22** #12704 Young]†.

224. Varga, R. S., Iterative methods for solving matrix equations, *Amer. Math. Monthly*, **72**, no. 2, pt. II, 67–74 (1965) [*MR* **30** #2677].

225. *Varga, R. S., *Matrix Iterative Analysis*, Prentice-Hall, 322 pp. (1962) [*CJ* **7** 88 Wilkinson]†, [*CR* **4** #4236 Hansen], [*MC* **17** 310 Bramble]†, [*MR* **28** #1725 Garder].

226. Varga, R. S., Price, H. S., and Warren, J. E., Application of oscillation matrices to diffusion-convection equations, *J. Math. and Phys.*, **45**, 301–311 (1966) [*MR* **34** #7046 (German) Collatz].

Varga, R. S., *see* Birkhoff, G.

Varga, R. S., *see* Cuthill, E. H.

Varga, R. S., *see* Douglas, J.

Varga, R. S., *see* Griffin, D. S.

Varga, R. S., *see* Hageman, L. A.

Vemuri, V., *see* Karplus, W. J.

227. Vichnevetsky, R., Application of hybrid computers to the integration of partial differential equations of the first and second order. In [248] *A*, 68–75 [*CR* **9** #15609].

Vitásek, E., *see* Babuška, I.

227a. *Voigt, S. J., *Bibliography on the Numerical Solution of Integral and Differential Equations and Related Topics*, U.S. Department of the Navy, Naval Ship Research and Development Center, Applied Mathematics Laboratory, Report **2423**, 526 pp.(1967) [*MC* **23** 207 Luke].

228. *Volynskii, B. A., and Bukhman, V. Ye., *Analogues for the Solution of Boundary-value Problems*, Pergamon, 460 pp. (1965) [*CJ* **9** 380 Gatehouse], [*CR* **7** #9508 Dames], [*MC* **21** 283 Fifer].

* Refers to books.

229. *Vorobyëv, Yu. V., *Method of Moments in Applied Mathematics*, Gordon and Breach, 168 pp. (1965) [*MC* **22** 218 Petryshyn]†, [*MR* **21** #7591 (review of 1958 Russian ed.) Householder].
230. *Wachspress, E. L., *Iterative Solution of Elliptic Systems and Application to the Neutron Diffusion Equations of Reactor Physics*, Prentice-Hall, 299 pp. (1966) [*CR* **8** #11605 Hoffman], [*Q* **25** 499 Varga]† = [*SR* **9** 756].
231. *Walsh, J., ed., *Numerical Analysis: an Introduction*, Thompson, 212 pp. (1967) [*CR* **8** #11967, 12419–12420], [*MC* **22** 458 Wrench].
      Warren, J. E., *see* Varga, R. S.
      Wasow, W. R., *see* Forsythe, G. E.
232. Weinstein, A., Some numerical results in intermediate problems for eigenvalue problems. In [23] 167–191 [*MR* **34** #5326 (German) Collatz].
233. *Wendroff, B., *Theoretical Numerical Analysis*, Academic Press, 239 pp. (1966) [*CR* **7** #10485 Fettis], [*MC* **21** 290 Isaacson], [*MR* **33** #5080 Evans], [*SR* **9** 758 Parter].
      Wendroff, B., *see* Birkhoff, G.
      Wetherald, R. T., *see* Carroll, A. B.
      Wheeler, M. F., *see* Young, D. M.
234. Widlund, O. B., On difference methods for parabolic equations and alternating direction implicit methods for elliptic equations, *IBM J. Res. Develop.*, **11**, 239–243 (1967) [*MR* **36** #7356].
235. Widlund, O. B., On the rate of convergence of an alternating direction implicit method in a noncommutative case, *Math. Comp.*, **20**, 500–515 (1966) [*CR* **8** #12056 Payne].
      Wilf, H. S., *see* Ralston, A.
236. Wrigley, H. E., Accelerating the Jacobi method for solving simultaneous equations by Chebyshev extrapolation when the eigenvalues of the iteration matrix are complex, *Comput. J.*, **6**, 169–176 (1963) [*CR* **5** #5070 Fettis], [*MR* **27** #2095].
      Yohe, M., *see* Greenspan, D.
237. Young, D., The numerical solution of elliptic and parabolic partial differential equations. In *Beckenbach, E. F., ed., *Modern Mathematics for the Engineer*, 2d ser., McGraw-Hill, 373–419 (1961) [*MR* **23** #B2205 Mullin].
238. Young, D., The numerical solution of elliptic and parabolic partial differential equations. In *Todd, J., ed., *Survey of Numerical Analysis*, McGraw-Hill, 380–438 (1962) [*MR* **24** #B2123 Laasonen].
239. Young, D., On the solution of linear systems by iteration. In [253] 283–298 [MR **18** 417 Forsythe]†.
240. Young, D. M., and Dauwalder, J. H., Discrete representations of partial differential equations. In [188] **2**, 181–217 [*CR* **8** #12045 Lynch], [*MR* **33** #894 Isaacson]†.
241. Young, D., and Ehrlich, L., Some numerical studies of iterative methods for solving elliptic difference equations. In [138] 143–162 [*MR* **22** #5127 Varga].
242. Young, D. M., and Frank, T. G., A survey of computer methods for solving elliptic and parabolic partial differential equations, *ICC Bull.*, **2**, 3–61 (1963) [*MR* **27** #952].

    * Refers to books.

243. Young, D. M., and Wheeler, M. F., Alternating direction methods for solving partial differential equations. In *Ames, W. F., ed., *Nonlinear Problems of Engineering*, Academic Press, 220–246 (1964) [*MR* **30** #4395 Lees].
244. Young, D. M., Wheeler, M. F., and Downing, J. A., On the use of the modified successive overrelaxation method with several relaxation factors. In [247] 177–182 [*CR* **7** #10115 Kolsky].
    Young, D., see Birkhoff, G.
    Zarantonello, E. H., see Birkhoff, G.
    Zhidkov, N. P., see Berezin, I. S.
    Žogolev, E. A., see Trifonov, N. P.

245. *Information Processing: Proceedings of the International Conference on Information Processing, Paris, 1959*, UNESCO, 520 pp. (1960).
246. *Information Processing 1962: Proceedings of IFIP Congress 62, Munich*, North-Holland, 780 pp. (1963).
247. *Information Processing 1965: Proceedings of IFIP Congress 65, New York*, Spartan, **1**, 1–304 (1965) [*CR* **6** #8418]; **2**, 305–688 (1966).
248. *Information Processing 1968: Proceedings of IFIP Congress 68, Edinburgh*, North-Holland, (1968) [*CR* **9** #15489].
249. *Modern Computing Methods*, 2d ed., Philosophical Library, 170 pp. (1961) [*MR* **19** 579 (review of 1957 ed.) Householder], [*MR* **22** #8637 Householder], [*SR* **1** 185 (review of 1957 ed.) Kelley], [*SR* **5** 172 Saltzer].
250. *Outlines of the Joint Soviet-American Symposium on Partial Differential Equations, Novosibirsk, 1963*, Acad. Sci. USSR, 378 pp. (1963).
251. *Proceedings of Symposia in Applied Mathematics, 1, Non-linear Problems in Mechanics of Continua*, Amer. Math. Soc., 219 pp. (1949).
252. *Proceedings of Symposia in Applied Mathematics, 4, Fluid Dynamics*, Amer. Math. Soc., 186 pp. (1953).
253. *Proceedings of Symposia in Applied Mathematics, 6, Numerical Analysis*, Amer. Math. Soc., 303 pp. (1956).
254. *Proceedings of Symposia in Applied Mathematics, 11, Nuclear Reactor Theory*, Amer. Math. Soc., 339 pp. (1961).
255. *Proceedings of Symposia in Applied Mathematics, 15, Experimental Arithmetic, High Speed Computing and Mathematics*, Amer. Math. Soc., 396 pp. (1963).
256. *Proceedings of Symposia in Applied Mathematics, 19, Mathematical Aspects of Computer Science*, Amer. Math. Soc., (1967).
257. *Proceedings of the International Symposium on Numerical Weather Prediction, Tokyo, 1960*, Meteorological Society of Japan (1962).
258. *Symposium on the Numerical Treatment of Partial Differential Equations with Real Characteristics*, Provisional International Computation Centre, Rome (1959).

* Refers to books.

BIBLIOGRAPHY

A short list of books and articles dealing with the topics of this text is furnished below in order to provide additional methods and illustrations. This list makes no pretense at completeness, but the recent literature is somewhat more fully represented both because it gives the latest developments and because the earlier literature is partially covered in the following two publications:

I. *Bull. Nat. Res. Council U.S.* **92**, Numerical integration of differential equations, by A. A. Bennett, W. E. Milne, Harry Bateman, National Academy of Sciences, Washington, D. C. (1933). (See [7].)

II. *A Manual of Operation for the Automatic Sequence Controlled Calculator*, by The Staff of the Computation Laboratory, Harvard University Press, Cambridge, Massachusetts (1946).

Adams, *see* Bashforth and Adams.

1. Aitken, A. C., Studies in practical mathematics V, On the iterative solution of a system of linear equations, *Proc. Roy. Soc. Edinburgh*, Sect. A, **63**, 52–60 (1950).

2. Aronszajn, N., The Rayleigh-Ritz and A. Weinstein methods for approximation of eigenvalues, I. Operators in Hilbert space, *Proc. Nat. Acad. Sci.*, **34**, 474–480 (1948); II. Differential operators, *ibid.*, 594–601 (1948).

3. Aronszajn, N., The Rayleigh-Ritz and A. Weinstein method for approximation of eigenvalues, Five technical reports issued by Department of Mathematics, Oklahoma A. and M.

Baggot, *see* Levy and Baggot.

4. Bashforth, F., and Adams, J. C., An attempt to test the theories of capillary action . . . with an explanation of the method of integration employed, Cambridge University Press, Cambridge, 80 pp. +59 pp. tables, pp. 15–62 (1883).

Bateman, *see* Bennett, Milne and Bateman.

5. *Bateman, H., *Differential Equations*, Longmans, London, Longmans Modern Math. Series, 287–295 (1918).

6. von Beck, E., Zwei Anwendungen der Obreschkoffschen Formel, *Z. Angew. Math. Mech.*, **30**, (3) 84–93 (1950).

7. Bennett, A. A., Milne, W. E., and Bateman, H., Numerical integration of differential equations, *Bull. Nat. Res. Council U.S.* **92**, 51–87 (1933).

8. Bickley, W. G. A simple method for the numerical solution of differential equations. *Phil. Mag.* (7) **13**, 1006–1114 (1932).

9. Bickley, W. G., Difference and associated operators, with some applications, *J. Math. Phys.*, **27**, 183–192 (1948).

10. Bickley, W. G., Finite difference formulae for the square lattice. *Quart. J. Mech. Appl. Math.*, **1**, 35–42 (1948).

* Refers to books.

11. Bieberbach, L., *Theorie der Differentialgleichungen*, Berlin (1946) (now Dover Publications, New York), p. 54.
12. Birkhoff, G., and Young, David, Numerical quadrature of analytic and harmonic functions, *J. Math. Phys.*, **29** (3), 217–221 (1950).
13. Bodewig, E., Bericht über die verschiedenen Methoden zur Lösung eines Systems linearer Gleichungen mit reellen Koeffizienten, I, II, III, IV, V, *Nederl. Akad. Wetensch. Proc.*, **50**, 930–941, 1104–1116, 1285–1295 (1947); **51**, 53-64, 211–219 (1948).
14. Bogoliouboff, N., and Kryloff, N., On Rayleigh's principle in the theory of differential equations and on Euler's method in the calculus of variations, *Ann. of Math.* (2), **29**, 255–275 (1928).
15. *Boole, George, *A Treatise on the Calculus of Finite Differences*, 2nd ed. (1872), 3rd ed. is reprint of 2nd ed., G. E. Stechert & Co. (1931). 2nd ed. reprinted by Dover.
16. Bowie, O. L., A least-square application to relaxation methods, *J. Appl. Physics*, **18**, 830–833 (1947).
17. Brauer, Alfred, Limits for the characteristic roots of a matrix, *Duke J.*, **13**, 387–395 (1946); **14**, 21–26 (1947).
    Brown, *see* Morris and Brown.
18. Bückner, H., Über eine Näherungslösung der gewöhnlichen linearen Differentialgleichung 1 Ordnung, *Z. Angew. Math. Mech.*, **22**, 143–152 (1942).
19. Bückner, H., Über ein unbeschränkt anwendbares Iterationsverfahren für Systeme linearer Gleichungen, *Arch. Math.*, **2**, 172–177 (1950).
20. Bukovics, E., Eine Verbesserung und Verallgemeinerung des Verfahrens von Blaess zur numerischen Integration gewöhnlicher Differentialgleichungen, *Osterreich. Ing. Arch.*, **4**, 338–349 (1950).
21. Carter, A. E., and Sadler, D. H., The Application of the national accounting machine to the solution of first-order differential equations, *Quart. J. Mech. Appl. Math.*, **1**, 433–441 (1948).
22. Chadaja, F. G., On the problem of numerical integration of ordinary differential equations (Russian, Georgian summary), *Bull. Acad. Sci. Georgian SSR* **2**, 601–608 (1941).
23. Chadaja, F. G., On the error in the numerical integration of ordinary differential equations by the method of finite differences (Russian), *Trav. I·'. Math. Tbilissi*, **11**, 97–108 (1942).
24. Chakrabarti, M. C., Remainders in quadrature formulas, *Bull. Calcutta Math. Soc.* **39**, 119–126 (1947).
25. *Charlier, C. L., *Die Mechanik des Himmels*, Bd. II, Leipzig (1907).
26. Collatz, L., Bemerkungen zur Fehlerabschätzung für des Differenzenverfahren bei partiellen Differentialgleichungen, *Z. Angew. Math. Mech.*, **13**, 56–57 (1933).
27. Collatz, L., Eine Verallgemeinerung des Differenzenverfahrens für Differentialgleichungen, *Z. Angew. Math. Mech.*, **14**, 350–351 (1934).
28. Collatz, L., Das Differenzenverfahren mit höherer Approximation für lineare Differentialgleichungen, *Schr. Math. Sem. Inst. Angew. Math. Univ. Berlin*, **3**, 1–34 (1935).
29. Collatz, L., Über das Differenzenverfahren bei Anfangswertproblemen partieller Differentialgleichungen, *Z. Angew. Math. Mech.*, **16**, 239–247 (1936).
30. Collatz, L., Konvergenzbeweis und Fehlerabschätzung für das Differenzenverfahren bei Eigenwertproblemen gewöhnlicher Differentialgleichungen zweiter und vierter Ordnung, *Deutsche Math.*, **2**, 189–215 (1937).

* Refers to books.

31. Collatz, L., Schranken für den ersten Eigenwert bei gewöhnlichen Differentialgleichungen zweiter Ordnung, *Ing. Arch.*, **8**, 325–331 (1937).

32. Collatz, L., Genäherte Berechnung von Eigenwerten, *Z. Angew. Math. Mech.*, **19**, 224–249, 297–318 (1939).

33. Collatz, L., Natürliche Schrittweite bei numerischer Integration von Differentialgleichungssystemen, *Z. Angew. Math. Mech.*, **22**, 216–225 (1942).

34. Collatz, L., Differenzenverfahren zur numerischen Integration von gewöhnlichen Differentialgleichungen n-ter Ordnung, *Z. Angew. Math. Mech.*, **29**, 199–209 (1949).

35. Collatz, L., Über die Konvergenzkriterien bei Iterationsverfahren für lineare Gleichungssysteme, *Math. Z.*, **53**, 149–161 (1950).

36. *Collatz, L., Numerische Behandlung von Differentialgleichungen, Springer Verlag (1951) (English ed. cited in Appendix I (no. 47)).

37. *Collatz, L., Eigenwertaufgaben mit technischen Anwendungen, Mathematik und ihre Anwendungen in Physik und Technik, Reihe A, Bd. **19**, Akademische Verlagsgesellschaft, Leipzig, xvii +466 pp. (1949).

38. Collatz, L., and Zurmühl, R., Beiträge zu den Interpolationsverfahren der numerischen Integration von Differentialgleichungen 1 & 2, Ordnung, *Z. Angew. Math. Mech.*, **22**, 42–55 (1942).

39. Collatz, L., and Zurmühl, R., Zur Genauigkeit verschiedener Integrationsverfahren bei gewöhnlichen Differentialgleichungen, *Ing. Arch.*, **13**, 34–36 (1942).

40. Courant, R., Friedrichs, K., and Lewy, H., Über die partiellen Differenzengleichungen der mathematischen Physik, *Math. Ann.*, **100**, 32–74 (1928).

41. Cowell, P. H., and Crommelin, A. D. C., Essay on the return of Halley's Comet, Greenwich Observations (1909).

42. Crank, J., and Nicolson, P., A practical method for numerical evaluation of solutions of partial differential equations of the heat-conduction type, *Proc. Cam. Philos. Soc.*, **43**, 50–67 (1947).

Crommelin, see Cowell and Crommelin.

43. Curtiss, John H., Sampling methods applied to differential and difference equations, *Proceedings, Seminar on Scientific Computation*, Nov. 1949, IBM, 87–109 (1950).

44. Cutkosky, R. E., A Monte Carlo method for solving a class of integral equations, *J. Res. Nat. Bur. Stand.*, **47**, 113–116 (1951)

Darby, see Shortley, Weller, Darby and Gamble.

45. *Dixon, W. J., and Massey, F. J., Introduction to Statistical Analysis, McGraw-Hill, New York (1951).

46. Donsker, M. D., and Kac, M., A sampling method for determining the lowest eigenvalue and the principal eigenfunction of Schrödinger's equation, *J. Res. Nat. Bur. Stand.*, **44**, 551–557 (1950).

47. Duncan, W. J., Galerkin's method in mechanics and differential equations, *Gt. Brit. Aeronaut. Res. Comm. Reports and Mem.* **1798**, 33 pp. (1937).

48. Duncan, W. J., The principles of the Galerkin method, *Gt. Brit. Aeronaut. Res. Comm. Reports and Mem.* **1848**, 24 pp. (1938).

49. Duncan, W. J., Assessment of error in approximate solution of differential equations. *Quart. J. Mech. Appl. Math.*, **1**, 470–476 (1948).

50. Duncan, W. J., Technique of the step-by-step integration of ordinary differential equations, *Phil. Mag.*, **39**, 493 (1948).

* Refers to books.

51. Emmons, Howard W., The numerical solution of partial differential equations, *Quart. Appl. Math.*, **2**, 173–195 (1944).

52. Epstein, Bernard, A method for the solution of the Dirichlet problem for certain types of domains, *Quart. Appl. Math.*, **6**, 301–317 (1948).

53. Faddeeva, V. N., The method of lines applied to some boundary problems, *Trudy Mat. Inst. Steklov*, **28**, 73–103 (1949).

54. Falkner, V. M., A method of numerical solution of differential equations, *Phil. Mag.* (7) **21**, 624–640 (1936).

55. Feinstein, L., and Schwarzschild, M., Automatic integration of linear second-order differential equations by means of punched card machines, *Rev. Sci. Instr.*, **12**, 405–408 (1941).

56. Fettis, Henry E., A method for obtaining the characteristic equation of a matrix and computing the associated modal columns, *Quart. Appl. Math.*, **8**, 206–212 (1950).

57. Flanders, D. A., and Shortley G., Numerical determination of fundamental modes, *J. Appl. Phys.*, **21**, 1326–1332 (1950).

58. Forsythe, G. E., Reprint of a note on rounding-off errors, *SIAM Rev.*, **1**, 66–67 (1959).

59. Forsythe, George E., Alternative derivations of Fox's escalator formulas for latent roots, *Quart. J. Mech. Appl. Math.*, **5**, 191–195 (1952).

60. Forsythe, George E., Annotated translation of a letter in which Gauss solved linear equations by relaxation, Nat. Bur. Stand. at Los Angeles, 7 pp., Prepublication copy (1951).

61. Forsythe, G. E., and Leibler, R. A., Matrix inversion by a Monte Carlo method, *Math. Tables and Other Aids to Computation*, **4**, 127–129 (1950).

62. Forsythe, G. E., and Motzkin, T. S., On a gradient method for solving linear equations, Nat. Bur. Stand. at Los Angeles, 5 pp. (1950).

63. Fortet, R., On the estimation of an eigenvalue by an additive functional of a stochastic process, with special reference to the Kac-Donsker method. *J. Res. Nat. Bur. Standards*, **48**, 68–75 (1952).

64. Fox, L., Solution by relaxation methods of plane potential problems with mixed boundary conditions, *Quart. Appl. Math.*, **2**, 251–257 (1944).

65. Fox, L., Some improvements in the use of relaxation methods for the solution of ordinary and partial differential equations, *Proc. Roy. Soc. London, Ser. A*, **190**, 31–59 (1947).

66. Fox, L., A short account of relaxation methods, *Quart. J. Mech. Appl. Math.*, **1**, 253–280 (1948).

67. Fox, L., The solution by relaxation methods of ordinary differential equations, *Proc. Camb. Philos. Soc.*, **45**, 50–68 (1949).

68. Fox, L., The numerical solution of elliptic differential equations when the boundary conditions involve a derivative, *Philos. Trans. Roy. Soc. London, Ser. A*, **242**, 345–378 (1950).

69. Fox, L., and Goodwin, E. T., Some new methods for the numerical integration of ordinary differential equations, *Proc. Camb. Philos. Soc.*, **45**, 373–388 (1949).

70 Fox, L., Huskey, H. D., and Wilkinson, J. H., Notes on the solution of algebraic linear simultaneous equations, *Quart. J. Mech. Appl. Math.*, **1**, 149–173 (1948).

71. Frankel, Stanley, Convergence rates of iterative treatments of partial differential equations, *Math. Tables and Other Aids to Computation*, **4**, 65–75 (1950).

72. *Frazer, R. A., Duncan W. J., and Collar, A. R., *Elementary Matrices*, Cambridge University Press, Cambridge (1947).
Friedrichs, *see* Courant, Friedrichs, and Lewy.

73. Frocht, Max M., The numerical solution of Laplace's equation in composite rectangular areas, *J. Appl. Phys.*, **17**, 730–742 (1946).

74. Frocht, M. M., and Leven, M. M., A rational approach to the numerical solution of Laplace's equation, *J. Appl. Phys.*, **12**, 596–604 (1941).

75. Galerkin, B. G., Series developments for some cases of equilibrium of plates and beams (Russian), Wjestnik Ingenerow Petrograd (1915).
Gamble, *see* Shortley, Weller, Darby, and Gamble.

76. *Gauss, K. V., Brief an Encke, *Gesammelte Werke*, **7**, 433 (1834).

77. *Geiringer, Hilda, On the solution of systems of linear equations by certain iterative methods, *Reissner Anniversary Volume Contributions to Applied Mech.*, J. W. Edwards, Ann Arbor, 365–393 (1948).

78. Geršgorin, S., Fehlerabschätzung für das Differenzenverfahren zur Lösung partieller Differentialgleichungen, *Z. Angew. Math. Mech.*, **10**, 373–382 (1930).

79. Geršgorin, S., Über die Abgrenzung der Eigenwerte einer Matrix, *Izv. Akad. Nauk. SSSR*, **7**, 749–754 (1931).

80. Gill, S., A process for the step-by-step integration of differential equations in an automatic digital computing machine, *Proc. Camb. Philos. Soc.*, **47**, 96–108 (1951).

81. Gilles, D. C., The use of interlacing nets for the application of relaxation methods to problems involving two dependent variables, *Proc. Roy. Soc. London, Ser. A.*, **193**, 407–433 (1948).
Goodwin, *see* Fox and Goodwin.

82. *Grinter, L. E., *Numerical Methods of Analysis in Engineering*, Macmillan, 207 pp. (1949).

83. Hartree, D. R., Notes on iterative processes, *Proc. Camb. Philos. Soc.*, **24**, 89 (1928).

84. Hartree, D. R., A method for the numerical integration of first-order differential equations, *Proc. Camb. Philos. Soc.*, **46**, 523–524 (1950).

85. Hausman, L. F., and Schwarzschild, M., Automatic integration of linear sixth-order differential equations by means of punched-card machines, *Rev. Sci. Inst.*, **18**, 877–883 (1947).

86. Herrick, Samuel, Step-by-step integration of $\ddot{x} = f(x, y, z, t)$ without a "corrector," *Math. Tables and Other Aids to Computation*, **5**, 61–67 (1951).
Hestenes, *see* Rosser, Lanczos, Hestenes, and Karush.

87. Hestenes, M. R., Numerical methods of obtaining solutions of fixed end point problems in the calculus of variations, Rand Corp., RM-102.

88. Hestenes, M. R., and Karush, W., The solutions of $Ax = \lambda Bx$, *J. Res. Nat. Bur. Stand.*, **47**, 471–478 (1951).

89. Heun, K., Neue Methode zur approximativen Integration der Differentialgleichungen einer unabhängigen Variable, *Z. Math. Phys.*, **45**, 23 (1900).

90. Hidaka, Koji, Stencils for integrating partial differential equations of mathematical physics, *Math. Japonicae*, **2**, 27–34 (1950).

* Refers to books.

346   BIBLIOGRAPHY

91. *Hotelling, Harold, Practical problems of matrix calculation, *Proc. Berkeley Symposium Math. Statistics and Probability, 1945, 1946*, University of California Press, Berkeley, Los Angeles, 275–293 (1949).

Huskey, *see* Fox, Huskey, and Wilkinson.

92. Huskey, Harry D., On the precision of a certain procedure of numerical integration, with an appendix by Douglas R. Hartree, *J. Res. Nat. Bur. Stand.*, **42**, 57–62 (1949).

Hyman, *see* O'Brien and Hyman.

93. *Ince, E. L., *Ordinary Differential Equations*, Longmans, London, 540–547 (1927). Dover reprint.

94. Jackson, D., The method of numerical integration in exterior ballistics, *War Dept. Document **984***, Govt. Print. Office, Washington D. C., 43 pp. (1921).

Kac, *see* Donsker and Kac.

95. Kac, M., On some connections between probability theory and differential and integral equations, *Proc. 2d Berkeley Symposium Math. Statistics and Probability 1950*, University of California Press, 189–215 (1951).

96. *Kantorovich, L. V., and Krylov, V. I., *Approximate Methods of Higher Analysis*; Leningrad, Moscow (1941). Interscience, 681 pp. (1958).

Karush, *see* Hestenes and Karush.

Karush, *see* Rosser, Lanczos, Hestenes, and Karush.

97. Karush, William, Determination of the extreme values of the spectrum of a bounded self-adjoint operator, *Proc. Amer. Math. Soc.*, **2**, 980–989 (1951).

98. Karush, W., An iterative method for finding characteristic vectors of a symmetric matrix, *Pac. J. Math.*, **1**, 233–248 (1951).

99. Kato, Tosio, On the upper and lower bounds of eigenvalues, *J. Phys. Soc. Japan*, **4**, 334–339 (1949).

100. Kormes, M., A note on the integration of linear second-order differential equations by means of punched cards, *Rev. Sci. Instr.*, **14**, 118 (1943).

101. *Kowalewski, G., *Interpolation und genäherte Quadratur*, Leipzig, Berlin (1932).

Kryloff, *see* Bogoliouboff and Kryloff.

102. Kryloff, N., Sur la solution approchée des problèmes de la physique mathématique et de la science d'ingénieur, *Bull. Acad. Sci. URSS Leningrad* (*Izvestia Akad. Nauk SSSR*) (7) 1089–1114 (1930). *Reprint Revista Mat. Hisp. Amer.* (2) **6**, 213–238 (1931).

Krylov, *see* Kantorovich and Krylov.

103. Kutta, W., Beitrag zur näherungsweisen Integration totaler Differentialgleichungen, *Z. Math. Phys.*, **46**, 435–453 (1901).

Lanczos, *see* Rosser, Lanczos, Hestenes, and Karush.

104. Lanczos, C., An iteration method for the solution of the eigenvalue problem of linear differential and integral operators, *J. Res. Nat. Bur. Stand.*, **45**, 255–281 (1950).

Leibler, *see* Forsythe and Leibler.

105. le Roux, J., Sur le problème de Dirichlet, *J. de Math.* (6) **10**, 189–230 (1914).

Leven, *see* Frocht and Leven.

106. *Levy, H., and Baggot, E. A., *Numerical Studies in Differential Equations*, Watts, London, vol. 1, viii + 238 pp. (1934).

Lewy, *see* Courant, Friedrichs, and Lewy.

* Refers to books.

107. Liebmann, L., Die angenährte Ermittelung harmonischer Funktionen und konformer Abbildungen, *Sitz. Bayer Akad. Wiss. Math.-Phys. Klasse*, 385–416 (1918).

108. Lindelöf, E., Remarques sur l'intégration numérique des équations différentielles ordinaires, *Acta Soc. Sci. Fennicae (Phys. Mat.)* (A) **2**, no. 13, 21 p. (1938).

109. Lonseth, A. T., The propagation of error in linear problems, *Trans. Amer. Math. Soc.*, **62** (2), 193–212 (1947).

110. Lyusternik, L. A., Remarks on the numerical solution of boundary problems for Laplace's equation and the calculation of characteristic values by the method of networks (Russian), *Trav. Inst. Math. Stekloff*, **20**, 49–64 (1947).

111. *MacDuffee, C. C., The Theory of Matrices, Springer, Berlin, 110 pp. (1933).

112. Marchant Methods, Milne method of step-by-step integration of ordinary differential equations when starting values are known, MM-216, Marchant Calc. Mach. Co., Oakland, Calif., 10 pp. (June 1942).

113. Marchant Methods, Milne method of step-by-step double integration of second order differential equations in which first derivatives are absent, MM-216A, Marchant Calc. Mach. Co., Oakland, Calif., 6 pp. (Jan. 1943).

114. Marchant Methods, Starting value for Milne-method integration of ordinary differential equations of the first order, The method of Milne, MM-260, Marchant Calc. Mach. Co., Oakland, Calif., 11 pp., (Jan. 1944).

115. Marchant Methods, Starting values for Milne-method integration or ordinary differential equations of first order, or of second order when first derivatives are absent, The method of Taylor's series, MM-261, Marchant Calc. Mach. Co., Oakland, Calif., 4 pp. (Oct. 1943).

116. Massera, Jose L., Formulae for finite differences with applications to the approximate integration of differential equations of first order (Spanish), *Publ. Inst. Mat. Univ. Nac. Litoral*, **4**, 99–166 (1943).

117. Mikeladze, Š. E., Über die numerische Lösung der Differentialgleichung $\partial^2 u/\partial x^2 + \partial^2 u/\partial y^2 + \partial^2 u/\partial z^2 = \phi(x, y, z)$, *C. R. (Doklady) Acad. Sci. URSS. (N.S.)* **14**, 177–179 (1937).

118. Mikeladze, Š. E., Über numerische Integration der Laplaceschen und Poissonschen Gleichungen, *C. R. (Doklady) Acad. Sci. URSS, (N.S.)* **14**, 181–182 (1937).

119. Mikeladze, Š. E., Über die numerische Lösung der Differentialgleichungen von Laplace und Poisson (Russian, German summary), *Bull. Acad. Sci. URSS, Ser. Math. (Izvestia Akad. Nauk SSSR)* 271–292(1938).

120. Mikeladze, Š. E., Über die Integration von Differentialgleichungen mit Hilfe der Differenzenmethode (Russian, German summary), *Bull. Acad. Sci. URSS Ser. Math. (Izvestia Akad. Nauk SSSR)* 627–642 (1939).

121. Mikeladze, Š. E., On the question of numerical integration of partial differential equations by means of nets (Russian), *Mitt. Georg. Abt. Akad. Wiss. USSR*, **1**, 249–254 (1940).

122. Mikeladze, Š. E., Über die Lösung von Randwertproblemen mit der Differenzenmethode, *C. R. (Doklady) Acad. Sci. URSS (N.S.)* **28**, 400–402 (1940).

123. Mikeladze, Š. E., Verallgemeinerung der Methode der numerischen Integration von Differentialgleichungen mit Hilfe der Formeln der Mechanischen Quadratur (Russian, German summary), *Trav. Inst. Math. Tbilissi (Trudy Tbilissi. Mat. Inst.)* **7**, 47–63 (1940).

* Refers to books.

348 BIBLIOGRAPHY

124. Mikeladze, Š. E., Numerische Integration der Gleichungen vom elliptischen
   und parabolischen Typus (Russian, German summary), *Bull. Acad. Sci.
   URSS, Ser. Math.* (*Izvestia Akad. Nauk SSSR*) **5**, 57–74 (1941).
125. Mikeladze, Š. E., On the approximate integration of linear differential equations with discontinuous coefficients (Russian), *Bull. Acad. Sci. Georgian
   SSR*, **3**, 633–639 (1942).
126. Mikeladze, Š. E., New formulas for the numerical integration of differential
   equations (Russian), *Bull. Acad. Sci. Georgian SSR*, **4**, 215–218 (1943).
127. Miller, J. C. P., Checking by differences I, *Math. Tables and Other Aids to
   Computation*, **4**, 3–11 (1950).
   Milne, see Bennett, Milne, and Bateman.
128. Milne, W. E., Numerical integration of ordinary differential equations,
   Amer. Math. Month., **33**, 455–460 (1926).
129. Milne, W. E., On the numerical solution of a boundary problem, *Amer. Math.
   Month.*, **38**, 14–17 (1931).
130. Milne, W. E., On the numerical integration of certain differential equations
   of the second order, *Amer. Math. Month.*, **40**, 322–327 (1933).
131. Milne, W. E., The numerical integration of $y'' + g(x)y = f(x)$, *Amer. Math.
   Month.*, **49**, 96–98 (1942).
132. Milne, W. E., A note on the numerical integration of differential equations,
   J. Res., Nat. Bur. Stand. **43**, 537–542 (1949).
133. *Milne, W. E., *Numerical calculus*, Princeton University Press, Princeton,
   N. J. (1949).
134. Milne, W. E., The remainder in linear methods of approximation, *J. Res.
   Nat. Bur. Stand.*, RP2401, **43**, 501–511 (1949).
135. Milne, W. E., Note on the Runge-Kutta method, *J. Res. Nat. Bur. Stand.*,
   RP2101, **44**, 549–550 (1950).
136. Milne, W. E., Numerical determination of characteristic numbers, *J. Res.
   Nat. Bur. Stand.*, RP2132, **45**, 245–254 (1950).
137. *Milne-Thompson, L. M., *Calculus of Finite Differences*, Macmillan and Co.,
   London (1933).
138. von Mises, R., Zur numerischen Integration von Differentialgleichungen,
   Z. Angew. Math. Mech., **10**, 81–92 (1930).
139. von Mises, R., and Pollaczek-Geiringer, H., Praktische Verfahren der Gleich-
   ungsauflösung, I, *Z. für Math. und Mech.*, **9**, 58–77 (1929).
140. *Morris, Joseph, *The Escalator Method in Engineering Vibration Problems*,
   Wiley, New York (1947).
141. *Morris, M., and Brown, O. E., *Differential Equations* (2nd revised ed.),
   Prentice Hall, New York, (1952).
   Motzkin, see Forsythe and Motzkin.
142. Moulton, F. R., Numerical solution of differential equations, Chap. X of
   Smithsonian Mathematical Formulae and Tables of Elliptic Functions
   (Edited by Adams, E. P., and Hippisley, R. L.), *Smithsonian Miscellaneous
   Collections*, **74**, no. 1, 220–242 (1922).
143. *Moulton, F. R., *New Methods in Exterior Ballistics*, University of Chicago
   Press, Chicago (1926). Dover reprint.
144. *Moulton, F. R., *Differential Equations*, Macmillan, New York, 179–231
   (1930). Dover reprint.
   Negoro, see Sunatani and Negoro.
   Nicolson, see Crank and Nicolson.

* Refers to books.

145. Nyström, E. J., Über die numerische Integration von Differentialgleichungen, *Acta Soc. Sci. Fennicae*, **50**, no. 13, 56 pp. (1926).

146. Nyström, E. J., Zur Praktischen Integration von linearen Differentialgleichungen, *Soc. Scient. Fennicae*, *Comm. Phys. Math.* XI, **14**, 14 pp. (1943).

147. Nyström, E. J., Zur numerischen Lösung von Randwertaufgaben bei gewöhnlichen differentialgleichungen, *Acta Mathematica*, **76**, 158–184 (1945).

148. Obrechkoff, N., Sur les quadratures mécaniques (Bulgarian, French summary), *Spisanie Bulgar. Akad. Nauk*, **65**, 191–289 (1942).

149. O'Brien, George G., Hyman, Morton A., and Kaplan, Sidney, A study of the numerical solution of partial differential equations, *J. Math. Phys.*, **29**, 223–251 (1951).

150. Piaggio, H. T. H., On the numerical integration of differential equations, *Phil. Mag.* (6) **37**, 596–600 (1919).

151. *Piaggio, H. T. H., An Elementary Treatise on Differential Equations and Their Applications, 2nd ed., Bell, London, 94–108, 224–228 (1928).

152. Picard, Émile., Mémoire sur la théorie des équations aux dérivées partielles et la méthode des approximations successives, *J. Math. Pures Appl.* (4) **6**, 145–210, 231 (1890).

153. Picard, É., Sur l'application des méthodes d'approximations successives à l'étude de certaines équations différentielles ordinaires, *J. Math. Pures Appl.* (4) **9**, 217–271 (1893).

154. *Picard, É., Traité d'analyse, 3rd ed., Gauthier-Villars, Paris, **2**, 368–394 (1926); **3**, 88–99 (1928).

Pollaczek-Geiringer, *see* von Mises and Pollaczek-Geiringer.

Posch, *see* Sauer and Posch.

155. Rabinovitch, Féodora, Sur une nouvelle méthode d'intégration approchée des équations différentielles du second ordre, *Ann. Radioélec.*, **1**, 134–151 (1945).

156. Rademacher, Hans A., On the accumulation of errors in processes of integration, *Annals of Comp. Lab. of Harvard U.*, Nat. Bur. Stand. at Los Angeles, 17 pp. (1950).

157. Rayleigh, Lord, On the calculation of the frequency of vibration of a system in its gravest mode with an example from hydrodynamics, *Phil. Mag.* (5) **47**, 566–572 (1899).

158. Richardson, L. F., The approximate arithmetical solution by finite differences of physical problems involving differential equations with an application to the stresses in a masonry dam, *Phil. Trans. Roy. Soc. Lond.*, **210A**, 307–357 (1910).

159. Richardson, L. F., How to solve differential equations approximately by arithmetic, *Math. Gaz.*, **12**, 415–421 (1925).

160. Richardson, L. F., A purification method for computing the latent columns of numerical matrices and some integrals of differential equations, *Philos. Trans. Roy. Soc. London, Ser. A*, **242**, 439–491 (1950).

161. Richardson, R. G. D., A new method in boundary problems for differential equations, *Trans. Amer. Math. Soc.*, **18**, 489–518 (1917).

162. Ritz, Walter, Über eine neue Methode zur Lösung gewisser Variationsprobleme der mathematischen Physik. *J. Reine Angew. Math.*, **135**, 1–61 (1908).

* Refers to books.

163. Ritz, W., Über eine neue Methode zur Lösung gewisser Randwertaufgaben, *Gött. Nachr.*, 236–248 (1908). or *Ges. Werke*, 251–264, (1911).

Robinson, *see* Whittaker and Robinson.

164. Rosser, J. Barkley, A method of computing exact inverses of matrices with integer coefficients, *J. Res. Nat. Bur. Standards*, **49**, 349–358 (1952). copy (1950).

165. Rosser, J. B., Lanczos, C., Hestenes, M. R., and Karush, W., The separation of close eigenvalues of a real symmetric matrix, *J. Res. Nat. Bur. Stand.*, **47**, 291–297 (1951).

166. Rubbert, F. K., Zur Praxis der numerischen Integration, *Astron. Nachr.* **277**, 161–166 (1949).

167. Runge, C., Über die numerische Auflösung von Differentialgleichungen, *Math. Ann.*, **46**, 167–178 (1895).

168. Runge, C., and Willers, F. A., Numerische und graphische Quadratur und Integration gewöhnlicher und partieller Differentialgleichungen, *Ency. Math. Wiss.*, **II**, 3, 1; article II C2, Teubner, Leipzig, 47–176 (1915).

Sadler, *see* Carter and Sadler.

169. Salzer, Herbert E., Formulas for the numerical integration of first and second order differential equations in the complex plane, *J. Math. Phys.*, **29**, 207–216 (1950).

170. Salzer, H. E., Coefficients for numerical differentiation with central differences, *J. Math. Phys.*, **22**, 115–135 (1943).

171. Salzer, H. E., Coefficients for numerical integration with central differences, *Phil. Mag.*, **35**, 262–264 (1944).

172. Salzer, H. E., Coefficients for mid-interval numerical integration with central differences, *Phil. Mag.*, **36**, 216–218 (1945).

173. Samuelson, Paul A., Efficient computation of the latent vectors of a matrix, *Proc. Nat. Acad. Sci. USA*, **29**, 391–397 (1943).

174. Samuelson, Paul A., A convergent iterative process, *J. Math. Phys. Mass. Inst. Tech.*, **24**, 131–134 (1945).

175. *von Sanden, Horst, *Practical Mathematical Analysis*, Methuen, London (1923).

176. von Sanden, H., Zur Berechnung des kleinsten Eigenwerts von $y'' + \lambda p(x)y = 0$, *Z. Angew. Math. Mech.*, **21**, 381–382 (1941).

177. *von Sanden, Horst, *Praxis der Differentialgleichungen, Eine Einführung*, 3rd ed., Walter de Gruyter & Co., Berlin, 105 pp. (1945).

178. Sard, Arthur, Remainders: functions of several variables, *Acta Math.*, **84**, 319–346 (1950).

179. Sauer, R., and Posch, H., Anwendungen des Adamsschen Integrationsverfahrens in der Ballistik, *Ing. Arch.*, **12**, 158–168 (1941).

180. *Scarborough, J. B., *Numerical Mathematical Analysis* (6th ed.), Johns Hopkins Press, Baltimore (1950).

181. *Schmeidler, Werner, *Vorträge über Determinanten und Matrizen*, Akad.-Verlag Berlin, 155 pp. (1949).

182. Schmidt, R. J., On the numerical solution of linear simultaneous equations by an iterative method, *Phil. Mag.*, **32**, 369–383 (1941).

183. Schulte, A. M., A slight improvement of Southwell's method for the approximative computation of the lowest frequency of a homogeneous membrane, *Appl. Sci. Res. Sect. A*, **2**, 93–96 (1949).

* Refers to books.

BIBLIOGRAPHY 351

184. Schulz, G., Interpolationsverfahren zur numerischen Integration gewöhnlicher Differentialgleichungen, *Z. Angew. Math. Mech.*, **12**, 44–59 (1932).
185. *Schulz, G., *Formelsammlung*, no. 1110, Berlin, Göschen, 1937.
186. Schulz, W., Über das Meissnersche Integrationsverfahren für Differentialgleichungen erster Ordnung, *Deut. Math.*, **6**, 271–276 (1941).
    Schwarzschild, *see* Feinstein and Schwarzschild.
    Schwarzschild, *see* Hausman and Schwarzschild.
187. Seidel, L. Über ein Verfahren, die Gleichungen, auf welche die Methode der kleinsten Quadrate führt, sowie lineäre Gleichungen überhaupt, durch successive Annäherung aufzulösen, *Abh. Akad. Wiss. München*, Kl. **2**, no. 3, 81–108 (1874).
188. Sheppard, W. F., Central-difference formulae, *Proc. Lond. Math. Soc.*, **31**, 449–488 (1899).
    Shortley, *see* Flanders and Shortley.
189. Shortley, Geo., Relaxation methods in theoretical physics, *Sci.*, **105**, 455 (1947).
190. Shortley, Geo., Weller, Royal, Darby, Paul, and Gamble, Edward H., Numerical solution of axisymmetrical problems with applications to electrostatics and torsion, *J. Appl. Phys.*, **18**, 116–129 (1947).
191. Shortley, G. H., and Weller, R., The numerical solution of Laplace's equation, *J. Appl. Phys.*, **9**, 334–348 (1938).
192. *Shortley, George, Weller, Royal, and Fried, Bernard, Numerical solution of Laplace's and Poisson's equations with applications to photoelasticity and torsion, *Ohio State Univ. Studies, Engineering ser.*, **11**, no. 5; *Engineering Exp. Station Bull.* **107**, Columbus, Ohio (1940).
193. Southwell, R. V., On the natural frequencies of vibrating systems, *Proc. Roy. Soc. London*, **174**, 433–457 (1940).
194. *Southwell, R. V., *Relaxation Methods in Theoretical Physics*, Oxford (1946).
195. *Steffensen, J. F., *Interpolation*, Williams & Wilkins, Baltimore (1927).
196. Stein, Marvin L., On methods for obtaining solutions of fixed end point problems in the calculus of variations, *J. Res. Nat. Bur. Standards*, **50**, 277–297 (1953).
197. Stein, P., The convergence of Seidel iterants of nearly symmetric matrices, *Math. Tables and other Aids to Computation*, **5**, 236–240 (1951).
198. Stohler, K., Eine Vereinfachung bei der numerischen Integration gewöhnlicher Differentialgleichungen, *Z. Angew. Math. Mech.*, **23**, 120–122 (1943).
199. Störmer, C., Sur les trajectoires des corpuscules électrisés dans l'espace sous l'action du magnétisme terrestre avec application aux aurores boréales, *Arch. Sci. Phys. Nat. Genève* (4) **24**, 5–18, 113–158, 221–247 (1907).
200. Störmer, C., Resultats des calculs numériques des trajectoires des corpuscules électriques dans le champ d'un aimant élémentaire, *Videnskaps-Selskabets Skrifter, Kristiania 1913:* **1**, no. 4, 74 pp.; **2**, no. 10, 58 pp.; **2**, no. 14, 64 pp. (1913).
201. Störmer, C., Méthode d'intégration numérique des équations différentielles ordinaires, *C. R. Congr. Intern. Math. Strasbourg 1920*, Toulouse, Privat, 243–257 (1921).
202. Sunatani, C., and Negoro, S., On a method of approximate solution of a plane harmonic function, *Tech. Reports of Tôhoku Imperial Univ.*, **12**, 339–360 (1937).

* Refers to books.

203. Taussky, Olga, Bounds for characteristic roots of matrices, *Duke Math. J.*, **15**: (4) 1043-1044 (1948).

204. Taussky, Olga, A recurring theorem on determinants, *Amer. Math. Monthly*, **56**, 672-675 (1949).

205. Taussky, Olga, Notes on numerical analysis II, Note on the condition of matrices, *Math. Tables and Other Aids to Computation*, **4**, 111-112 (1950).

206. Thomas, L. H., Stability of solution of partial differential equations, Naval Ordnance Laboratory, White Oak, Md., *NOLR* **1132**, 83-94 (1950).

207. Tifford, Arthur N., On the solution of total differential boundary value problems, *J. Aeronaut. Sci.*, **18**, 65-66 (1951).

208. Todd, John, The condition of a certain matrix, *Proc. Camb. Phil. Soc.*, **46** (1), 116-118 (1949).

209. Todd, John, The condition of certain matrices, *Quart. J. Mech. Appl. Math.*, **2**, 469-472 (1949).

210. Todd, John, Notes on modern numerical analysis, I. Solution of differential equations by recurrence relations, *Math. Tables and Other Aids to Computation*, **4**, 39-44 (1950).

211. Tollmien, W., Über die Fehlerabschätzung beim Adamsschen Verfahren zur Integration gewöhnlicher Differentialgleichungen, *Z. Angew. Math. Mech.*, **18**, 83-90 (1938).

212. Tranter, C. J., The combined use of relaxation methods and Fourier transforms in the solution of some three-dimensional boundary value problems, *Quart. J. Mech. Appl. Math.*, **1**, 281-286 (1948).

213. *Turnbull, H. W., and Aitken, A. C., *An Introduction to the Theory of Canonical Matrices*, Blackie, London (1932). Dover reprint.

214. Turton, F. J., Two notes on the numerical solution of differential equations, *Philos. Mag.* (7) **28**, 381-384 (1939).

215. Turton, F. J., The errors in the numerical solution of differential equations, *Philos. Mag.* (7) **28**, 359-363 (1939).

216. van Wijngaarden, A., Rounding-off errors (Dutch), Math. Centrum, Amsterdam Rapport ZW-1950-001, 13 pp. (1950).

217. War Department, United States of America, A course in exterior ballistics, *War Dept. Document* **1051**, Govt. Print. Office, Washington, D.C., 127 pp. (1921).

218. *Wedderburn, J. H. M., Lectures on matrices. *Am. Math. Soc. Colloquium Pub.* 17 (1934). Dover reprint.

Weller, *see* Shortley and Weller.

Weller, *see* Shortley, Weller, Darby, and Gamble.

219. *Whittaker, E. T., and Robinson, G., *The Calculus of Observations*, Blackie and Son, London, Glasgow (1932). Dover reprint.

Wilkinson, *see* Fox, Huskey, and Wilkinson.

Willers, *see* Runge and Willers.

220. Wilson, E. M., A note on the numerical integration of differential equations, *Quart. J. Mech. Appl. Math.*, **2**, 208-211 (1949).

Young, *see* Birkhoff and Young.

221. Young, Jr., David M., Iterative methods for solving partial differential equations of elliptic type (Ph.D. thesis), Harvard University, Cambridge, Mass. (May 1950) (published in *Trans. Amer. Math. Soc.*, **76**, 92-111 (1954).

222. Yowell, Everett C., A Monte Carlo method of solving Laplace's equation, *Proc. Computation Seminar, Dec. 1949*, IBM, 89-91 (1951).

* Refers to books.

223. Yowell, Everett C., Numerical solution of partial differential equations, *Proc. Computation Seminar*, Dec. *1949*, IBM, 24–28 (1951).

224. Yu\u0161kov, P. P., On the application of triangular nets to the numerical integration of the equation of heat conduction, *Prikl. Math. Meh.*, **12**, 223–226 (1948) (Russian). Zurmühl, *see* Collatz and Zurmühl.

225. Zurmühl, R., Zur numerischen Integration gewöhnlicher Differentialgleichungen zweiter und höherer Ordnung, Untersuchungen zu den Verfahren von Blaess und Runge-Kutta-Nyström, *Z. Angew. Math. Mech.*, **20**, 104–116 (1940).

226. Zurmühl, Rudolf, Runge-Kutta Verfahren zur numerischen Integration von Differentialgleichungen n-ter Ordnung, *Z. Angew. Math. Mech.*, **28**, 173–182 (1948).

227. *Zurmuhl, Rudolf, *Matrizen*, Springer Verlag, Berlin (1950).

SUPPLEMENTARY REFERENCES

228. Allen, D. N. de G., and Severn, R. T., The application of relaxation methods to the solution of non-elliptic partial differential equations, I. The heat-conduction equation, *Quart. J. Mech. Appl. Math.*, **4**, 209–222 (1951).

229. Blanch, Gertrude, On the numerical solution of parabolic partial differential equations, *J. Res. Nat. Bur. Standards*, **50**, 343–356 (1953).

230. *Bodewig, E., *Bericht über die Methoden zur numerischen Lösung von algebraischen Eigenwertproblemen*, The Hague, 36 pp. (1950).

231. Brock, J. E., An iterative numerical method for non-linear vibrations, *J. Appl. Mech.*. **18**. 1–11 (1951).

232. Couffignal, Louis, Sur la résolution numérique des systèmes d'équations linéaires, II, *Revue Sci.*, **89**, 3–10 (1951).

233. Fehlberg, Erwin, Bemerkungen zur Entwicklung gegebener Funktionen nach Legendreschen Polynomen mit Anwendung auf die numerische Integration gewöhnlicher linearer Differentialgleichungen, (German, English, French, and Russian summaries) *Z. Angew. Math. Mech.*, **31**, 104–114 (1951).

234. Jenne, W., Zur Auflösung linearer Gleichungssysteme, *Astr. Nachr.*, **278**, 73–95 (1949).

235. Mohr, Ernst, Über das Verfahren von Adams zur Integration gewöhnlicher Differentialgleichungen, *Math. Nachr.*, **5**, 209–218 (1951).

236. Polachek, H., On the solution of systems of linear equations of high order, Report NOLM-9522, Naval Ordnance Laboratory, White Oaks, Md., 8 pp. (1948).

237. Sponder, Erich, Ein zeichnerisches Lösungsverfahren für Differentialgleichungen zweiter Ordnung, *Elemente der Math.*, **6**, 53–58 (1951).

238. Vlasov, I. O., and Čarnyĭ, I. A., On a method of numerical integration of ordinary differential equations (Russian), *Akad. Nauk SSSR, Inženernyĭ Sbornik*, **8**, 181–186 (1950).

239. Weissinger, Johannes, Eine verscharfte Fehlerabschätzung zum Extrapolationsverfahren von Adams (German, English, French, and Russian summaries), *Z. Angew. Math. Mech.*, **30**, 356–363 (1950).

240. *Kendall, M. G., and Smith, B. Babington, *Tracts for Computers, No. XXIV, Tables of Random Sampling Numbers*, Cambridge University Press, London, 60 pp. (1946).

* Refers to books.

354 BIBLIOGRAPHY

241. *Feller, William, *An Introduction to Probability Theory and Its Applications*, John Wiley, New York, **1**, 419 pp. + xii.

242. *Goursat, E., *Cours d'analyse mathématique*, Gauthier-Villars, Paris. Dover reprint (Eng. trans.).

243. *Jordan, M. C., *Cours d'analyse*, Gauthier-Villars, Paris (1887).

244. *Murnaghan, Francis D., *Introduction to Applied Mathematics*, John Wiley, New York, 389 pp. + ix (1948). Dover reprint.

245. *Webster, Arthur Gordon, *Partial Differential Equations of Mathematical Physics*, B. G. Teubner, Leipzig, 440 pp. + vii (1927). Dover reprint.

246. *High-Speed Computing Devices*, Engineering Research Associates, McGraw-Hill, New York (1950).

247. *Wilkes, M. V., Wheeler, D. J., and Gill, S., *Preparation of Programs for an Electronic Digital Computer*, Addison Wesley Press, Cambridge, Mass. (1951).

248. *Hartree, Douglas R., *Calculating Instruments and Machines*, University of Illinois Press (1949).

249. *Synthesis of electronic computing and control circuits, *Annals of Comp. Lab. of Harvard U.*, **27**, Staff of Computation Lab., Harvard University Press, Cambridge, Mass. (1951).

250. Huskey, H. D., Characteristics of the Institute for Numerical Analysis Computer, *Math. Tables and Other Aids to Computation*, **4**, pp. 103–108 (1950).

251. Huskey, H. D., Semi-automatic instruction on the Zephyr, *Proc. of Second Symposium on Large-Scale Digital Calculating Machinery, Annals of Comp. Lab. of Harvard U.*, **26**, 83–90 (1951).

252. Richter, Willy, Sur l'erreur commise dans la méthode d'intégration de Milne, *Comptes rendus de l'Acad. Sci.*, **233**, 1342–1344 (1951).

253. Forsythe, George E., Tentative classification of methods and bibliography on solving systems of linear equations. In Paige, L. J., and Taussky, O., eds., *Simultaneous Linear Equations and the Determination of Eigenvalues*, Nat. Bur. Standards Appl. Math. Ser., **29**, U. S Gov't. Printing Off., 1–28 (1953).

254. Milne, W. E., Numerical methods associated with Laplace's equation, *Proc. of Second Symposium on Large-Scale Digital Calculating Machinery, Annals of Comp. Lab. of Harvard U.*, **26**, 152–163 (1951).

255. John, Fritz, On integration of parabolic equations by difference methods, *Comm. Pure Appl. Math.*, **5**, 155–211 (1952).

256. Lanczos, Cornelius, Solution of systems of linear equations by minimized iterations, *J. Res. Nat. Bur. Standards*, **49**, 33–53 (1952).

257. Hyman, Morton A., Non-iterative numerical solutions of boundary-value problems, *Appl. Sci. Res. Sect. B*, **2**, 325–351 (1952).

258. Wasow, Wolfgang, On the truncation error in the solution of Laplace's equation by finite differences, *J. Res. Nat. Bur. Stand.*, **48**, 345–348 (1952).

* Refers to books.

NAME INDEX

SUBJECT INDEX

SOME DOVER SCIENCE BOOKS

SOME DOVER SCIENCE BOOKS

WHAT IS SCIENCE?,
Norman Campbell
This excellent introduction explains scientific method, role of mathematics, types of scientific laws. Contents: 2 aspects of science, science & nature, laws of science, discovery of laws, explanation of laws, measurement & numerical laws, applications of science. 192pp. 5⅜ x 8. 60043-2 Paperbound $1.25

FADS AND FALLACIES IN THE NAME OF SCIENCE,
Martin Gardner
Examines various cults, quack systems, frauds, delusions which at various times have masqueraded as science. Accounts of hollow-earth fanatics like Symmes; Velikovsky and wandering planets; Hoerbiger; Bellamy and the theory of multiple moons; Charles Fort; dowsing, pseudoscientific methods for finding water, ores, oil. Sections on naturopathy, iridiagnosis, zone therapy, food fads, etc. Analytical accounts of Wilhelm Reich and orgone sex energy; L. Ron Hubbard and Dianetics; A. Korzybski and General Semantics; many others. Brought up to date to include Bridey Murphy, others. Not just a collection of anecdotes, but a fair, reasoned appraisal of eccentric theory. Formerly titled *In the Name of Science*. Preface. Index. x + 384pp. 5⅜ x 8.
20394-8 Paperbound $2.00

PHYSICS, THE PIONEER SCIENCE,
L. W. Taylor
First thorough text to place all important physical phenomena in cultural-historical framework; remains best work of its kind. Exposition of physical laws, theories developed chronologically, with great historical, illustrative experiments diagrammed, described, worked out mathematically. Excellent physics text for self-study as well as class work. Vol. 1: Heat, Sound: motion, acceleration, gravitation, conservation of energy, heat engines, rotation, heat, mechanical energy, etc. 211 illus. 407pp. 5⅜ x 8. Vol. 2: Light, Electricity: images, lenses, prisms, magnetism, Ohm's law, dynamos, telegraph, quantum theory, decline of mechanical view of nature, etc. Bibliography. 13 table appendix. Index. 551 illus. 2 color plates. 508pp. 5⅜ x 8.
60565-5, 60566-3 Two volume set, paperbound $5.50

THE EVOLUTION OF SCIENTIFIC THOUGHT FROM NEWTON TO EINSTEIN,
A. d'Abro
Einstein's special and general theories of relativity, with their historical implications, are analyzed in non-technical terms. Excellent accounts of the contributions of Newton, Riemann, Weyl, Planck, Eddington, Maxwell, Lorentz and others are treated in terms of space and time, equations of electromagnetics, finiteness of the universe, methodology of science. 21 diagrams. 482pp. 5⅜ x 8.
20002-7 Paperbound $2.50

CHANCE, LUCK AND STATISTICS: THE SCIENCE OF CHANCE,
Horace C. Levinson
Theory of probability and science of statistics in simple, non-technical language. Part I deals with theory of probability, covering odd superstitions in regard to "luck," the meaning of betting odds, the law of mathematical expectation, gambling, and applications in poker, roulette, lotteries, dice, bridge, and other games of chance. Part II discusses the misuse of statistics, the concept of statistical probabilities, normal and skew frequency distributions, and statistics applied to various fields—birth rates, stock speculation, insurance rates, advertising, etc. "Presented in an easy humorous style which I consider the best kind of expository writing," Prof. A. C. Cohen, Industry Quality Control. Enlarged revised edition. Formerly titled *The Science of Chance*. Preface and two new appendices by the author. xiv + 365pp. 5⅜ x 8. 21007-3 Paperbound $2.00

BASIC ELECTRONICS,
prepared by the U.S. Navy Training Publications Center
A thorough and comprehensive manual on the fundamentals of electronics. Written clearly, it is equally useful for self-study or course work for those with a knowledge of the principles of basic electricity. Partial contents: Operating Principles of the Electron Tube; Introduction to Transistors; Power Supplies for Electronic Equipment; Tuned Circuits; Electron-Tube Amplifiers; Audio Power Amplifiers; Oscillators; Transmitters; Transmission Lines; Antennas and Propagation; Introduction to Computers; and related topics. Appendix. Index. Hundreds of illustrations and diagrams. vi + 471pp. 6½ x 9¼.
61076-4 Paperbound $2.95

BASIC THEORY AND APPLICATION OF TRANSISTORS,
prepared by the U.S. Department of the Army
An introductory manual prepared for an army training program. One of the finest available surveys of theory and application of transistor design and operation. Minimal knowledge of physics and theory of electron tubes required. Suitable for textbook use, course supplement, or home study. Chapters: Introduction; fundamental theory of transistors; transistor amplifier fundamentals; parameters, equivalent circuits, and characteristic curves; bias stabilization; transistor analysis and comparison using characteristic curves and charts; audio amplifiers; tuned amplifiers; wide-band amplifiers; oscillators; pulse and switching circuits; modulation, mixing, and demodulation; and additional semiconductor devices. Unabridged, corrected edition. 240 schematic drawings, photographs, wiring diagrams, etc. 2 Appendices. Glossary. Index. 263pp. 6½ x 9¼. 60380-6 Paperbound $1.75

GUIDE TO THE LITERATURE OF MATHEMATICS AND PHYSICS,
N. G. Parke III
Over 5000 entries included under approximately 120 major subject headings of selected most important books, monographs, periodicals, articles in English, plus important works in German, French, Italian, Spanish, Russian (many recently available works). Covers every branch of physics, math, related engineering. Includes author, title, edition, publisher, place, date, number of volumes, number of pages. A 40-page introduction on the basic problems of research and study provides useful information on the organization and use of libraries, the psychology of learning, etc. This reference work will save you hours of time. 2nd revised edition. Indices of authors, subjects, 464pp. 5⅜ x 8.
60447-0 Paperbound $2.75

THE RISE OF THE NEW PHYSICS (formerly THE DECLINE OF MECHANISM),
A. d'Abro
This authoritative and comprehensive 2-volume exposition is unique in scientific publishing. Written for intelligent readers not familiar with higher mathematics, it is the only thorough explanation in non-technical language of modern mathematical-physical theory. Combining both history and exposition, it ranges from classical Newtonian concepts up through the electronic theories of Dirac and Heisenberg, the statistical mechanics of Fermi, and Einstein's relativity theories. "A must for anyone doing serious study in the physical sciences," *J. of Franklin Inst.* 97 illustrations. 991pp. 2 volumes.
20003-5, 20004-3 Two volume set, paperbound $5.50

THE STRANGE STORY OF THE QUANTUM, AN ACCOUNT FOR THE GENERAL
READER OF THE GROWTH OF IDEAS UNDERLYING OUR PRESENT ATOMIC
KNOWLEDGE, *B. Hoffmann*
Presents lucidly and expertly, with barest amount of mathematics, the problems and theories which led to modern quantum physics. Dr. Hoffmann begins with the closing years of the 19th century, when certain trifling discrepancies were noticed, and with illuminating analogies and examples takes you through the brilliant concepts of Planck, Einstein, Pauli, de Broglie, Bohr, Schroedinger, Heisenberg, Dirac, Sommerfeld, Feynman, etc. This edition includes a new, long postscript carrying the story through 1958. "Of the books attempting an account of the history and contents of our modern atomic physics which have come to my attention, this is the best," H. Margenau, Yale University, in *American Journal of Physics.* 32 tables and line illustrations. Index. 275pp. 5⅜ x 8.
20518-5 Paperbound $2.00

GREAT IDEAS AND THEORIES OF MODERN COSMOLOGY,
Jagjit Singh
The theories of Jeans, Eddington, Milne, Kant, Bondi, Gold, Newton, Einstein, Gamow, Hoyle, Dirac, Kuiper, Hubble, Weizsäcker and many others on such cosmological questions as the origin of the universe, space and time, planet formation, "continuous creation," the birth, life, and death of the stars, the origin of the galaxies, etc. By the author of the popular *Great Ideas of Modern Mathematics.* A gifted popularizer of science, he makes the most difficult abstractions crystal-clear even to the most non-mathematical reader. Index.
xii + 276pp. 5⅜ x 8½.   20925-3 Paperbound $2.50

GREAT IDEAS OF MODERN MATHEMATICS: THEIR NATURE AND USE,
Jagjit Singh
Reader with only high school math will understand main mathematical ideas of modern physics, astronomy, genetics, psychology, evolution, etc., better than many who use them as tools, but comprehend little of their basic structure. Author uses his wide knowledge of non-mathematical fields in brilliant exposition of differential equations, matrices, group theory, logic, statistics, problems of mathematical foundations, imaginary numbers, vectors, etc. Original publications, appendices. indexes. 65 illustr. 322pp. 5⅜ x 8. 20587-8 Paperbound $2.25

THE MATHEMATICS OF GREAT AMATEURS, *Julian L. Coolidge*
Great discoveries made by poets, theologians, philosophers, artists and other non-mathematicians: Omar Khayyam, Leonardo da Vinci, Albrecht Dürer, John Napier, Pascal, Diderot, Bolzano, etc. Surprising accounts of what can result from a non-professional preoccupation with the oldest of sciences. 56 figures. viii + 211pp. 5⅜ x 8½.   61009-8 Paperbound $2.00

COLLEGE ALGEBRA, H. B. Fine
Standard college text that gives a systematic and deductive structure to algebra;
comprehensive, connected, with emphasis on theory. Discusses the commutative,
associative, and distributive laws of number in unusual detail, and goes on
with undetermined coefficients, quadratic equations, progressions, logarithms,
permutations, probability, power series, and much more. Still most valuable
elementary-intermediate text on the science and structure of algebra. Index.
1560 problems, all with answers. x + 631pp. 5⅜ x 8. 60211-7 Paperbound $2.75

HIGHER MATHEMATICS FOR STUDENTS OF CHEMISTRY AND PHYSICS,
J. W. Mellor
Not abstract, but practical, building its problems out of familiar laboratory
material, this covers differential calculus, coordinate, analytical geometry,
functions, integral calculus, infinite series, numerical equations, differential
equations, Fourier's theorem, probability, theory of errors, calculus of variations, determinants. "If the reader is not familiar with this book, it will repay
him to examine it," Chem. & Engineering News. 800 problems. 189 figures.
Bibliography. xxi + 641pp. 5⅜ x 8.   60193-5 Paperbound $3.50

TRIGONOMETRY REFRESHER FOR TECHNICAL MEN,
A. A. Klaf
A modern question and answer text on plane and spherical trigonometry. Part I
covers plane trigonometry: angles, quadrants, trigonometrical functions, graphical representation, interpolation, equations, logarithms, solution of triangles,
slide rules, etc. Part II discusses applications to navigation, surveying, elasticity,
architecture, and engineering. Small angles, periodic functions, vectors, polar
coordinates, De Moivre's theorem, fully covered. Part III is devoted to spherical
trigonometry and the solution of spherical triangles, with applications to
terrestrial and astronomical problems. Special time-savers for numerical calculation. 913 questions answered for you! 1738 problems; answers to odd numbers.
494 figures. 14 pages of functions, formulae. Index. x + 629pp. 5⅜ x 8.
                                                      20371-9 Paperbound $3.00

CALCULUS REFRESHER FOR TECHNICAL MEN,
A. A. Klaf
Not an ordinary textbook but a unique refresher for engineers, technicians,
and students. An examination of the most important aspects of differential and
integral calculus by means of 756 key questions. Part I covers simple differential
calculus: constants, variables, functions, increments, derivatives, logarithms,
curvature, etc. Part II treats fundamental concepts of integration: inspection,
substitution, transformation, reduction, areas and volumes, mean value, successive and partial integration, double and triple integration. Stresses practical
aspects! A 50 page section gives applications to civil and nautical engineering,
electricity, stress and strain, elasticity, industrial engineering, and similar fields.
756 questions answered. 556 problems; solutions to odd numbers. 36 pages of
constants, formulae. Index. v + 431pp. 5⅜ x 8.   20370-0 Paperbound $2.25

INTRODUCTION TO THE THEORY OF GROUPS OF FINITE ORDER,
R. Carmichael
Examines fundamental theorems and their application. Beginning with sets,
systems, permutations, etc., it progresses in easy stages through important types
of groups: Abelian, prime power, permutation, etc. Except 1 chapter where
matrices are desirable, no higher math needed. 783 exercises, problems. Index.
xvi + 447pp. 5⅜ x 8.   60300-8 Paperbound $3.00

FIVE VOLUME "THEORY OF FUNCTIONS" SET BY KONRAD KNOPP

This five-volume set, prepared by Konrad Knopp, provides a complete and readily followed account of theory of functions. Proofs are given concisely, yet without sacrifice of completeness or rigor. These volumes are used as texts by such universities as M.I.T., University of Chicago, N. Y. City College, and many others. "Excellent introduction . . . remarkably readable, concise, clear, rigorous," *Journal of the American Statistical Association*.

ELEMENTS OF THE THEORY OF FUNCTIONS,
Konrad Knopp
This book provides the student with background for further volumes in this set, or texts on a similar level. Partial contents: foundations, system of complex numbers and the Gaussian plane of numbers, Riemann sphere of numbers, mapping by linear functions, normal forms, the logarithm, the cyclometric functions and binomial series. "Not only for the young student, but also for the student who knows all about what is in it," *Mathematical Journal*. Bibliography. Index. 140pp. 5⅜ x 8.   60154-4 Paperbound $1.50

THEORY OF FUNCTIONS, PART I,
Konrad Knopp
With volume II, this book provides coverage of basic concepts and theorems. Partial contents: numbers and points, functions of a complex variable, integral of a continuous function, Cauchy's integral theorem, Cauchy's integral formulae, series with variable terms, expansion of analytic functions in power series, analytic continuation and complete definition of analytic functions, entire transcendental functions, Laurent expansion, types of singularities. Bibliography. Index. vii + 146pp. 5⅜ x 8.   60156-0 Paperbound $1.50

THEORY OF FUNCTIONS, PART II,
Konrad Knopp
Application and further development of general theory, special topics. Single valued functions. Entire, Weierstrass, Meromorphic functions. Riemann surfaces. Algebraic functions. Analytical configuration, Riemann surface. Bibliography. Index. x + 150pp. 5⅜ x 8.   60157-9 Paperbound $1.50

PROBLEM BOOK IN THE THEORY OF FUNCTIONS, VOLUME 1.
Konrad Knopp
Problems in elementary theory, for use with Knopp's *Theory of Functions*, or any other text, arranged according to increasing difficulty. Fundamental concepts, sequences of numbers and infinite series, complex variable, integral theorems, development in series, conformal mapping. 182 problems. Answers. viii + 126pp. 5⅜ x 8.   60158-7 Paperbound $1.50

PROBLEM BOOK IN THE THEORY OF FUNCTIONS, VOLUME 2,
Konrad Knopp
Advanced theory of functions, to be used either with Knopp's *Theory of Functions*, or any other comparable text. Singularities, entire & meromorphic functions, periodic, analytic, continuation, multiple-valued functions, Riemann surfaces, conformal mapping. Includes a section of additional elementary problems. "The difficult task of selecting from the immense material of the modern theory of functions the problems just within the reach of the beginner is here masterfully accomplished," *Am. Math. Soc*. Answers. 138pp. 5⅜ x 8.
60159-5 Paperbound $1.50

NUMERICAL SOLUTIONS OF DIFFERENTIAL EQUATIONS,
H. Levy & E. A. Baggott
Comprehensive collection of methods for solving ordinary differential equations of first and higher order. All must pass 2 requirements: easy to grasp and practical, more rapid than school methods. Partial contents: graphical integration of differential equations, graphical methods for detailed solution. Numerical solution. Simultaneous equations and equations of 2nd and higher orders. "Should be in the hands of all in research in applied mathematics, teaching," *Nature.* 21 figures. viii + 238pp. 5⅜ x 8. 60168-4 Paperbound $1.85

ELEMENTARY STATISTICS, WITH APPLICATIONS IN MEDICINE AND THE BIOLOGICAL SCIENCES, *F. E. Croxton*
A sound introduction to statistics for anyone in the physical sciences, assuming no prior acquaintance and requiring only a modest knowledge of math. All basic formulas carefully explained and illustrated; all necessary reference tables included. From basic terms and concepts, the study proceeds to frequency distribution, linear, non-linear, and multiple correlation, skewness, kurtosis, etc. A large section deals with reliability and significance of statistical methods. Containing concrete examples from medicine and biology, this book will prove unusually helpful to workers in those fields who increasingly must evaluate, check, and interpret statistics. Formerly titled "Elementary Statistics with Applications in Medicine." 101 charts. 57 tables. 14 appendices. Index. vi + 376pp. 5⅜ x 8. 60506-X Paperbound $2.25

INTRODUCTION TO SYMBOLIC LOGIC,
S. Langer
No special knowledge of math required — probably the clearest book ever written on symbolic logic, suitable for the layman, general scientist, and philosopher. You start with simple symbols and advance to a knowledge of the Boole-Schroeder and Russell-Whitehead systems. Forms, logical structure, classes, the calculus of propositions, logic of the syllogism, etc. are all covered. "One of the clearest and simplest introductions," *Mathematics Gazette.* Second enlarged, revised edition. 368pp. 5⅜ x 8. 60164-1 Paperbound $2.25

A SHORT ACCOUNT OF THE HISTORY OF MATHEMATICS,
W. W. R. Ball
Most readable non-technical history of mathematics treats lives, discoveries of every important figure from Egyptian, Phoenician, mathematicians to late 19th century. Discusses schools of Ionia, Pythagoras, Athens, Cyzicus, Alexandria, Byzantium, systems of numeration; primitive arithmetic; Middle Ages, Renaissance, including Arabs, Bacon, Regiomontanus, Tartaglia, Cardan, Stevinus, Galileo, Kepler; modern mathematics of Descartes, Pascal, Wallis, Huygens, Newton, Leibnitz, d'Alembert, Euler, Lambert, Laplace, Legendre, Gauss, Hermite, Weierstrass, scores more. Index. 25 figures. 546pp. 5⅜ x 8. 20630-0 Paperbound $2.75

INTRODUCTION TO NONLINEAR DIFFERENTIAL AND INTEGRAL EQUATIONS,
Harold T. Davis
Aspects of the problem of nonlinear equations, transformations that lead to equations solvable by classical means, results in special cases, and useful generalizations. Thorough, but easily followed by mathematically sophisticated reader who knows little about non-linear equations. 137 problems for student to solve. xv + 566pp. 5⅜ x 8½. 60971-5 Paperbound $2.75

AN INTRODUCTION TO THE GEOMETRY OF N DIMENSIONS,
D. H. Y. Sommerville
An introduction presupposing no prior knowledge of the field, the only book in English devoted exclusively to higher dimensional geometry. Discusses fundamental ideas of incidence, parallelism, perpendicularity, angles between linear space; enumerative geometry; analytical geometry from projective and metric points of view; polytopes; elementary ideas in analysis situs; content of hyper-spacial figures. Bibliography. Index. 60 diagrams. 196pp. 5⅜ x 8.
60494-2 Paperbound $1.50

ELEMENTARY CONCEPTS OF TOPOLOGY, *P. Alexandroff*
First English translation of the famous brief introduction to topology for the beginner or for the mathematician not undertaking extensive study. This unusually useful intuitive approach deals primarily with the concepts of complex, cycle, and homology, and is wholly consistent with current investigations. Ranges from basic concepts of set-theoretic topology to the concept of Betti groups. "Glowing example of harmony between intuition and thought," David Hilbert. Translated by A. E. Farley. Introduction by D. Hilbert. Index. 25 figures. 73pp. 5⅜ x 8.
60747-X Paperbound $1.25

ELEMENTS OF NON-EUCLIDEAN GEOMETRY,
D. M. Y. Sommerville
Unique in proceeding step-by-step, in the manner of traditional geometry. Enables the student with only a good knowledge of high school algebra and geometry to grasp elementary hyperbolic, elliptic, analytic non-Euclidean geometries; space curvature and its philosophical implications; theory of radical axes; homothetic centres and systems of circles; parataxy and parallelism; absolute measure; Gauss' proof of the defect area theorem; geodesic representation; much more, all with exceptional clarity. 126 problems at chapter endings provide progressive practice and familiarity. 133 figures. Index. xvi + 274pp. 5⅜ x 8.
60460-8 Paperbound $2.00

INTRODUCTION TO THE THEORY OF NUMBERS, *L. E. Dickson*
Thorough, comprehensive approach with adequate coverage of classical literature, an introductory volume beginners can follow. Chapters on divisibility, congruences, quadratic residues & reciprocity. Diophantine equations, etc. Full treatment of binary quadratic forms without usual restriction to integral coefficients. Covers infinitude of primes, least residues. Fermat's theorem. Euler's phi function, Legendre's symbol, Gauss's lemma, automorphs, reduced forms, recent theorems of Thue & Siegel, many more. Much material not readily available elsewhere. 239 problems. Index. I figure. viii + 183pp. 5⅜ x 8.
60342-3 Paperbound $1.75

MATHEMATICAL TABLES AND FORMULAS,
compiled by Robert D. Carmichael and Edwin R. Smith
Valuable collection for students, etc. Contains all tables necessary in college algebra and trigonometry, such as five-place common logarithms, logarithmic sines and tangents of small angles, logarithmic trigonometric functions, natural trigonometric functions, four-place antilogarithms, tables for changing from sexagesimal to circular and from circular to sexagesimal measure of angles, etc. Also many tables and formulas not ordinarily accessible, including powers, roots, and reciprocals, exponential and hyperbolic functions, ten-place logarithms of prime numbers, and formulas and theorems from analytical and elementary geometry and from calculus. Explanatory introduction. viii + 269pp. 5⅜ x 8½.
60111-0 Paperbound $1.50

APPLIED OPTICS AND OPTICAL DESIGN,
A. E. Conrady
With publication of vol. 2, standard work for designers in optics is now complete for first time. Only work of its kind in English; only detailed work for practical designer and self-taught. Requires, for bulk of work, no math above trig. Step-by-step exposition, from fundamental concepts of geometrical, physical optics, to systematic study, design, of almost all types of optical systems. Vol. 1: all ordinary ray-tracing methods; primary aberrations; necessary higher aberration for design of telescopes, low-power microscopes, photographic equipment. Vol. 2: (Completed from author's notes by R. Kingslake, Dir. Optical Design, Eastman Kodak.) Special attention to high-power microscope, anastigmatic photographic objectives. "An indispensable work," *J., Optical Soc. of Amer.* Index. Bibliography. 193 diagrams. 852pp. 6⅛ x 9¼.
60611-2, 60612-0 Two volume set, paperbound $8.00

MECHANICS OF THE GYROSCOPE, THE DYNAMICS OF ROTATION,
R. F. Deimel, Professor of Mechanical Engineering at Stevens Institute of Technology
Elementary general treatment of dynamics of rotation, with special application of gyroscopic phenomena. No knowledge of vectors needed. Velocity of a moving curve, acceleration to a point, general equations of motion, gyroscopic horizon, free gyro, motion of discs, the damped gyro, 103 similar topics. Exercises. 75 figures. 208pp. 5⅜ x 8. 60066-1 Paperbound $1.75

STRENGTH OF MATERIALS,
J. P. Den Hartog
Full, clear treatment of elementary material (tension, torsion, bending, compound stresses, deflection of beams, etc.), plus much advanced material on engineering methods of great practical value: full treatment of the Mohr circle, lucid elementary discussions of the theory of the center of shear and the "Myosotis" method of calculating beam deflections, reinforced concrete, plastic deformations, photoelasticity, etc. In all sections, both general principles and concrete applications are given. Index. 186 figures (160 others in problem section). 350 problems, all with answers. List of formulas. viii + 323pp. 5⅜ x 8.
60755-0 Paperbound $2.50

HYDRAULIC TRANSIENTS,
G. R. Rich
The best text in hydraulics ever printed in English . . . by former Chief Design Engineer for T.V.A. Provides a transition from the basic differential equations of hydraulic transient theory to the arithmetic integration computation required by practicing engineers. Sections cover Water Hammer, Turbine Speed Regulation, Stability of Governing, Water-Hammer Pressures in Pump Discharge Lines, The Differential and Restricted Orifice Surge Tanks, The Normalized Surge Tank Charts of Calame and Gaden, Navigation Locks, Surges in Power Canals—Tidal Harmonics, etc. Revised and enlarged. Author's prefaces. Index. xiv + 409pp. 5⅜ x 8½. 60116-1 Paperbound $2.50